MW00563927

CRC Handbook of Immunoblotting of Proteins

Volume I
Technical Descriptions

Editors

Ole J. Bjerrum, Ph.D., M.D.
Associate Professor
The Protein Laboratory
University of Copenhagen
Copenhagen, Denmark

Niels H. H. Heegaard, M.D.
The Protein Laboratory
University of Copenhagen
Copenhagen, Denmark

CRC Press, Inc.
Boca Raton, Florida

Library of Congress Cataloging-in-Publication Data

CRC handbook of immunoblotting of proteins.

Includes bibliographies and index.
Contents: v. 1. Technical descriptions -- v.2.
Experimental and clinical applications.
1. Immunoblotting--Handbooks, manuals, etc.
2. Proteins--Analysis--Handbooks, manuals, etc.
I. Bjerrum, Ole J. II. Heegaard, Niels H. H.
III. Title: Handbook of immunoblotting of proteins.
[DNLM: 1. Membrane Proteins--analysis--handbooks.
2. Membrane Proteins--immunology--handbooks.
QU 39 C911]
QP519.9.I43C73 1988 574.19′245 87-24238
ISBN 0-8493-0548-9 (set)
ISBN 0-8493-0549-7 (v. 1)
ISBN 0-8493-0550-0 (v.2)

This book represents information obtained from authentic and highly regarded sources. Reprinted material is quoted with permission, and sources are indicated. A wide variety of references are listed. Every reasonable effort has been made to give reliable data and information, but the author and the publisher cannot assume responsibility for the validity of all materials or for the consequences of their use.

All rights reserved. This book, or any parts thereof, may not be reproduced in any form without written consent from the publisher.

Direct all inquiries to CRC Press, Inc., 2000 Corporate Blvd., N.W., Boca Raton, Florida, 33431.

© 1988 by CRC Press, Inc.

International Standard Book Number 0-8493-0548-9 (set)
International Standard Book Number 0-8493-0549-7 (Volume I)
International Standard Book Number 0-8493-0550-0 (Volume II)

Library of Congress Card Number 87-24238
Printed in the United States

PREFACE

The development of the immunoblotting technique has been an event of major importance in protein chemistry. The combination of the high resolving powers of electrophoretic separation techniques with the specificity of antibody detection has brought a new dimension into the molecular characterization of proteins.

In spite of a landslide of methodological papers and reviews, no books devoted to immunoblotting of proteins from a practical point of view have been published until now. We have, therefore, found it expedient to try to collect the sum of our present knowledge about applications and techniques of immunoblotting in this monograph.

Volume I focuses on technical descriptions (Sections 1 to 7). Volume II focuses on experimental applications (Section 8), and Clinical applications (Section 9). Volume I is built up like the chronological progress of an immunoblotting experiment, starting with chapters about separation methods and ending with chapters on detection principles and artifacts. Much attention is given to detailed descriptions and recipes, so that the book can be of direct use for bench work in the laboratory. In the application chapters (Volume II), we have tried to select some representative topics. A chapter on future aspects closes the book.

The volumes deal with immunoblotting but not with DNA, RNA, or ligand blotting. There are two exceptions: a chapter on lectin blotting and a chapter on cell blotting, which have been included to exemplify the versatility of the technique. No further attempts to discuss the heterogenous group of ligand blotting have been made.

Immunoblotting is now an established technique and part of the standard technology of biological sciences, but it has been used in such a diversity of connections that no "authorized" version of the procedure exists. We have chosen to let specialists in different fields write about their own part of the spectrum. The inclusion of so many authors, all describing their personal variant of immunoblotting, we think especially increases the value of the book despite some unavoidable repetitions. This approach should give the researcher seeking a solution to a particular problem in immunoblotting a chance to find useful information. We also hope that the combination of techniques and applications in one handbook, besides illustrating the applicability of the immunoblotting technique, will be a source of inspiration for researchers working in other fields. Therefore, abbreviations have been avoided as much as possible to make the chapters appear intelligible and readable.

We have the sole responsibility for the edition of the chapters, as no editorial board has been involved. We are grateful to our fellow authors for their fine contributions and for granting us editorial license to obtain, we hope, uniformity in style and composition.

Cophenhagen, July 1986
Ole J. Bjerrum
Niels H. H. Heegaard

THE EDITORS

Ole J. Bjerrum, Ph.D., is Director, Research and Development, Department of Immunotechnology, Novo Industry A/S, Bagsvaerd, Denmark.

Dr. Bjerrum was born 1944 in Copenhagen and obtained his medical degree from the University of Copenhagen in 1969. After finishing his medical internship at the University Hospital, he studied protein chemistry at the Institute of Biochemistry, University of Uppsala, Sweden. At the Protein Laboratory, University of Copenhagen, he was appointed in 1971 as assistant professor and in 1974 as associate professor. Dr. Bjerrum has been director of the institution for 4 years. The Ph.D. degree from University of Copenhagen was obtained in 1978. In 1980-81 Dr. Bjerrum spent a year as visiting professor at the Department of Biochemistry and Molecular Biology, Northwestern University, Illinois. In 1987 he moved to Novo Industry A/S.

Dr. Bjerrum is a member of The Danish Academy of Natural Sciences, Danish Biochemical Society (vice-president 1983-84 and president 1984-86), Scandinavian Electrophoresis Society (council member since 1986), and International Electrophoresis Society (council member since 1986). He is on the editorial boards for *Journal of Biochemical and Biophysical Methods* and for *Electrophoresis*.

Dr. Bjerrum has presented over 30 invited lectures at International and National Meetings, approximately 40 guest lectures at Universities and Institutes, and organized 10 international training courses and workshops. He has published more than 110 research papers and edited handbooks on immuno-biochemical methodology. His current major research interest is the structure and function of membrane proteins and immunoassay technology.

Niels H. H. Heegaard, M.D., was born 1959 in Copenhagen and obtained the medical degree from the University of Copenhagen in 1986.

He is now appointed to the University Hospital for his clinical internship. Simultaneously, he is doing research at the Protein Laboratory, University of Copenhagen, where he has been associated since 1981.

He has presented lectures at International Meetings, acted as a referee for International Journals, and participated in arrangement of and teaching of postgraduate courses in membrane protein chemistry. The focus of his current research is on methodology development, and on the influence of ageing processes and hematological disorders on red cell membrane proteins.

CONTRIBUTORS — VOLUME I

Byron E. Batteiger, M.D.
Department of Medicine
Indiana University School of Medicine
Indianapolis, Indiana

Donna Mia Benko
Department of Pharmacology
Johns Hopkins University
School of Medicine
Baltimore, Maryland

Patricia A. Billing, M.S.
Department of Clinical Immunology
Abbott Laboratories
North Chicago, Illinois

Michael Bittner, Ph.D.
Amoco Research Center
Naperville, Illinois

Ole J. Bjerrum, Ph.D.
Protein Laboratory
University of Copenhagen
Copenhagen, Denmark

Guy Daneels
Laboratory of Biochemical Cytology
Pharmaceutical Research Laboratories
Beerse, Belgium

Jan R. De Mey, Ph.D.
Laboratory of Biochemical Cytology
Janssen Pharmaceutica
Beerse, Belgium

Marc De Raeymaeker
Laboratory of Biochemical Cytology
Janssen Pharmaceutica
Beerse, Belgium

Keith B. Elkon, M.B.B.Ch.
Department of Rheumatic Disease
Hospital for Special Surgery
Cornell University Medical Center
New York, New York

Ruth Feldborg
Institute of Medical Microbiology
University of Copenhagen
Copenhagen, Denmark

Steffen Ulrik Friis, M.D.
Department of Biochemistry
University of Copenhagen
Copenhagen, Denmark

Johan Geysen
Laboratory for Developmental
Physiology
Zoological Institute
K.U. Leuven
Leuven, Belgium

Wade Gibson, Ph.D.
Department of Pharmacology
Johns Hopkins University
School of Medicine
Baltimore, Maryland

Julian Gordon, Ph.D.
Diagnostics Division
Abbott Laboratories
North Chicago, Illinois

Kathy M. Hancock, M.Sc.
Division of Parasitic Diseases
Parasitic Diseases Branch
Centers of Disease Control
Atlanta, Georgia

Niels H. H. Heegaard, M.D.
Protein Laboratory
University of Copenhagen
Copenhagen, Denmark

Peter M. H. Heegaard
Protein Laboratory
University of Copenhagen
Copenhagen, Denmark

Karl-Erik Johansson, Ph.D.
The National Veterinary Institute
Uppsala, Sweden

Jason M. Kittler, Ph.D.
Department of Biochemistry
University of Connecticut School of Medicine
Farmington, Connecticut

Jan Kyhse-Andersen, Ph.D.
Department of Biochemistry and Nutrition
Technical University of Denmark
Lyngby, Denmark

Kurt Pii Larsen
Protein Laboratory
University of Copenhagen
Copenhagen, Denmark

Natalie T. Meisler, Ph.D.
Department of Biochemistry
College of Medicine
University of Vermont
Burlington, Vermont

Marc Moeremans
Laboratory for Biochemical Cytology
Janssen Pharmaceutica
Beerse, Belgium

Bodil Norrild, Ph.D.
Department of Virology
Institute for Medical Microbiology
Copenhagen, Denmark

Lise Nyholm, Ph.D.
Harboes Laboratorium Aps
Copenhagen, Denmark

Kenji Ogata, Ph.D.
Autoimmune Disease Center
Scripps Clinic and Research Foundation
La Jolla, California

Marnix Peferoen, Ph.D.
Department of Medical Research
Catholic University of Leuven
Leuven, Belgium

Jakob Ramlau, M.D.
Harboes Laboratorium Aps
Copenhagen, Denmark

Karl-Johan Pluzek
Dakopatts a/s
Glostrup, Denmark

Edwin Rowold, Jr., B.A.
Department of Biological Sciences
Monsanto Company
St. Louis, Missouri

John W. Thanassi, Ph.D.
Department of Biochemistry
College of Medicine
University of Vermont
Burlington, Vermont

Victor C. W. Tsang, Ph.D.
Division of Parasitic Diseases
Parasitic Diseases Branch
Centers for Disease Control
Atlanta, Georgia

Kai-Chung Leonard Yuen, Ph.D.
Laboratory of Viral Diseases
National Institute of Allergy and Infectious Disease
National Institutes of Health
Bethesda, Maryland

TABLE OF CONTENTS — VOLUME I

TABLE OF CONTENTS — VOLUME II

Section 1

IMMUNOBLOTTING — GENERAL PRINCIPLES AND PROCEDURES

Niels H. H. Heegaard and Ole J. Bjerrum

INTRODUCTION

A specific technique and the term "blotting" were first joined in 1975 when E. M. Southern[1] described the transfer of electrophoretically separated single-stranded DNA from gels to an immobilized state on a membrane. The approach was soon applied to RNA by Alwine et al.[2] and in 1979 Renart et al.[3] and Towbin et al.[4] introduced the blotting of proteins by means of capillary and electrophoretic transfer, respectively. The concept of immobilization added a new dimension to analytical electrophoresis. Prior to this, direct biochemical characterization of individual proteins buried within the separation support had been difficult and unsatisfactory, particularly in tight gels, such as polyacrylamide.

Earlier, immunochemical characterization of gel-separated proteins had attracted special interest. An array of immunoelectrophoretic techniques was developed for the purpose of identification and characterization such as crossed immunoelectrophoresis, crossed sodium dodecyl sulfate polyacrylamide gel immunoelectrophoresis (SDS-PAGE IE), crossed immunoelectrofocusing, and immunofixation (procedures described extensively in References 5 to 7). A common characteristic is that they work in agarose. However, for immunochemical analysis of proteins separated by high resolution electrophoresis in polyacrylamide gels new tricks had to be invented (see the following text and Table 1). Thus, analysis by (SDS-PAGE) of an antigen in a mixture is possible after immunoprecipitation with a monospecific antibody. It does not necessarily require a precipitating antibody,[8] but coprecipitation and cross-reactions or aggregation of proteins make it difficult to identify unambiguously the unique antigen relating to the antibody preparation. Other methods employ analysis by crossed immunoelectrophoresis of excised bands from SDS-PAGE,[9] or the reverse where excised bands from crossed immunoelectrophoresis are investigated by SDS-PAGE.[10] Both need precipitating antibodies. Diffusion of antibodies into the gel by applying the antibodies directly onto the gel surface followed by detection with radioactively labeled anti-antibodies (primary binding assays, Burridge gel technique),[11] does not demand precipitating antibodies. The procedure is time consuming (days), needs relatively large amounts of protein (micrograms), and is hampered by the restrictions of antigen accessibility determined by the pore size of the gel matrix (i.e., the acrylamide concentration and crosslinking). Moreover, the long diffusion time reduces the resolution originally achieved by the gel electrophoresis. Better resolution was gained when SDS-PAGE was combined with electrophoresis or diffusion into an antibody-containing overlayed agarose gel (immunoreplica electrophoresis),[12] but these techniques also require precipitating antibodies and the resolution of the SDS-PAGE is not preserved during the different procedures.

The majority of the drawbacks mentioned above are eliminated when the proteins of the separation gel are transferred to a protein binding membrane (immobilizing matrix) followed by antibody incubation. In this way, the analytical potential of the separation techniques is expanded with the specificity and sensitivity of immunodetection. This idea was first practically employed for proteins by Towbin et al.[4] who introduced the technique of electroblotting from SDS-polyacrylamide gels to nitrocellulose sheets followed by immunodetection. An attempt to show the relationship between immunoblotting and other immunochemical techniques used for identification, characterization, and quantification of antigens and antibodies is shown in Figure 1. For a description of the present state of the technique, three recent reviews can also be consulted.[13,14]

Table 1
COMBINATIONS OF IMMUNOCHEMICAL DETECTION AND SDS-PAGE ANALYSIS BEFORE THE ERA OF IMMUNOBLOTTING

Modification	Ref.
Material from immunoprecipitation in solution analyzed by SDS-PAGE	8
SDS-PAGE followed by incubation with antibodies	11
SDS-PAGE followed by electrophoresis into a second dimension antibody-containing agarose gel	15
SDS-PAGE/2D-IEF-SDS-PAGE followed by diffusion into an overlayed agarose gel containing antibodies	12
Excised bands from SDS-PAGE analyzed by crossed immunoelectrophoresis	9
Excised immunoprecipitates from crossed immunoelectrophoresis subjected to SDS-PAGE and 2D-IEF-SDS-PAGE	10

Abbreviations: SDS-PAGE, sodium dodecyl sulfate-polyacrylamide gel electrophoresis; 2D-IEF-SDS-PAGE; two-dimensional isoelectric focusing-SDS-PAGE.

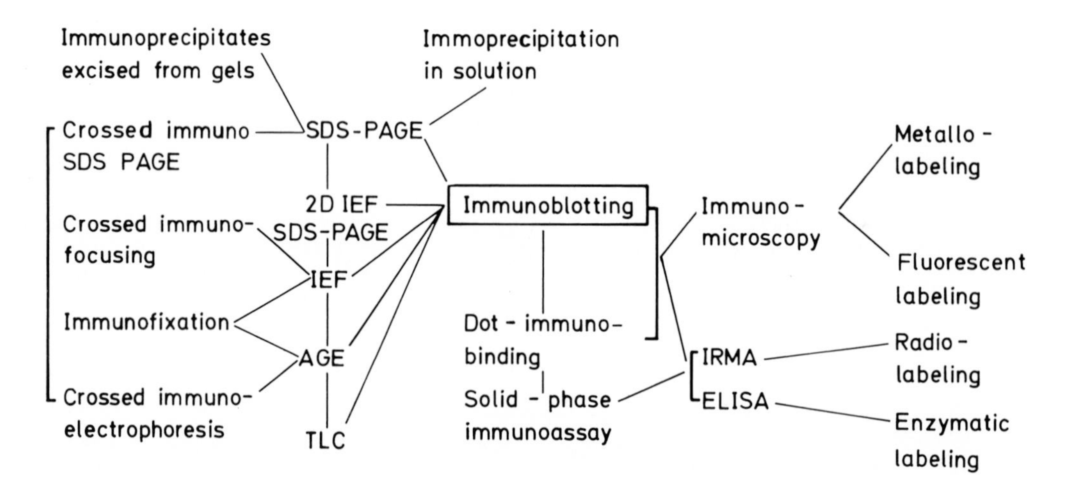

FIGURE 1. Relationship between immunoblotting and other techniques for immunochemical identification, quantification, and characterization of antigens. Abbreviations: AGE, Agarose gel electophoresis; ELISA, enzyme-linked immunosorbent assay; IEF, isoelectric focusing; IRMA, immunoradiometric assay; RIA, radioimmunoassay; SDS-PAGE, sodium dodecyl sulfate-polyacrylamide gel electrophoresis; TLC, thin layer chromatography; 2D, two-dimensional.

PRINCIPLES

A blotting experiment can, in principle, be delineated as a method where affinity interactions between ligands and acceptors take place on the spot and where molecular species have been rendered accessible and maintained in a pattern depending on their physicochemical properties. Biological material, e.g., proteins, can be immobilized on a membrane by wetting (dotting or spotting) or transferred by capillary, diffusional, or electrical forces. The material might be separated before the transfer. The binding forces of the membrane can be of a covalent and noncovalent nature. The main advantages of the blotting principle are due to:

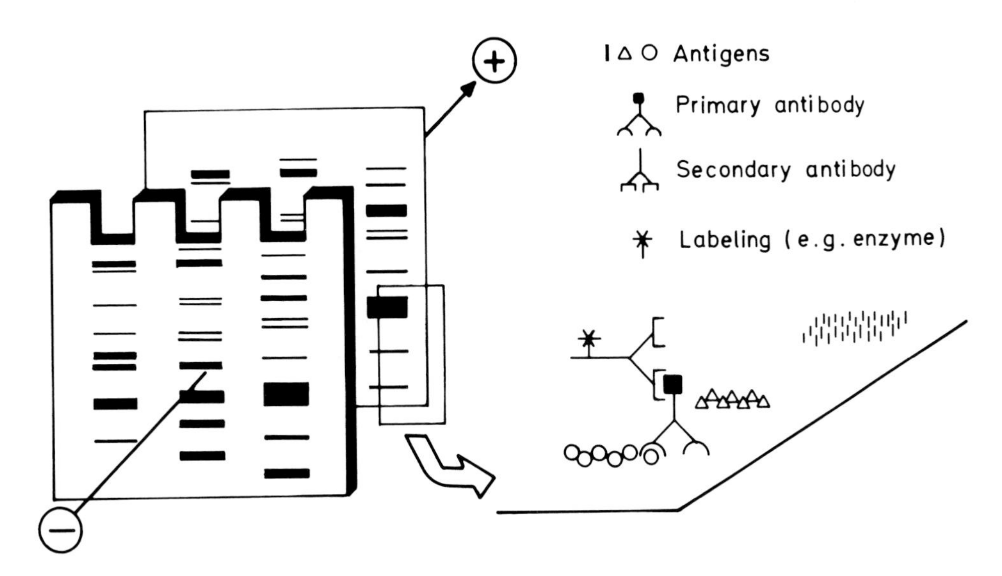

FIGURE 2. General principle of electroimmunoblotting.

1. Accessibility — The antigens are concentrated at the surface of the membrane texture.
2. Immobilization — Diffusion is avoided. The original resolution is thus preserved.
3. Time — Electroblotting can be performed in 1 hr (see Chapter 4.3.2).[16] The accessibility of the transferred proteins further shortens the time required for reactions with ligands. (In addition, the procedure normally does not involve especially difficult steps and most blotting membranes are easily handled.)
4. Economics — Small amounts of proteins are required (10 to 100 ng) as almost all of it is recovered on the blotting membrane ready for the ligand binding assays.
5. Flexibility — The blots might be stored and reused by probing with different ligands. The principle is applicable to all types of ligands (e.g., antigens/antibodies, receptors/hormones, glycoproteins/lectins) and electrophoresis principles (see Chapter 3) and other separation methods such as thin-layer chromatography.

The topic of these volumes is blotting of proteins and the detection by antibodies which is termed immunoblotting. Figure 2 shows the general principle of immunoblotting which can be modified in numerous ways.

TERMINOLOGY

We find that the term "immunoblotting" should be a common denominator for all blotting methods from all kinds of analytical supports with subsequent immunochemical recognition of the transferred species. Less used synonymous expressions would be immunoprinting, immunoreplicas, or immunotransfer. Hence, the words blot, replica, transfer, and imprint are also considered synonymous. When immunoblotting is connected with enzyme-based marker systems it might be called enzyme-linked immunotransfer blot (EITB).[18] The somewhat semigeographical terminology used in naming DNA-blotting Southern blotting after its inventor and followed by calling RNA-blotting Northern blotting and the electrotransfer of proteins Western blotting is not recommended. First, the names are confusing; second, they are not logical; and finally, the use of geographical terms is at the risk of inducing political controversies.[19] For precise descriptions it is preferable to define the blotting method by stating (1) the class of molecules transferred, (2) the method of transfer, and (3) the type of immobilizing matrix. For example: protein transfer by electrophoresis to nitrocellulose, protein replica by vacuum-blotting to Zetaprobe, or protein blotting by diffusion to diazobenzyloxymethyl membranes. Direct application (spotting) of a protein onto a blotting membrane followed by probing with antibodies should be called dot immunobinding (see Chapter 2); [20] synonymous terms would be dot blot immunoassay,[21] spot immunodetection,[22] immunodot,[23] and antigen spot detection.[2] The term dotblotting, though it is very idiomatic, is not adequately descriptive since the transfer from a solution to a solid phase by pipetting is not considered a blotting method. Special modifications have their own names. Examples are filter affinity transfer (FAT), [25,26] a reverse blotting method, and lectin and cell blotting, where the detection steps are not based on immunological recognition (Chapters 8. 6, and 8.7, respectively).

EQUIPMENT

The basic equipment differs depending on the method of blotting employed. Consult the special chapters covering the different blotting methods. Some examples are shown in Figures 3, 4, and 5.

REAGENTS

Detailed descriptions of the sources of chemicals and other reagents are given in the individual methodology chapters in the form of laboratory-ready recipes. The appendix to this chapter contains a thorough description of buffers and chemicals used in our laboratory for immunoblotting of SDS-PAGE separated proteins onto nitrocellulose.

PROCEDURES

The main steps of the technique are briefly considered under the headings: transfer, membranes, and visualization.

Look for the technical descriptions and modifications of these steps in the relevant chapters. A detailed description of our routine procedure for electroimmunoblotting of proteins to nitrocellulose membranes is found in the Appendix.

Transfer

Fundamentally, all kinds of driving forces might be used to transport proteins from analytical supports to blotting membranes: diffusion, osmosis, electrical potential, centrifugal forces, vacuum pressure. The choice of transfer principle generally depends on the class of molecules involved, on the composition of the support nesting these molecules, on the available equipment, and on economic and time considerations. Almost all blotting experiments so far have been performed with capillary, diffusional, vacuum, or electrical forces, the latter being the transfer principle most used for blotting of proteins. The transfer principles are presented schematically in Table 2 and the build-ups of the apparatus are sketched in Figures 3A to E.

The optimal transfer step gives a quantitative elution in the shortest possible time with retention of biological and antigenic characteristics. This is achieved to varying degrees with the different techniques. Blotting by diffusion, capillary forces, or by vacuum suction have no restrictions concerning the choice of buffer, and if the blotting membrane itself has no special requirements the buffer chosen should provide the best conditions for binding, renaturation, and solubility of the proteins. However, elution is not easily controlled or reproduced with any of these methods, and resolution might be lost during the diffusion and capillary blotting processes as a consequence of the long elution times. The poor elution forces restrict the use of diffusion and capillary blotting methods to agarose gels.

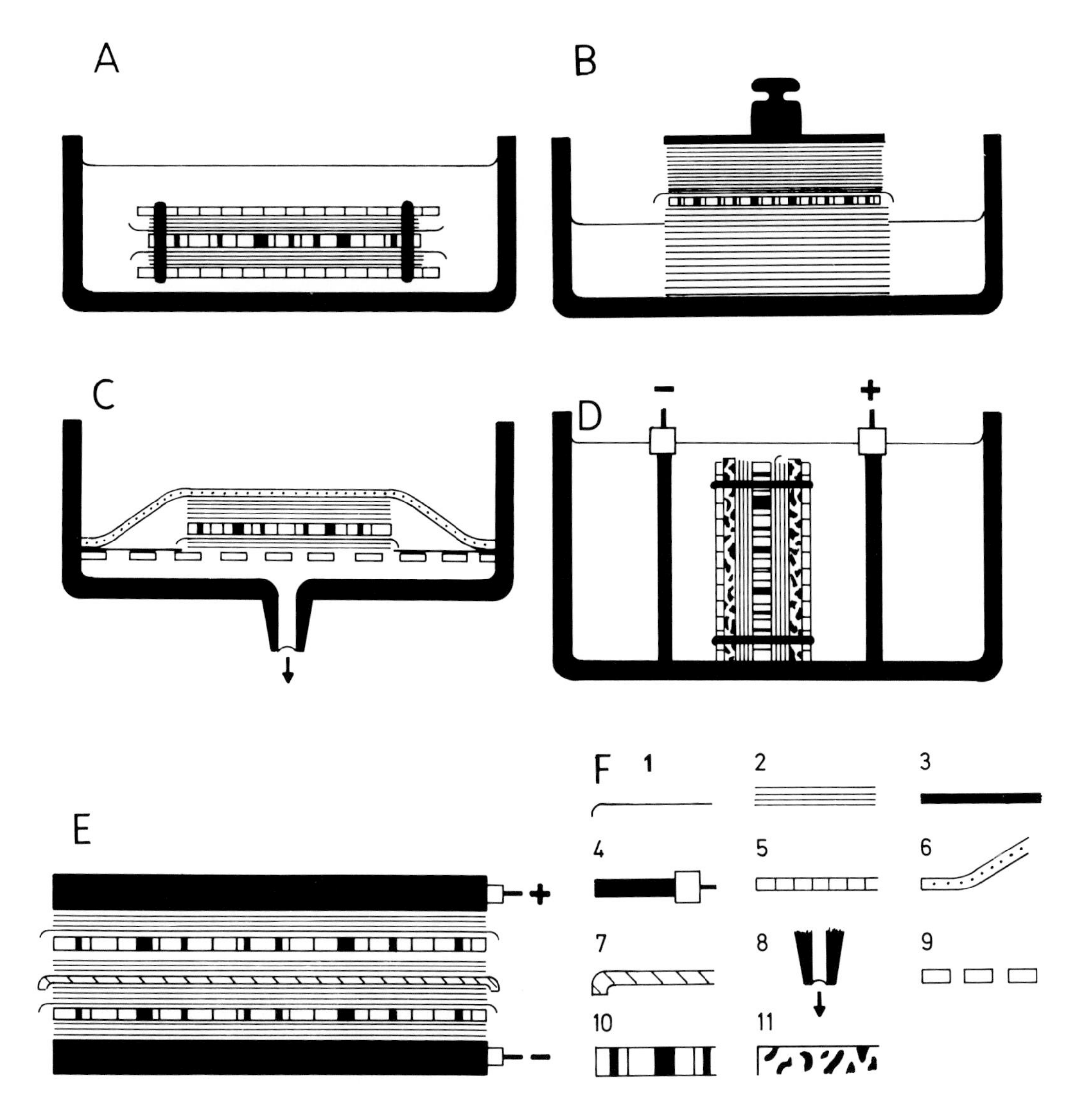

FIGURE 3. Transfer techniques in immunoblotting. A, diffusion; B, capillary flow-convection; C, vacuum-convection; D, electroblotting in buffer tank; E, semidry electroblotting; F, explanation of symbols (1, nitrocellulose sheet; 2, filter paper; 3, electrode; 4, electrode terminal; 5, plexiglass grid; 6, rubber sheet; 7, dialysis film; 8, outlet; 9, porous support; 10, gel; 11, sponge).

In electroelution, the direction and rate of transfer depend on the pH of the transfer buffer and placement of electrodes. The efficiency of elution depends on the composition of the separation medium (e.g., a gel), transfer time, size of the proteins, field strength, and on the presence of detergents. Application of special techniques like *in situ* digestion by pronase (see Chapter 4.3.4),[27] partial chemical breaking of the gel before blotting,[3] or widening of the gel pores by employing aqueous or urea-containing buffers for preequilibration of the gel also make elution more efficient. The binding of proteins to the blotting membranes depends on the characteristics of the proteins and membrane (see Membranes), presence of detergents, substances (like methanol in the case of nitrocellulose) which enhance the binding capacity of the membrane,[4] and on elements of the buffer interacting with the binding sites on the blotting matrix.

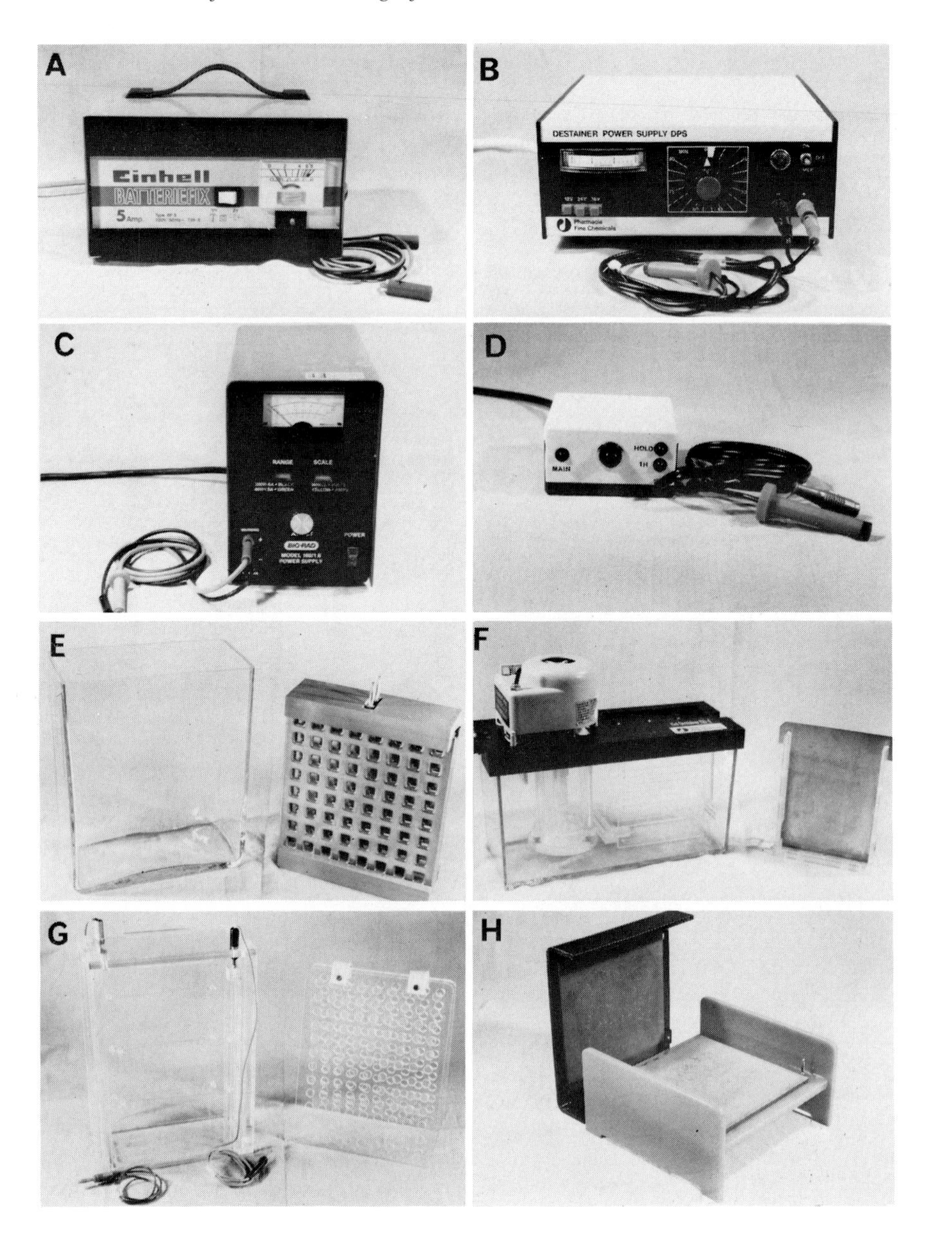

FIGURE 4. Equipment for electroblotting experiments. Power supplies (A to D) and transfer apparatus (E to H). (A) Battery charger 6 V/12 V, 5 A (Einhell). (B) Destainer power supply 12 V/24 V/36 V, equipped with timer (Pharmacia Fine Chemicals). (C) Power supply for electroblotting in buffer vessels 200 V, 0.6 A/60 V, 1.8 A (Bio-Rad). (D) Power supply for semidry electroblotting, variable voltage, 200 mA with 1 hr timer (Kem-En-Tec). (E) Simple electroblotting apparatus. Buffer vessel: glass jar, electrodes, and sandwich holders; stainless steel bathroom gratings (courtesy of Dr. Ib Rode Pedersen). (F) Gel destainer (Pharmacia Fine Chemicals). Plexiglass vessel with platinum anode and stainless steel cathode. (G) Electroblotting apparatus (Bio-Rad) with polycarbonate vessel, platinum electrodes, and locking gel cassette with hinged plexiglass sheets. (H) Semidry electroblotting apparatus (JKA-Biotech); graphite electrodes, plexiglass mounted.

At a given pH a more rapid and efficient transfer is obtained by use of higher field strengths for which reason the power supply should have a variable voltage and a timer. The upper limit depends on the heat dissipating abilities of the system in question and might be increased by using buffers of low ionic strength which generate less heat.[28] Large buffer volumes, cooling, and stirring can also reduce the temperature increase.

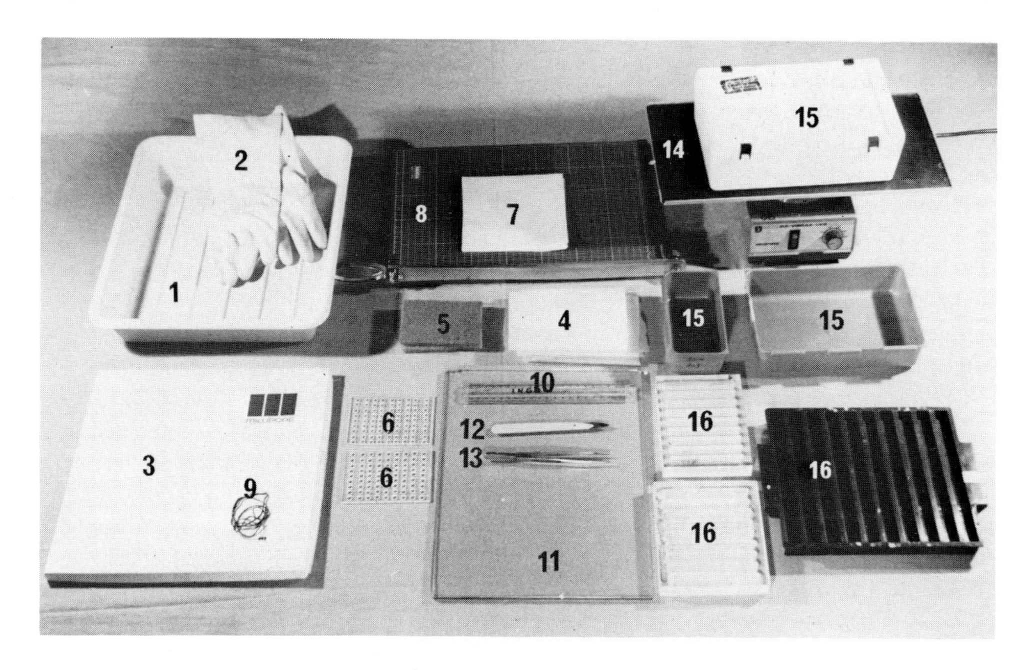

FIGURE 5. Acessories for immunoblotting. 1, bath for assembly of sandwich; 2, gloves; 3, nitrocellulose sheets; 4, 5 packing material (4, rubber foam, 5, scouring pads); 6, perforated plexiglass holders; 7, Whatman no. 1 filter paper; 8, paper cutter; 9, rubber bands for clamping the sandwich together; 10, ruler; 11, glassplate; 12, scalpel; 13, tweezers; 14, rocking table; 15, various sized incubation vessels; 16, trays with ditches of various sizes.

Table 2
METHODS FOR TRANSFER OF PROTEINS TO BLOTTING MEMBRANES

Transfer principle	Driving force	Described in chapter	Original ref.
Pipetting	Wetting	2	20
Capillary	Buffer flow	4.1	3
Vacuum	Buffer flow	4.2	19
Diffusion	Diffusion	—	50
Electrophoretic with buffer tank	Electric potential	4.3.1	4
Electrophoretic without buffer tank	Electric potential	4.3.2	16, 38

The transfer also depends on the type of gel and the size and charge of the protein molecules. Generally, electroelution from electrofocusing and nondenaturing gels is faster than from SDS-PAGE gels.[28] Elution is also more efficient from homogeneous gels than from gradient gels. Large molecules are more slowly eluted than small molecules, and the nearer the pH is to the pI of the protein the slower it is transferred even if it is from polyacrylamide gels containing SDS. Thus, basic proteins might advantageously be eluted at an acidic pH (see Chapter 4.3.3). Elution seems to be more dependent on molecular size alone when SDS is added to the transfer buffer.[28,29] It is also possible to blot from gels containing nonionic detergents as demonstrated for agarose gels with Triton® X-100 in Chapter 8.5. Transfer of immunoprecipitates by electroblotting after a brief treatment of the gel with SDS is another approach to blotting from agarose gels.[30] Characterization of precipitates with nonprecipitating antibodies is then possible. With time, elution is more complete and prolonged transfer times do not seem to have deleterious effects themselves, but small proteins (less than 20,000) might be lost if nitrocellulose membranes with big pores (0.45 μm pore size) are used probably because they have less available surface for binding.[31] This loss is partially avoided by using membranes with smaller pore sizes. To obtain an even elution of big and small molecules from polyacrylamide gels a gradient electric field can be employed.[32]

Electrophoretic transfer from SDS-PAGE is often performed in 25 m*M* Tris and 192 m*M* glycine, pH 8.0 to 8.3 with 20% methanol which counteracts the swelling of gels during transfer and increases the binding capacity of nitrocellulose, but slows down the elution.[33] Another way of retaining the gel size during transfer is to preequilibrate the gel in transfer buffer before elution.[34] Tris and glycine-containing buffers cannot be used in connection with diazo-paper membranes because amino groups bind to the paper (see Chapter 5.2). Usually acetate or phosphate buffer, pH 6.5 or borate, pH 9.2 are used instead.[35,36]

The vertical blotting apparatus of Towbin et al.[4] (Figure 3C), which is submerged in a bath with transfer buffer, usually gives an efficient electroblotting of proteins by an overnight run at 4°C using a power supply generating 25 to 100 V at 0.5 to 1.5 A.([37]) Electrophoretic transfer in the horizontal semidry blotting apparatus (Figure 3D)[16] normally yields a complete blotting at room temperature in 1 hr employing 200 mA at a voltage gradient of 12 V/cm for a normal sized gel (15 × 15 cm) (see Chapter 4.3.2).

Conclusion

If low cost in equipment is demanded, blotting by capillary forces might be chosen. This works best with agarose gels. It has no special buffer requirements. In other cases, electroblotting will be the best choice as it is fast, efficient, reproducible, and allows semi-quantitative considerations. Many instruments using a sandwich assembly of gel and membrane for electroblotting are available (see Section IV). Easy assembly, multiple, and rapid transfers are offered by the semidry blotting apparatus.

Membranes

The solid phase of immunoblotting is the blotting membrane which functions as an immobilizing matrix capable of adsorbing and retaining the transferred molecules in the original separation pattern during the subsequent detection steps. Membranes with different characteristics exist, and as outlined in Table 3, they can be grouped on the basis of the nature of the binding forces between membrane and the transferred molecules.

The rather broad spectrum of matrices available illustrates that a matrix suitable for all blotting experiments does not exist. Nitrocellulose, which was introduced in 1967,[40] is very useful for immobilization of transferred proteins as it is easy to handle, stable, durable, and has a high binding capacity. If multiple assays on the same imprint are needed, the covalently binding diazo-papers can be used (see Chapter 5.2),[2] but they require activation before use, are unstable, and not very durable. For special purposes where nitrocellulose cannot be used, nylon-based membranes such as Genescreen or Hybond N or the charge-modified (cationized) Zetaprobe can be employed (see Chapter 5.1.2).[33,41] These membranes have binding capacities four to five times higher than nitrocellulose, but extensive blocking procedures are necessary and they cannot be stained with the common anionic protein dyes. An anion-exchange membrane such as DEAE-paper binds acidic or amphoteric molecules above their pI and is merely used for preparative purposes since it is structurally weak.[42]

Nitrocellulose is the matrix of choice for blotting of proteins if the blots are not going to be subjected to extensive treatment with detergents or other agents displacing the bound proteins.[31,43] The binding involves hydrophobic forces,[44] and thus it is not disturbed by agents such as 6 *M* urea, 6 *M* guanidinium hydrochloride, or ampholines.[43] Fixing the proteins *in situ* to the paper may be performed with glutardialdehyde,[44] or *N*-hydroxysuccinimidyl-*p*-azidobenzoate.[45] The binding capacity is reported to be from 15 to 80 μg/cm^2

Table 3
SOME CHARACTERISTICS OF DIFFERENT BLOTTING MEMBRANES

Feature	Nitrocellulose	Zetabind/GeneScreen nylon membranes	Diazobenzyloxymethyl/ diazophenylthioether-cellulose	Diethylaminoethyl paper
Bond	"Hydrophobic"	Ionic + electrostatic	Covalent	Ionic
Activation necessary	No	No	Yes	No
Buffer requirements	Any buffer but nonionic detergent block binding	—	Amino groups bind to paper	—
Electroblotting of proteins	Yes	Yes	Yes	Yes
Electroblotting of nucleic acids	No	Yes	No	Yes
Direct staining	Possible	Not possible with anionic stains	—	—
Capacity	20—80 μg/cm^2	200—400 μg/cm^2	High	—
Handling/durability	Easy/durable	Easy/durable	Easy/unstable	Low wet strength
Reprobing	Possible	Possible	Easy	Not possible
Preparative purposes	Yes	—	No	Yes
Described in chapter	5.1.1	5.1.2	5.2	—

since nitrocellulose containing cellulose acetate has a lower capacity.[4,33,46] The capacity is not only correlated with the available binding surface depending on pore sizes since low molecular weight proteins are lost first when the capacity limits are approached. Thus, loss of protein in blotting procedures besides the previously mentioned incomplete elution and disturbances of binding might occur as a result of transport of molecules through the membrane.

Blocking

The blocking procedure (also called quenching) is essential if the detection step involves specific recognition with proteins (e.g., antibodies, lectins, or protein A; see Visualization). Quenching saturates additional binding sites on the matrix and it might help renaturation of the bound proteins. The blocking method is highly dependent on the type of membrane, but also on the detection system, the nature of the transferred molecules, and the reagents available. The two major classes of blocking agents are proteins and detergents. The proteins employed for blocking of nitrocellulose membranes include bovine serum albumin,[4] fetal calf serum, hemoglobin,[33] casein, and animal serum.[47] Here, the possibility of cross reactions between the detecting antibodies and the components of the blocking solution should always be considered. When visualization involves peroxidase-dependent color reactions, hemoglobin should not be used since it has intrinsic peroxidase activity. If detergents such as polyoxyethylene alcohols (e.g., Nonidet P-40[31,48]) or polyoxyethylene sorbitan alcohols (e.g., Tween® 20[43,49]) are used it should be borne in mind that the detergent which is able to disturb hydrophobic interactions competes for the same binding sites as the protein, so that prolonged incubation or high concentration might result in displacement of the primary bound protein. Thus, displacement problems with Tween® 20 have been reported (see Chapter 7). The activated papers (diazobenzyloxymethyl/diazophenylthioether cellulose) are usually inactivated by means of Tris/ethanolamine buffers containing gelatin.[3]

The blocking procedure is considered in Chapter 6.2. In our laboratory, Tween® 20 for quenching of nitrocellulose membranes is used and routinely incorporated in a concentration of 0.05% (v/v) into all washing solutions to facilitate removal of nonspecifically adsorbed proteins throughout the detection procedures. Consult the Appendix for details.

Conclusion

For protein blotting, nitrocellulose is normally the membrane of choice. The other membranes are for purposes such as when high binding capacity (e.g., Zetaprobe or Hybond N) is required or to avoid the risk of loss is antigen during the assay procedure (covalent binding). Useful and cheap blocking agents for nitrocellulose appear to be nonionic detergents (e.g., Tween® 20).

Visualization

In our laboratory, Pyronin G is transferred together with the proteins to the blotting membrane as a marker of the positions of the lanes. Detection of the retained molecules on the replica can then proceed generally or specifically. In both instances the visualization achieved by deposits of color, radioactivity, or fluorescence is linked either directly or through some amplification cascade to the recognizing molecules. Many of the designs have originated in the field of histochemistry, and especially for the specific, multistep enzyme and isotope-based visualization systems new variations continue to emerge. Some possibilities divided on the basis of the detection principle and visualization method are mentioned in Tables 4 and 5. The following comments are given with a bias toward detection on nitrocellulose.

General Detection

The transfer efficiency and the degree of preservation of original gel geometry and size are assessed by performing a general staining of the blot. This also helps to correlate the

Table 4
GENERAL DETECTION PRINCIPLES FOR VISUALIZATION OF PROTEINS

Technique	Chapter
Conventional protein stains (e.g., Amido Black, Coomassie brilliant blue)	1 (Append.)
India ink	6.1.2
Silver staining	6.1.1
Colloidal metal staining	6.1.3
Derivatization with hapten followed by incubation with antihapten antibody	6.4.5

Table 5
SPECIFIC DETECTION PRINCIPLES IN PROTEIN BLOTTING AND POSSIBLE COMBINATIONS WITH VISUALIZATION SYSTEMS

Primary recognition	Possible secondary ligands	Linkage of primary or secondary ligands with visualization systems	Visualization systems	Chapter
Ab	a-Ab, protein A, Ag	Direct conjugation; bio-	Enzymatic,	6.4.1, 6.4.2
Ag	Ab, ligand	tin-avidin, biotin-strep-	autoradiographic,	6.4.3
Protein A	Unspecific IgG	tavidin coupling; PAP-	fluorescent,	—
Lectin	a-Lectin Ab	systems, or other sys-	colloidal metal/	6.4.4
Ligand	a-Ligand Ab, receptor	tems with repetitive	silver development	6.4.4
		layers.		

Abbreviations: Ab, antibody; Ag, antigen; a-, anti-; PAP, peroxidase-antiperoxidase.

specific staining patterns with the picture of the total separation. The methods listed in Table 4 differ with respect to sensitivity, applicability, and complexity. As a rule, the sensitivity of general staining reagents depends on the degree of background coloration and varies directly with the strength of the bonds formed between the stain and the protein in question. As the absolute sensitivity varies from one group of proteins to another it will be sufficient for a survey to consider the methods relative to each other.

The common protein dyes are easy and quick staining reagents. They include Amido Black 10B,[4] Coomassie brilliant blue R-250,[34] aniline blue black,[50] Ponceau S which can be washed off after use,[51] and Fast green.[52] Immunoassays performed after staining might be impeded by the binding of dyes,[31] therefore reversible staining methods with heparin toluidine blue[53] or phenol red[54] have been devised. Sufficient antigenicity, however, seemed to be preserved in a modification where a gel was stained with Coomassie brilliant blue and proteins and dye directly blotted onto nitrocellulose, where an immunoperoxidase detection was performed.[55] The anionic stains are not very sensitive and they cannot be used on nylon membranes. On nitrocellulose, a disturbing background staining may appear as seen with Coomassie brilliant blue.[56] Amido Black 10B is more easily applied, but fast green appears to be the most sensitive of the general protein stains.[57]

India ink is not compatible with nylon membranes,[56] but it is easy to work with and claimed to be more sensitive than the ordinary protein stains (see Chapter 6.1.2). For higher sensitivity a special silver stain technique has been developed for nitrocellulose,[58] (see Chapter 6.1.1). At best, the sensitivity matches the detection limits obtained for silver staining of polyacrylamide gels, making the technique approximately 10 and 50 to 100 times more sensitive than India ink and ordinary protein staining, respectively.[58] Since the method is based on deposits of silver in areas not occupied by protein maintenance of antigenicity is ensured.[54,58] A more straightforward approach is the staining with FerriDye® described in Chapter 6.1.3, which is applicable for all types of membranes, or the the colloidal gold (or colloidal silver) staining procedure treated in Chapter 6.1.3,[57] which is of about equal sensitivity.

The derivatization methods (Chapter 6.4.5) are very sensitive (about 100 times the ordinary protein stains) and comparable with silver staining. Unlike the other procedures mentioned, they are equally suitable for all types of matrices. The amino groups of the proteins are derivatized *in situ* on the membrane with, e.g., 2,4-dinitrofluorobenzene (2,4 DNB),[59] or with pyridoxal-5′-phosphate and sodium borohydride (P5P),[60] followed by immunochemical detection with antidinitrophenol or anti-5′-phosphopyridoxyl antibodies connected with some visualization system *(vide infra)*. Conceptually, the different derivatization methods are alike.[60]

Conclusion

For a quick, rough estimate of transfers on nitrocellulose, Amido Black 10B is easily used. More sensitive general detection systems include colloidal metal or silver staining. Hapten derivatization of the blots followed by staining with antihapten antibodies is more sensitive, but also more complicated.

Specific Detection

The large diversity of specific detection principles in immunoblotting is indicated in Table 5. Their common feature is a specific binding of a macromolecule to the immobilized proteins. Any ligand with a dissociation constant in the range of 10^{-3} to 10^{-10} M^{-1} (the range comprised of lectins and antibodies) should be applicable.[61] With the purpose of rendering this primary interaction visible, a connection which may have an amplifying function between the ligand and an imaging principle is established. The choice of system depends on several considerations: the primary ligand, sensitivity and time consumption, acquaintance with the different procedures, and the availability, durability, and costs of the reagents in addition to safety considerations.

All incubations with specific reagents must be performed in the presence of the blocking agent and separated by washing steps. Interaction of antibodies with antigens constitute the detection principle in classical immunoblotting. The dilutions and incubation times required depend on the antigen-antibody system in question, and on the characteristics of the visualization system employed. The extent of binding between antibody and antigen also depends on the possible renaturation of the antigen taking place after blotting, but the interplay between antibody concentration, avidity, and shaking and incubation time seem to be more important.[43,54] It is useful to distinguish between detection with polyclonal antibody preparations (Chapter 6.3.1) and monoclonal antibody detection (Chapter 6.3.2). The former is a mixture of antibodies of variable affinity recognizing different epitopes on the immunogenic protein, of which some nearly always are preserved even in a partially SDS-denatured molecule. With polyspecific antibodies, however, several bands may appear after staining. The significance of such extra bands is difficult to evaluate. Some specificity tests are mentioned in Table 1 of Chapter 7. Absorbing the antibody preparation by various procedures should make it possible to obtain a monospecific reaction (see Table 2 of Chapter 7). Unspecific interactions may be counteracted by changing the incubation conditions, allowing only antibodies of high affinity to be bound. Figure 6 shows that an increase of pH during incubation with the primary antibody counteracts unspecific bands (see also Chapter 7).[37]

The monoclonal antibody is dependent on a single epitope which may have been denaturated by SDS and not restored during the transfer step.[62] In fact, a characterization of monoclonal antibodies with respect to their specificity toward conformational determinants can be achieved by examining their reactivity in dot immunobinding toward SDS-treated

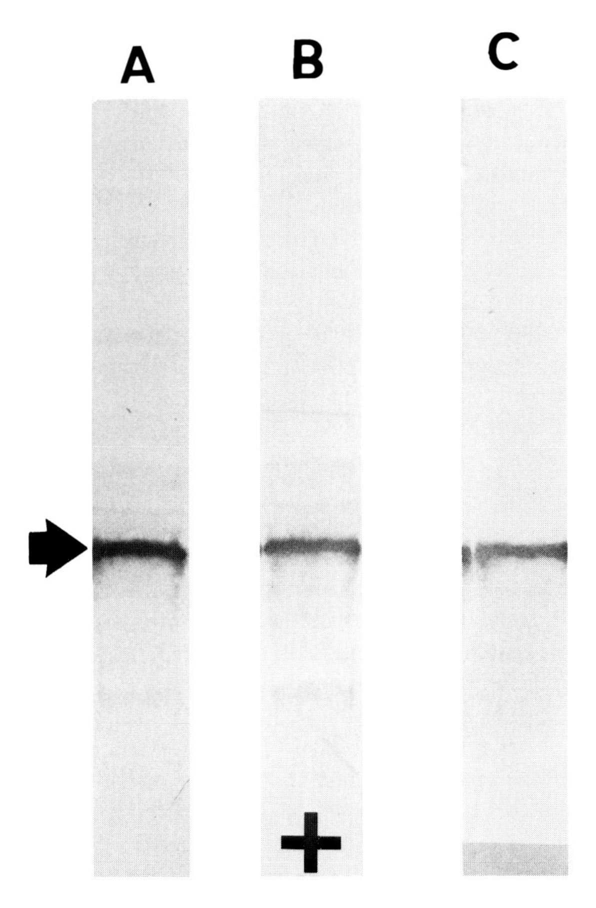

FIGURE 6. Depression of unspecific binding in immunoblotting of human transferrin by incubating the primary antitransferrin antibody at an elevated pH. A, pH 7.2; B, pH 8.7; and C, pH 10.2. Arrow points to transferrin. Experimental: SDS-PAGE of 70 μg thiol-reduced human serum proteins was performed in a 5% (w/v) stacking gel combined with a 9—15 % (w/v) linear separation gel. Both were crosslinked with 0.5% (w/v) bisacrylamide. After blotting to nitrocellulose (pore size: 0.45 μm) in a destaining apparatus (see Figures 4B and F) for 16 hr at 36 V the lanes were blocked for 5 min in 2% (v/v) Tween® 20, pH 10.2. Lanes A—C were cut out and incubated for 18 hr at 22°C on a rocking table with 10 mℓ of the following buffers as indicated: (A) 0.116 *M* phosphate, pH 7.4, (B) 0.1 *M* glycine, 0.038 *M* Tris, pH 8.6, and (C) 0.154 *M* NaCl, 0.05 *M* Tris/HCl, pH 10.2 each containing 0.05% (v/v) Tween® 20 and 1.5 μℓ rabbit antitransferrin antibodies (Dakopatts, Copenhagen). After 5 washings in the respective buffers for 10 min, the blots were incubated with peroxidase-conjugated swine antirabbit IgG antibodies (Dakopatts) at 1 μℓ/mℓ in buffer (C). After additional 5 washings in buffer (C) the lanes were stained with 3-amino-9-ethyl carbazole. [35]

proteins.[63] Depending on the antigen, 20 to 80% reactivity can be expected.[64] Rather sensitive detection systems are often required for immunoblotting with monoclonal antibodies due to a low affinity or scarcely distributed epitopes.[62]

Staphylococcal protein A is rarely used for the first layer detection but often as the second layer binding to antibodies of the primary layer.[34] It must always be taken into consideration that the affinity of protein A depends on the species origin and the isotype of the antibody.[65,66]

Detection of glycoproteins on blots by means of lectins (lectin blotting, see Chapter 8.6) is accomplished with antilectin antibodies coupled with a visualization system,[67] or by means of directly labeled lectins.[68] A general detection principle comparable with the immunochemical detection of derivatized proteins has also been worked out for sialic acid containing glycoproteins. Here, the blots are subjected to periodate oxidation and reductive phenylamination and reacted with labeled wheat germ agglutinin.[68] Lectin blotting often requires buffers of special composition and modified blocking procedures.[69,70]

Strictly speaking, the term "ligand blotting" covers all blotting methods, but currently it is used for blotting techniques involving specific recognition by means of other ligands than antibodies, antigens, protein A, or lectins (see examples in Table 3 or Chapter 8.5). An advantage is that there is no need for an antireceptor antibody provided the ligand is obtainable, but a prerequisite is that the transfer step (e.g., after an SDS-PAGE) and the subsequent immobilization allow a recovery of the original binding activity.[61,71] Therefore, better conditions for renaturation[72] or separation in nondenaturing gels (native blots),[73] might be considered. A special type of ligand blotting is cell blotting where interactions located on the cell surface are studied (Chapter 8.7).[74]

Among the many possible linkage systems (Table 5), secondary antibodies labeled with an enzyme or with an isotope are most commonly used. Enzyme and isotopic labeling are employed about equally often, while fluorescent labeling is only used in about 3% of published blotting experiments.[75] Direct covalent conjugation of the visualization system with the primary detecting ligand is rarely used, because it requires a purified first layer reactant which should be specifically labeled, and because the sensitivity is lower than the indirect techniques.

Attachment through biotin-avidin-peroxidase coupling,[76] peroxidase-antiperoxidase (PAP) systems (both in Chapter 6.4.1) as known from immunohistochemical procedures, or through lectin-avidin[77] renders detection very sensitive. Problems with nonspecific binding of avidin to protein might be encountered,[78] and the PAP method is rather complicated (two antibody layers between the detecting primary antibody and the PAP complex increase the risk of unspecific binding).[79] The high sensitivity of these multilayer techniques is a result of binding of several antibodies to one primary antibody with a further increase in the following layers. In the PAP technique, where secondary antibodies are in excess, the existence of free antigen binding sites is exploited. This is also the background for special modifications where inactivated enzyme antigens are detected on blots by means of an excess of anti-enzyme antibodies capable of catching with their unoccupied binding sites the native form of the enzyme introduced in a second layer incubation. Binding is then detected by assaying for enzyme activity (see Chapter 6.6).[51,79] The same principle is used in the double antigen detection techniques.[80] Other special modifications are mentioned in Chapter 10.

The extent to which a blot can be reused (reprobed) (Chapter 6.5) depends on the type of membrane and types of ligands involved. Generally all blotting membranes can be subjected to reprobing by applying much the same desorption conditions as used in affinity chromatography procedures,[54,81] i.e., dissociation by lowering the pH, [82] incubation with chaotropic agents such as urea([29]) or ammonium thiocyanate,[36,83] or with detergents (SDS).[84] Elution might also constitute a simple way of purifying small amounts of monospecific antibodies from a crude antiserum preparation (see Chapter 8.8).[85,86] Bound lectins can be removed by incubations with the appropiate sugar in presence of nonionic detergent.[68]

For preparative purposes, the transferred proteins themselves can be recovered from the blotting matrix. Such a preparative recovery from nitrocellulose can be achieved by dissolving the blot in acetone.[87] It is also possible to elute the proteins by treatment with solvents such as 50% pyridine, 40% acetonitrile in 0.1 *M* ammonium acetate, or with detergent mixtures. It should be noted, however, that small proteins are most easily eluted.[88]

The visualization systems of importance are the enzymatic, radioactive, and the colloidal metal/silver development methods.[89,90] The visualization methods are treated in five chapters (6.4.1 to 6.4.5). Sensitivity, safety, speed, simplicity, expense, availability, and durability of reagents and results are the main features distinguishing the methods from each other. The detection sensitivity of a visualization method is, of course, dependent on the previous detecting layers employed, but for the visualization per se it has been found that a radioactive label (^{125}I) and autoradiography was about twice as sensitive as a peroxidase-based imaging with 4-chloro-2-naphthol as a substrate, but then the background staining of the former was more pronounced.[91] In absolute figures the detection limit for a peroxidase-coupled antiimmunoglobulin can be as little as 0.1 ng immunoglobulin applied as a spot.[20] The detection limits for dot immunobinding experiments are not necessarily valid for other blotting experiments, but for evaluation of different detection systems it constitutes a rapid and simple test.

The enzyme-based systems are preferred by many as being fast, easy, and safe methods, and especially when combined with multilayer techniques as mentioned above, sensitive (50 to 100 times that of the techniques with one secondary layer).[91] The substrates *o*-dianisidine and 3,3-diaminobenzidine for peroxidase-dependent reactions give a higher background staining than 4-chloro-2-naphthol.[92] The latter does not work in the presence of Tween® 20 (see Chapter 7). Here, 3-amino-9-ethylcarbazole is convenient, but it gives less contrast and has a pronounced tendency to fade.

Enzyme detection with alkaline phosphatase is treated in Chapter 6.4.2. A useful substrate is 5-bromo-4-chloro-3-indolyl phosphate, which after reacting with Nitroblue tetrazolium, gives a blue-colored, nonfading deposit.[93] The procedure is about equally sensitive as peroxidase visualization. Consult the Appendix for details.

The autoradiographic methods imply handling and disposal problems as well. With ^{125}I they are very sensitive but time consuming. Also weak β-emitters such as ^{14}C and ^{3}H can be used, in which case the film detection can be enhanced by fluorographic measures.[94]

An attractive visualization design is offered by metal particles conjugated with an antibody (immunogold method, Chapter 6.4.4)[89] or protein A.[90] The procedure is safe, the color is inherent, and not dependent on some additional imaging step with chromogenic substrates or with autoradiographic procedures. The rather expensive immunogold method works on nitrocellulose and nylon membranes provided gelatin is included in the buffers. It gives a dark red staining of bands, equals the sensitivity of indirect immunoperoxidase methods, and the sensitivity is further increased about 5 to 10 times with a photochemical silver staining.[89,90]

Conclusion

A lot of specific detection approaches for blotted proteins exist. They must be adapted with respect to the antigen-antibody system investigated and to the sensitivity and specificity requirements. In each case, proper control experiments must be considered. Techniques working with additional layers are more sensitive. The enzyme-based visualization systems seem to be the most generally suitable, while the autoradiographic systems might offer better sensitivities. The colloidal gold conjugation/silver development methods are simple and quick but await further evaluation before their applicability can be delineated.

CONCLUDING REMARKS

This introductory survey of principles, potentialities, and procedures of the immunoblotting technique was intended to be a support for the absorbed reading of the following chapters where the details of the various technical steps and concepts of various applications are uncovered.

In the Appendix to this chapter, step-by-step descriptions of our standard electroblotting procedures are given as a reference point for laboratory work and for the following chapters.

ACKNOWLEDGMENTS

The financial support from Weimann's Foundation (NHHH), the Danish Medical Research Council, grant no. 12-5257 (OJB), and Harboefonden (NHHH, OJB) is greatly appreciated. Dr. S. U. Friis, Dr. Peter M. H. Heegaard, and Mr. Glenn R. Williams have read the text and helped with valuable criticism and suggestions. We thank Mrs. Inga Vaarst Andersen and Mr. Kurt Pii Larsen for expert technical help and Mr. Bjarne Eriksen for the drawings.

APPENDIX: A PRACTICAL DESCRIPTION OF TWO BASIC VERSIONS OF IMMUNOBLOTTING

I. INTRODUCTION

The immunoblotting technique consists of several individual steps. Many small modifications automatically develop in a laboratory working seriously with the procedure. Thus, each research team has its own modifications and this, in fact, is the background for the appearance of this book.

In the following we describe in practical terms our current routine immunoblotting techniques, which have developed as a synthesis of elements from different approaches. It must be emphasized that they are not necessarily better than the other techniques described in this book but mentioned here as reference procedures for the following chapters and to give detailed descriptions of all steps not expected to be found in the chapters on special modifications.

Two approaches which differ with respect to the transfer steps, viz., the electroblotting in a buffer tank (Section II) and the semidry electroblotting (Section III) are described. Also, two sets of staining techniques based on peroxidase and alkaline phosphatase activity and on Amido Black and gold staining are mentioned. The techniques are simple and robust. They have been applied for many types of antigens such as serum proteins, erythrocyte membrane proteins, platelet proteins, endothelial cell proteins, spermatozoal proteins, adipocyte membrane proteins, and proteins from *Escherichia coli*. Antibodies from various sources, for example, from rabbits, mice, swine, and humans of both poly- and monoclonal origin, have also been employed.

II. ELECTROBLOTTING IN A BUFFER TANK

A. Equipment and Reagents

1. Apparatus

Trans-blot cell and power supply 250/2.5 from Biorad, Richmond, California (see Figures 4C and G). Bath (30 × 25 × 5 cm) for assembly of sandwich. Incubation tray with 10 ditches of 10 × 15 × 100 mm suitable for 10 mℓ solutions, rocking table (IKA-VIBRAX-VXR, Janke & Kunkel, West Germany) plastic boxes, glass plate, scalpel, ruler, gloves, and tweezers (see Figure 5).

2. Materials

Packing material (rubber foam sheets), heavily perforated plexiglass grids (14 × 14 cm, perforations with a diameter of approximately 6 mm corresponding to a perforated area of 25 cm^2), clamps, or rubber bands. Nitrocellulose membrane, GSWP 293 25, pore size 0.22 μm, Millipore, Bedford, Mass. Magic Tape® 810, 3M Company.

3. Chemicals

Acetone, Amido Black 10B, glacial acetic acid, glycine, hydrogen peroxide (30%), methanol, sodium acetate, sodium azide, sodium disulfite, and Tween® 20 were obtained from Merck, Schuchardt, West Germany. 3-Amino-9-ethyl carbazole and Tris 7-9 (Tris(hydroxymethyl)aminomethane), from Sigma, St. Louis.

4. Antibodies

Primary antibody examples: rabbit, mouse, or human antibody. Secondary antibodies: Horseradish peroxidase-conjugated swine antirabbit immunoglobulins, rabbit antimouse IgG, rabbit antihuman IgG, IgM, or IgA were delivered by Dakopatts, Glostrup, Denmark.

5. Solutions

Transfer buffer (pH 8.4)	
25 m*M* Tris	9.1 g
192 m*M* glycine	43.2 g
20% (v/v) methanol	600mℓ
Distilled water	ad 3000mℓ
Washing and incubation buffers	
Washing buffer (pH 10.2):	
50 m*M* Tris	30.3 g
150 m*M* sodium chloride	43.8 g
5 m*M* sodium azide	1.6 g
Distilled water	ad 5000mℓ
Incubation buffer (pH 10.2):	
0.05% (w/v) Tween® 20	0.5 g
Washing buffer	ad 1000mℓ
Blocking buffer (pH 10.2):	
2% (w/v) Tween® 20	20.0 g
Washing buffer	ad 1000mℓ
Staining solutions	
Amido Black solution:	
0.5% (w/v) Amido Black 10B	5.0 g
Distilled water	450 mℓ
Methanol	450 mℓ
Glacial acetic acid	100 mℓ
Amido Black destaining solution:	
Same as above without Amido Black	
Peroxidase staining buffer (pH 5.5):	
50 m*M* sodium acetate	6.8 g
Glacial acetic acid	1.8 mℓ
Distilled water	ad 1000mℓ
Carbazole stock solution:	
1% (w/v)	
3-Amino-9-ethyl carbazole	500 mg
Acetone	ad 50mℓ
Store at 4°C in an amber bottle.	
Staining solution (mix just before use):	
Carbazole stock solution	2.0 mℓ
Peroxidase staining buffer	50.0 mℓ
Hydrogen peroxide, 30% (w/v)	25 μℓ

No filtering is necessary in spite of some precipitation.

B. Procedures

1. Gel Electrophoresis

In principle, any gel and buffer system can be used. Pyronin G (final concentration: 0.04% (w/v)) is electrophoresed together with the sample. The position of the top of the lane is marked by applying 5 μℓ Pyronin G solution (0.02% (w/v), 10% (w/v) sucrose, 1% (v/v) SDS) in each well just before the electrophoresis is terminated. Wear gloves when nitrocellulose, filter paper, gels, and staining solutions are handled. All procedures take place at room temperature.

2. Assembly of Blotting Cell

The cell (also called "the sandwich") is assembled in a large bath containing transfer buffer. It consists of: (1) plexiglass grid, (2) three layers of foam rubber sheets, (3) gel slab, (4) nitrocellulose sheet trimmed to the size of the gel slab (air bubbles between the gel and the nitrocellulose membrane must be avoided; the wet nitrocellulose can be marked with a soft pencil), (5) three layers of foam rubber sheets, and plexiglass grid. The sandwich is clamped together with rubber bands.

3. Transfer

The transfer is performed at a constant voltage of 6 to 8 V/cm (distance measured between electrodes) for 3 to 18 hr. Stirring and cooling is recommended. If the blot is going to be stained specifically, transfer is followed by blocking and washing and forward. All these procedures take place on a rocking table if not otherwise mentioned. Routinely the SDS-PAGE gel should be stained to control the transfer.

4. Amidoblack Staining

For control of transfer and staining of molecular weight markers: immerse in Amido Black staining solution for 10 min. Remove excess of stain in water. Destain until a light blue background is obtained. Do not block in Tween® 20 before this staining. It becomes lighter upon drying.

5. Blocking and Washing

Immerse in a plastic box with blocking buffer for 2 to 5 min. Incubations for longer periods should be avoided as they may displace a fraction of the primary bound antigen. Now the individual lanes can be out of the nitrocellulose sheet with a ruler and a scalpel. When Pyronin G marker stain is co-transferred from the separation gel, it is possible to produce strips corresponding to the lanes of the separation gel. The strips are transferred to the incubation tray and washed for 10 min in incubation buffer.

6. Incubation with Primary Antibody

It is advisable to use diluted antibodies with concomitant long incubation times which result in lower background staining. Overnight incubations fit into the time schedule. Unspecific antibody binding is also reduced by means of the high pH of the incubation solution. If primary antibodies of low avidity are employed, the incubation can take place with a buffer of a neutral pH; for monoclonal antibodies the pH must always be 7.4. The antibody concentration should be optimized by performing titration experiments as it is possible to dilute away additional antibody specificities and/or unspecific binding. The load of antigen can also be titrated. As a rule-of-thumb, a load of about 10% of the load giving a good Coomassie brilliant blue-stained pattern on the PAGE gel will be suitable for immunoblotting.

7. Washings

3 × 10 min with incubation buffer.

8. Incubation with Secondary Antibody

The peroxidase-conjugated anti-antibody preparation is diluted 1:1000 in incubation buffer. Incubate for 2 hr. The incubation time can be decreased by a proportional increase in the antibody concentration.

9. Washings

3 × 10 min with incubation buffer.

10. Staining for Peroxidase Activity

The staining reagents are mixed immediately before use. The incubations do not necessarily take place on a rocking table. The staining is complete within 5 min. In case of faint bands the procedure can be repeated. Finally, the blots are washed in distilled water for 10 min.

11. Conservation

The fading process is delayed if the blots are incubated for 5 min in 50 m*M* sodium disulfite. They are then dried and finally the strips can be mounted and protected with a piece of Magic Tape®. Store protected from light. As the color might fade within days the wet blots should be photographed the same day as the processing.

The schedule for the electrophoresis and the immunoblotting procedure occupies 4 days, for example as follows: Day 1, casting of polyacrylamide gel; Day 2, electrophoresis, blotting; Day 3, incubation with primary antibody; Day 4, incubation with secondary antibodies, staining, drying, and mounting.

III. SEMIDRY ELECTROBLOTTING

A. Equipment and Reagents

1. Apparatus

Semidry blotting apparatus from JKA Biotech, Brønshøj; Ancos, Lstykke, or Kem-En-Tec, Hellerup, Denmark with corresponding power supply, giving a constant power and equipped with timer (1 hr) (see Figures 4D and H).

2. Materials

Filter paper, Whatman no. 1 cut into pieces 140 × 140 mm. Dialysis membrane, PT 300, 140 × 140 mm, Dansk Emballage Industri, Søborg, Denmark. Otherwise as described in Section II.

3. Chemicals

Tetrachlorogold (III) from Degussa AG Hanau, West Germany; sodium dodecylsulfate from Serva, Heidelberg, West Germany; citric acid, ethanolamine, magnesium chloride, and trisodium citrate from Merck; and 6-amino-*n*-hexanoic acid purchased from BDH, Poole, England. Nitroblue tetrazolium (N-6876), and 5-bromo-4-chloro-3-indolyl phosphate (B-8503) from Sigma. Otherwise as in Section II.

4. Antibodies

As for Section II, but the secondary antibodies are alkaline phosphatase conjugated. Also delivered by Dakopatts.

5. Solutions

Transfer buffer[39]

48 m*M* Tris	5.83 g
39 m*M* glycine	4.15 g
1.3 m*M* SDS	0.38 g
20% (v/v) methanol	200 mℓ
Distilled water	ad 1000 mℓ

Washing and incubation buffers

As in Section II.

Staining solutions

Gold stock solution (30 nm particles):

0.01% (w/v) $HAuCl_4$	25 mg
Distilled water	ad 250 mℓ
When boiling add trisodium citrate (10 mg/mℓ)	7.5 mℓ

Boil with reflux for 30 min, the color of the solution changes in approximately 10 min from weak yellow over dark blue to wine red. Avoid protein contaminations which consume the gold. In such cases the solution stays blue and must be discharged. Store in an amber bottle at 4°C.

Citrate buffer (pH 3.0):

50 m*M* citric acid	1.05 g
Adjust to pH 3.0 with NaOH	
Distilled water	ad 100 mℓ

Gold staining solution:

Gold stock solution	75 mℓ

Citrate buffer (pH 3.0)	25 mℓ
0.1% (v/v) Tween® 20	100 μℓ
Ethanolamine buffer (pH 9.6):	
0.1 *M* ethanolamine buffer	
Titrated with 4 *M* HCl to pH 9.6	6.3 mℓ
Distilled water	ad 1000 mℓ
Nitroblue tetrazolium stock solution:	
Nitroblue tetrazolium	50.0 mg
Ethanolamine buffer	ad 50.0 mℓ
Store at 4°C	
Substrate stock solution:	
66% (v/v) methanol	8.0 mℓ
33% (v/v) acetone	4.0 mℓ
5-Bromo-4-chloro-3-indolyl phosphate	48.0 mg
Store at −20°C	
Magnesium chloride solution:	
1 *M* $MgCl_2$	0.95 g
Distilled water	ad 10 mℓ

B. Procedures

1. Gel Electrophoresis (See Section II).

2. Assembly of Blotting Cell

The technique is described in detail in Chapter 4.3.2. The version given here is simplified so that the transfer buffer is a 1/40 dilution of the anodic buffer of Laemmli.[95] Briefly, filter papers and nitrocellulose sheets are trimmed to the size of the separation gel. Air bubbles in and between the various layers must be avoided, for example, by rolling a glass rod several times over the paper when immersed in buffer. The upper graphite plate of the blotting cell is the cathode, the lower the anode (see Figure 3).[14] The electrodes are rinsed in distilled water before use and the following layers are built up.

1. Six layers of filter paper soaked in transfer buffer (drip off well)
2. Nitrocellulose sheet wetted in distilled water
3. Separation gel
4. Six layers of filter paper soaked in transfer buffer

The cover containing the cathodic graphite plate finishes the stacking.

3. Transfer

A constant current density of 0.8 mA/cm² of the gel (i.e., 200 mA for a 14 × 14 cm gel) is applied for 1 hr. Then the cell is disconnected, dismounted, and the nitrocellulose processed as described below.

4. Gold Staining[50]

The nitrocellulose is blocked in 2% (v/v) Tween® 20 (pH 10.2) for 2 to 5 min, washed in 0.05% (v/v) Tween® 20, and rinsed for 3 min in distilled water. After one wash in citrate/NaOH (pH 3.0), the sheet is incubated for at least 30 min in this buffer. It is important that the pH throughout the paper is 3.0. The nitrocellulose is then incubated on a rocking table at room temperature with the gold solution (0.2 mℓ/cm²) for 4 to 16 hr.

5 to 7. As for Section II.

8. Incubation with Secondary Antibody

The phosphatase-conjugated anti-antibody preparation is diluted 1:2000 in incubation buffer. Incubate for 2 hr.

9. *Washings*
   As in Electroblotting Section.

10. *Staining for Phosphatase Activity*[89]
   Not necessarily on a rocking table. The staining reagent is freshly prepared:

Nitroblue tetrazolium stock solution	5.0 mℓ
Substrate stock solution	0.75 mℓ
1 *M* $MgCl_2$	0.10 mℓ
Ethanolamine buffer	ad 50.0 mℓ

Normally, bands are visible after about 5 to 15 min. The incubation should be carried on a little further though, because there is a slight tendency to fade. Prolonged incubation is possible, but then the background will also be stained. The reaction is terminated by washing the blots in distilled water. After drying they can be mounted. Only very little fading upon storage is observed with this stain.

The schedule for this procedure is shorter than that outlined in Section II and may be as follows: Day 1, casting of polyacrylamide gel; Day 2, electrophoresis, blotting, incubation with primary antibody; Day 3, incubation with secondary antibody, staining, drying, and mounting.

REFERENCES

1. **Southern, E. M.,** Detection of specific sequences among DNA fragments separated by gel electrophoresis, *J. Mol. Biol.,* 98, 503, 1975.
2. **Alwine, J. C., Kemp, D. J., Parker, B. A., Reiser, J., Renart, J., Stark, G. R., and Wahl, G. M.,** Detection of specific RNAs or specific fragments of DNA by fractionation in gels and transfer to diazobenzyloxymethyl paper, *Methods Enzymol.,* 68, 220, 1979.
3. **Renart, J., Reiser, J., and Stark, G. R.,** Transfer of proteins from gels to diazobenzyloxymethyl-paper and detection with antisera: a method for studying antibody specificity and antigen structure, *Proc. Natl. Acad. Sci. U.S.A.,* 76, 3116, 1979.
4. **Towbin, H., Staehelin, T., and Gordon, J.,** Electrophoretic transfer of proteins from polyacrylamide gels to nitrocellulose sheets: procedure and some applications, *Proc. Natl. Acad. Sci. U.S.A.,* 76, 4350, 1979.
5. **Axelsen, N. H., Ed.,** Handbook of immunoprecipitation-in-gel techniques, *Scand. J. Immunol.,* 17 (Suppl. 10), Blackwell Scientific, Oxford, 1983.
6. **Bjerrum, O. J. Ed.,** *Electroimmunochemical Analysis of Membrane Proteins,* Elsevier, Amsterdam, 1983.
7. **Heegaard, P. M. H. and Bøg-Hansen, T. C.,** Immunoelectrophoretic methods, in *Analytical Gel Electrophoresis of Proteins,* Dunn, M. J., Ed., Adam Hilger Ltd., London, 1985.
8. **Palmiter, R. D., Palacios, R., and Schimke, R. T.,** Identification and isolation of ovalbumin-synthesizing polysomes, *J. Biol. Chem.,* 247, 3296, 1972.
9. **Bjerrum, O. J., Bhakdi, S., Bøg-Hansen, T. C., Knufermann, H., and Wallach, D. F. H.,** Quantitative immunoelectrophoresis of proteins in human erythrocyte membranes. Analysis of protein bands obtained by sodium dodecylsulfate polyacrylamide gel electrophoresis, *Biochim. Biophys. Acta,* 406, 489, 1975.
10. **Norrild, B., Bjerrum, O. J., and Vestergaard, B. F.,** Polypeptide analysis of individual immunoprecipitates from crossed immunoelectrophoresis, *Anal. Biochem.,* 81, 432, 1977.
11. **Burridge, K.,** Changes in cellular glycoproteins after transformation: identification of specific glycoproteins and antigens in sodium dodecyl sulphate gels, *Proc. Natl. Acad. Sci. U.S.A.,* 79, 4457, 1978.
12. **Showe, M. K., Isobe, E., and Onorato, L.,** Bacteriophage T4 prehead proteinase. II. Its cleavage from the product of gene *21* and regulation in phage-infected cells, *J. Mol. Biol.,* 107, 55, 1976.
13. **Beisiegel, U.,** Protein blotting, *Electrophoresis,* 7, 1, 1986.
14. **Gershoni, J. M.,** Protein blotting: developments and perspectives, *TIBS,* 10, 103, 1985.

15. **Converse, C. A. and Papermaster, D. S.,** Membrane protein analysis by two-dimensional immunophoresis, *Science,* 189, 469, 1986.
16. **Kyhse-Andersen, J.,** A simple horizontal apparatus without buffer tank for electrophoretic transfer of proteins from polyacrylamide to nitrocellulose, *J. Biochem. Biophys. Methods,* 10, 203, 1984.
17. **Towbin, H., Schoenenberger, C., Ball, R., Braun, D. G., and Rosenfelder, G.,** Glycosphingolipid-blotting: an immunological detection procedure after separation by thin-layer chromatography, *J. Immunol. Methods,* 72, 471, 1984.
18. **Tsang, V. C. W., Peralta, J. M., and Simons, A. R.,** Enzyme-linked immunoelectrotransfer blot techniques (EITB) for studying the specificities of antigens and antibodies separated by gel electrophoresis, *Methods Enzymol.,* 92, 377, 1983.
19. **Peferoen, M., Huybrechts, R., and de Loof, A.,** Vacuum-blotting: a new simple and efficient transfer of proteins from sodium dodecyl sulfate-polyacrylamide gels to nitrocellulose, *FEBS Lett.,* 145, 369, 1982.
20. **Hawkes, R., Niday, E., and Gordon, J.,** A dot-immunobinding assay for monoclonal and other antibodies, *Anal. Biochem.,* 119, 142, 1982.
21. **Yen, T. S. B. and Webster, R. E.,** Translational control of bacteriophage f1 gene II and gene X proteins by gene V protein, *Cell,* 29, 337, 1982.
22. **Huet, J., Sentenac, A., and Fromageot, P.,** Spot-immunodetection of conserved determinants in eukaryotic RNA polymerases, *J. Biol. Chem.,* 257, 2613, 1982.
23. **McDougal, J. S., Browning, S. W., Kennedy, S., and Moore, D. D.,** Immunodot assay for determining the isotype and light chain type of murine monoclonal antibodies in unconcentrated hybridoma culture supernates, *J. Immunol. Methods,* 63, 281, 1983.
24. **Herbrink, P., van Bussel, F. J., and Warnar, S. O.,** The antigen spot test (AST): a highly sensitive assay for the detection of antibodies, *J. Immunol. Methods,* 48, 293, 1982.
25. **Erlich, H. A., Levinson, J. R., Cohen, S. N., and McDevitt, H. O.,** Filter affinity transfer — a new technique for the *in situ* identification of proteins in gels, *J. Biol. Chem.,* 254, 12240, 1979.
26. **Ahmed, F., Moore, M. A., and Dunlap, R. B.,** A nitrocellulose-filter assay for the binary complex of 5-fluorodeoxyuridylate and *Lactobacillus casei* thymidylate synthetase, *Anal. Biochem.,* 145, 151, 1985.
27. **Gibson, W..,** Protease-facilitated transfer of high-molecular-weight proteins during electrotransfer to nitrocellulose, *Anal. Biochem.,* 118, 1, 1981.
28. **Sutton, R., Wrigley, C. W., and Baldo, B. A.,** Detection of IgE- and IgG-binding proteins after electrophoretic transfer from polyacrylamide gels, *J. Immunol. Methods,* 52, 183, 1982.
29. **Erickson, P. F., Minier, L. N., and Lasher, R. S.,** Quantitative electrophoretic transfer of polypeptides from SDS polyacrylamide gels to nitrocellulose sheets: a method for their re-use in immunoautoradiographic detection of antigens, *J. Immunol. Methods,* 51, 241, 1982.
30. **Bhakdi, S., Dieter, J., and Hugo, F.,** Electroimmunoassay-immunoblotting (EIA-IB) for the utilization of monoclonal antibodies in quantitative immunoelectrophoresis: the method and its applications, *J. Immunol. Methods,* 80, 25, 1985.
31. **Lin, W. and Kasamatsu, H.,** On the electrotransfer of polypeptides from gels to nitrocellulose membranes, *Anal. Biochem.,* 128, 302, 1983.
32. **Gershoni, J. M., Davis, F. E., and Palade, G. E.,** Protein blotting in uniform or gradient electric fields, *Anal. Biochem.,* 144, 32, 1985.
33. **Gershoni, J. M. and Palade, G. E.,** Electrophoretic transfer of proteins from sodium dodecyl sulfate polyacrylamide gels to a positively charged membrane filter, *Anal. Biochem.,* 124, 396, 1982.
34. **Burnette, W. N.,** "Western blotting": electrophoretic transfer of proteins from sodium dodecyl sulfate-polyacrylamide gels to unmodified nitrocellulose and radiographic detection with antibody and radioiodinated protein A, *Anal. Biochem.,* 112, 195, 1981.
35. **Bittner, M., Kupferer, P., and Morris, C. F.,** Electrophoretic transfer of proteins and nucleic acids from slab gels to diazobenzyloxymethyl cellulose or nitrocellulose sheets, *Anal. Biochem.,* 102, 459, 1980.
36. **Reiser, J. and Wardale, J.,** Immunological detection of specific proteins in total cell extracts by fractionation in gels and transfer to diazophenylthioether paper, *Eur. J. Biochem.,* 114, 569, 1981.
37. **Naaby-Hansen, S. and Bjerrum, O. J.,** Auto- and isoantigens of the human spermatozoa detected by immunoblotting with human sera after SDS-PAGE, *J. Reprod. Immunol.,* 7, 41, 1985.
38. **Svoboda, M., Meuris, S., Robyn, C., and Christophe, J.,** Rapid electrotransfer of proteins from polyacrylamide gels to nitrocellulose membrane using surface-conductive glass as anode, *Anal. Biochem.,* 151, 16, 1985.
39. **Bjerrum, O. J. and Schafer-Nielsen, C.,** Buffer systems and transfer parameters for semidry electroblotting with a horizontal apparatus, in *Electrophoresis '86,* Dunn, M. J. Ed., VCH Publishers, Weinheim, 1986, 315.
40. **Kuno, H. and Kihara, H. K.,** Simple microassay of protein with membrane filter, *Nature (London),* 215, 974, 1967.
41. **NEN's Application Laboratory,** Electrophoretic transfer and detection of proteins using genescreen, *NEN Res TIPS,* 1M783P-2207, 1984.

42. **Winberg, G. and Hammarskjold, M. L.,** Isolation of DNA from agarose gels using DEAE-paper. Application to restriction site mapping of adenovirus type 16 DNA, *Nucleic Acids Res.,* 8, 253, 1980.
43. **Bjerrum, O. J., Larsen, K. P., and Wilken, M.,** Some recent developments of the electroimmunochemical analysis of membrane proteins. Application of Zwittergent, Triton X-114 and Western blotting technique, in *Modern Methods in Protein Chemistry,* Tschesche, H., Ed., Walter de Gruyter, Berlin, 1983, 79.
44. **Palfree, R. G. E. and Elliott, B. E.,** An enzyme-linked immunosorbent assay (ELISA) for detergent solubilized Ia glycoproteins using nitrocellulose membrane discs, *J. Immunol. Methods,* 52, 395, 1982.
45. **Kakita, K., O'Conell, K., and Permutt, M. A.,** Immunodetection of insulin after transfer from gels to nitrocellulose filters. A method of analysis in tissue extracts, *Diabetes,* 31, 648, 1982.
46. **St. John, T. P. and Davis, R. W.,** Isolation of galactose-inducible DNA sequences from *Saccharomyces cerevisiae* by differential plaque filter hybridization, *Cell,* 16, 443, 1979.
47. **Gordon, J., Towbin, H., and Rosenthal, M.,** Antibodies against ribosomal protein determinants in the sera of patients with connective tissue diseases, *J. Rheumatol.,* 9, 247, 1982.
48. **Petit, C., Sauron, M. E., Gilbert, M., and Theze, J.,** Direct detection of idiotypic determinants on blotted monoclonal antibodies, *Ann. Immunol.,* 133, 77, 1982.
49. **Batteiger, B., Newhall, W. J., and Jones, R. B.,** The use of Tween 20 as a blocking agent in the immunological detection of proteins transferred to nitrocellulose membranes, *J. Immunol. Methods,* 55, 297, 1982.
50. **Bowen, B., Steinberg, J., Laemmli, U. K., and Weintraub, H.,** The detection of DNA-binding proteins by protein blotting, *Nucleic Acids Res.,* 8, 1, 1980.
51. **Muilerman, H. G., ter Hart, H. G., and Van Dijk, W.,** Specific detection of inactive enzyme protein after polyacrylamide gel electrophoresis by a new enzyme-immunoassay method using unspecific antiserum and partially purified active enzyme: application to rat liver phosphodiesterase I, *Anal. Biochem.,* 120, 46, 1982.
52. **Parchment, R. E., Ewing, C. M., and Shaper, J. H.,** The use of galactosyltransferase to probe nitrocellulose-immobilized glycoproteins for nonreducing terminal *N*-acetylglucosamine residues, *Anal. Biochem.,* 154, 460, 1986.
53. **Towbin, H., Ramjoue, H.-P., Huster, H., Liverani, D., and Gordon, J.,** Monoclonal antibodies against eukaryotic ribosomes, *J. Biol. Chem.,* 257, 12709, 1982.
54. **Towbin, H. and Gordon, J.,** Immunoblotting and dot immunobinding — current status and outlook, *J. Immunol. Methods,* 72, 313, 1984.
55. **Jackson, P. and Thompson, R. J.,** The immunodetection of brain proteins blotted onto nitrocellulose from fixed and stained two-dimensional polyacrylamide gels, *Electrophoresis,* 5, 35, 1984.
56. **Hancock, K. and Tsang, V. C. W.,** India ink staining of proteins on nitrocellulose paper, *Anal. Biochem.,* 133, 157, 1983.
57. **Moeremans, M., Daneels, G., and De Mey, J.,** Sensitive colloid metal (gold or silver) staining of protein blots on nitrocellulose membranes, *Anal. Biochem.,* 145, 315, 1985.
58. **Yen, K. C. L., Johnson, T. K., Denell, R. E., and Consigli, R. A.,** A silver-staining technique for detecting minute quantities of proteins on nitrocellulose paper: retention of antigenicity of stained proteins, *Anal. Biochem.* 126, 398, 1982.
59. **Wojtkowiak, Z., Briggs, R. C., and Hnilica, L. S.,** A sensitive method for staining proteins transferred to nitrocellulose sheets, *Anal. Biochem.,* 129, 486, 1983.
60. **Kittler, J. M., Meisler, N. T., Viceps-Madore, D., Cidlowski, J. A., and Thanassi, J. W.,** A general immunochemical method for detecting proteins on blots, *Anal. Biochem.,* 137, 210, 1984.
61. **Gershoni, J. M. and Palade, G. E.,** Review: protein blotting principles and applications, *Anal. Biochem.,* 131, 1, 1983.
62. **de Blas, A. L. and Cherwinski, H. M.,** Detection of antigens on nitrocellulose paper immunoblots with monoclonal antibodies, *Anal. Biochem.,* 133, 214, 1983.
63. **Vartio, T., Zardi, L., Balza, E., Towbin, H. and Vaheri, A.,** Monoclonal antibodies in analysis of cathespsin G-digested proteolytic fragments of human plasma fibronectin, *J. Immunol. Methods,* 55, 309, 1982.
64. **Bjerrum, O. J., Selmer, J., Larsen, F. S., and Naaby-Hansen, S.,** Exploitation of antibodies in the study of cell membranes, in *Investigation and Exploitation of Antibody Combining Sites,* Reid, E., Ed., Plenum Press, London, 1985, 231.
65. **Goudswaard, J., van der Donk, J. A., Noordzig, A., van Dam, R. H., and Vaerman, J. P.,** Protein A reactivity of various mammalian immunoglobulins, *Scand. J. Immunol.,* 8, 21, 1978.
66. **Richman, D. G., Cleveland, P. H., Oxman, M. N., and Johnson, K. M.,** The binding of staphylococcal protein A by the sera of different animal species, *J. Immunol.,* 128, 2300, 1982.
67. **Glass, W. F., Briggs, R. C., and Hnilica, L. S.,** Use of lectins for detection of electrophoretically separated glycoproteins transferred onto nitrocellulose sheets, *Anal. Biochem.,* 115, 219, 1981.
68. **Bartles, J. R. and Hubbard, A. L.,** 125-I Wheat germ agglutinin Blotting: Increased Sensitivity with polyvinylpyrrolidone quenching and periodate oxidation reductive phenylamination, *Anal. Biochem.,* 140, 284, 1984.

69. **Hawkes, R.,** Identification of concanavalin A-binding proteins after sodium dodecyl sulfate-gel electrophoresis and protein blotting, *Anal. Biochem.*, 123, 143, 1982.
70. **Clegg, J. C. S.,** Glycoprotein detection in nitrocellulose transfers of electrophoretically separated protein mixtures using concanavalin A and peroxidase: application to arena-virus and flavivirus proteins, *Anal. Biochem.*, 127, 389, 1982.
71. **Daniel, T. O., Schneider, W.J., Goldstein, J. L., and Brown, M. S.,** Visualization of lipoprotein receptors by ligand blotting, *J. Biol. Chem.*, 258, 4606, 1983.
72. **Flanagan, S. D. and Yost, B.,** Calmodulin-binding proteins: visualization by ^{125}I-calmodulin overlay on blots quenched with Tween 20 or bovine serum albumin and poly(ethylene oxide), *Anal. Biochem.*, 140, 510, 1984.
73. **Reinhardt, M. P. and Malamud, D.,** Protein transfer from isoelectric focusing gels: the native blot, *Anal. Biochem.*, 123, 229, 1982.
74. **Hayman, E. G., Engvall, E., A'Hearn, E., Barnes, D., Pierschbacher, M., and Ruoslahti, E.,** Cell attachment on replicas for SDS polyacrylamide gels reveals two adhesive plasma proteins, *J. Cell Biol.*, 95, 20, 1982.
75. **Beisiegel, U., Weber, W., and Utermann, G.,** Proteinblotting, in *Electrophoresis '84*, Neuhoff, V., Ed., Walter de Gruyter, Berlin, 1984, 49.
76. **Lanzillo, J. J., Stevens, J., Tumas, J., and Faburg, B. L.,** Avidin-biotin amplified immunoperoxidase staining of angiotensin-1-converting enzyme transferred to nitrocellulose after agarose isoelectric focusing, *Electrophoresis*, 4, 313, 1983.
77. **Rohringer, R. and Holden, D. W.,** Protein blotting: detection of proteins with colloidal gold, and of glycoproteins and lectins with biotin-conjugated and enzyme probes, *Anal. Biochem.*, 144, 118, 1985.
78. **Ogata, K., Arakawa, M., Kasahara, T., Shioiri-Nakanao, K., and Hiraoka, K.,** Detection of toxoplasma membrane antigens transferred from SDS-polyacrylamide gel to nitrocellulose with monoclonal antibody and avidin-biotin, peroxidase antiperoxidase and immunoperoxidase methods, *J. Immunol. Methods*, 65, 75, 1983.
79. **Van der Meer, J., Dorssers, L., and Zabel, P.,** Antibody-linked polymerase assay on protein blots: a novel method for identifying polymerases following SDS-polyacrylamide gel electrophoresis, *EMBO J.*, 2, 233, 1983.
80. **Larsson, L. -I.,** Simultaneous ultrastructural demonstration of multiple peptides in endocrine cells by a novel immunocytchemical method, *Nature (London)*, 282, 743, 1979.
81. **Geysen, J., de Loof, A., and Vandesande, F.,** How to perform subsequent or "double" immunostaining of two different antigens on a single nitrocellulose blot within one day with an immunoperoxidase technique, *Electrophoresis*, 5, 129, 1984.
82. **Legocki, R. P. and Verma, D. P. S.,** Multiple immunoreplica technique: screening for specific proteins with a series of different antibodies using one polyacrylamide gel, *Anal. Biochem.*, 111, 385, 1981.
83. **Anderson, N. L., Nance, S. L., Pearson, T. W., and Anderson, N. G.,** Specific antiserum staining of two-dimensional electrophoretic patterns of human plasma proteins immobilized on nitrocellulose, *Electrophoresis*, 3, 135, 1982.
84. **Symington, J. Green, M., and Brackmann, K.,** Immunoautoradiographic detection of proteins after electrophoretic transfer from gels to diazo-paper: analysis of adenovirus encoded proteins, *Proc. Natl. Acad. Sci. U.S.A.* 78, 177, 1981.
85. **Olmsted, J. B.,** Affinity purification of antibodies from diazotized paper blots of heterogeneous protein samples, *J. Biol. Chem.*, 256, 11955, 1981.
86. **D'Angelo Siciliano, J. and Craig, S. W.,** Meta-vinculin — a vinculin-related protein with solubility properties of a membrane protein, *Nature (London)*, 300, 533, 1982.
87. **Anderson, P. J.,** The recovery of nitrocellulose-bound protein, *Anal. Biochem.*, 148, 105, 1985.
88. **Parekh, B. S., Metha, H. B., West, M. D., and Montelaro, R. C.,** Preparative elution of proteins from nitrocellulose membranes after separation by sodium dodecyl sulfate-polyacrylamide gel electrophoresis, *Anal. Biochem.*, 148, 87, 1985.
89. **Moeremans, M., Daneels, G., Van Dijck, A., Langanger, G., and De May, J.,** Sensitive visualization of antigen-antibody reactions in dot and blot immuneoverlay assays with immunogold and immunogold/silver staining, *J. Immunol. Methods*, 74, 353, 1984.
90. **Brada, D. and Roth, J.,** "Golden blot" — Detection of polyclonal and monoclonal antibodies bound to antigens by protein A-gold complexes, *Anal. Biochem.*, 142, 79, 1984.
91. **Vissing, H. and Madsen, O. D.,** Comparison of detection limits for various nitrocellulose binding immunoassays using beta-2-microglobulin as a model antigen, *Electrophoresis*, 5, 313, 1984.
92. DNA, RNA and Protein Electrophoretic Blotting, No. 1080 EG, bio-rad Laboratories, 1981.
93. **Knecht, D. A. and Dimond, R. L.,** Visualization of antigenic proteins on western blots, *Anal. Biochem.*, 136, 180, 1984.
94. **Heegaard, N. H. H., Hebsgaard, K. P., and Bjerrum, O. J.,** Sodium salicylate for flurographical detection of immunoprecipitated proteins in agarose gels, *Electrophoresis*, 5, 263, 1984.

95. **Laemmli, U. K.,** Cleavage of structural proteins during the assembly of the head of bacteriophage T4, *Nature (London),* 227, 680, 1970.
96. **Blake, M. S., Johnston, K. H., Russell-Jones, G. J., and Gotschlich, E. C.,** A rapid, sensitive method for detection of alkaline phosphatase-conjugated anti-antibody on western blots, *Anal. Biochem.,* 136, 175, 1984.

Section 2

DOT IMMUNOBINDING — GENERAL PRINCIPLES AND PROCEDURES

Julian Gordon and Pat Billing

INTRODUCTION AND BASIC PRINCIPLE

A prerequisite of hybridoma methodology is the ability to screen large numbers of clones in order to select the desired specificity. Dot immunobinding was developed as a simplification of the original immunoblotting procedure to fulfill this need. In 1982, four publications appeared, apparently independently dealing with this procedure.[1-4] The methodology was anticipated by two prescient publications in the early 1960s.[5,6]

The dot immunobinding assay takes advantage of the same immunoassay protocols as immunoblotting, and of the extreme versatility of nitrocellulose as a solid phase for the binding of a wide variety of macromolecules. The substance is applied directly onto the nitrocellulose in a small volume. Since multiple dots can be placed on a strip, it permits multiple simultaneous assays on the same analyte. In the simplest form, an antigen is bound to the solid phase, incubated with unknown antibody, and then probed with a labeled anti-immunoglobulin.[1-4] More elaborate sandwich assays are possible, where the antibody is applied as a dot to the solid phase, incubated with unknown antigen, and probed with a second layer of antibody.[7]

The original assay provided a convenient method for screening monoclonal antibodies. It was also shown to be applicable to serological screening of clinical samples, and to be a potentially quantitative assay.[1] Subsequent developments have demonstrated the applicability as a spot test in which the analyte is spotted onto the nitrocellulose surface, and then probed with a labeled antibody.[8-12] Further developments of the technique were the additional use of the nitrocellulose as a filtration medium to concentrate virus from relatively large volumes,[13] and blotting of bacterial-secreted products from agar stab culture to nitrocellulose, followed by immunoprobing.[14]

Here, we described a generic procedure which has been optimized for multiantibody screening of human sera.

EQUIPMENT AND REAGENTS

Equipment

Automatic pipettor, dilutor (Microlab M, Hamilton, Bonaduz, Switzerland), 5 $\mu\ell$ Microdispenser (Drummond, Broomall, Pa.), or 5 $\mu\ell$ Transferpettor (Brand, Wertheim, West Germany); micropipettes (e.g., Gilson, Middleton, Wisc.,) Pipetman for >5 $\mu\ell$, and Nichiryo, Tokyo, with disposable plastic capillary tips for the range 1 to 5 $\mu\ell$); rotary or reciprocating horizontal shaker; multireservoir trays (Dynatec, VI) or transfer trays (Inotech, Wohlen, Switzerland); razor blades or scalpels and steel ruler, or rotary-blade paper cutter (Alvin, Windsor, Conn.); gridded nitrocellulose-acetate mixed ester, 0.45 μm, 10 ft rolls (Millipore Corp., Bedford, Mass.); drafting pen with India ink (e.g., Rapidograph type, 00 or 000 size); wash bottles; glass vials with Teflon cap liner; index cards; paper towel.

The Transferpettor or Microdispenser, originally designed to deliver five 1 $\mu\ell$ aliquots, is modified by machining additional grooves midway between those in the shaft of the original, thus providing for ten 0.5 $\mu\ell$ aliquots. The orifice of the wash bottle is cut off to provide high volume, low velocity flow. Otherwise, only standard items are needed.

Reagents

Tris (Trizma base and Trizma hydrochloride, Sigma, St. Louis); NaCl (analytical grade); inactivated horse serum (Flow, McLean, Va.); deionized fish gelatin (30%, w/v, solution LB gelatin, Inotech, Wohlen, Switzerland); Nonidet P40 (Sigma); bovine serum albumin (BSA) (Serva, Heidelberg, West Germany); NaN_3 (Fluka, Buchs, Switzerland); peroxidase-conjugated goat antihuman IgG antibodies, affinity purified (Kirkegaard-Perry, Gaithersburg, Md.), also anti-IgM, -IgE, etc., as appropriate; H_2O_2 (30%, w/v, aqueous solution, Sigma); 4-chloronaphthol (Sigma or Merck, Darmstadt, West Germany); methanol (analytical grade); rubber cement (for mounting photographs); Mylar® plastic sheets (3M, Minneapolis, Minn.) as for overhead projector.

Stock and Working Solutions

1. Tris buffered saline (TBS) — 0.15 *M* NaCl, 0.01 *M* Tris-Cl, pH 7.6. It is convenient to stock a 10x concentrate of 1.5 *M* NaCl, 0.1 *M* Trizma chloride and 1.5 *M* NaCl, and 0.1 *M* Trizma base at 4°C. The appropriate ratio to obtain pH 7.6 need only be titrated once. Thereafter, the same mixture can be used.
2. TBS — gelatin (1 mg/mℓ) using the 30% (w/v) gelatin stock, made up as needed.
3. TBS — bovine serum albumin (1 mg/mℓ), can be stored at 4°C with 0.02% (w/v) NaN_3 as preservative.
4. TBS — horse serum 10% (v/v), made up as needed from frozen aliquots.
5. Peroxidase conjugated anti-immunoglobulin — stored at −20°C in 50% (v/v) glycerol, diluted in solution 6 as needed.
6. TBS — gelatin (10 mg/mℓ), 0.1% Nonidet P40, 0.1% (v/v), made up as needed.
7. Chloronaphthol — 3 mg/mℓ in methanol, stored at 4°C in glass vial for up to 1 month.
8. Peroxidase substrate mixture — rapidly mix 10 vol TBS with 0.6 vol of chloronaphthol solution and 0.004 vol 30% (w/v) H_2O_2, immediately before use.

PROCEDURES

Preparation of Nitrocellulose

Matrices of nitrocellulose are cut according to the number of dots/assay and the number of assays anticipitated, plus additional rows and columns for labeling. A generous estimate of the number of tests anticipated is made, as once dotted, the antigens are invariably extremely stable. In this way, repeated dotting can be avoided, and many experiments can be performed with one batch. Nitrocellulose can be cut with a sharp edge provided by the razor blade or scalpel and steel ruler. It handles best when kept between the layers of blue paper used for protection in shipping. Direct manual contact should be avoided at all times.

The top row of grids are enumerated using a very light touch with the drafting pen, so that individual strips can be identified later. The ink remains in place during the entire process, and does not interfere.

Application of the Dots

Antigen dilutions are made up in the TBS-1 mg/mℓ BSA. The most appropriate dilution is determined by titration experiments; the upper limit is not much more than 1 mg/mℓ of protein. The BSA in the diluent is not essential, but prevents loss at very high dilutions of very pure antigen. In some cases, sera may contain antibodies against BSA. This interaction is blocked by inclusion of 1 mg/mℓ BSA in the sample diluent. Each antigen is dispensed across the rows repetitively, using 0.5 μℓ with the microdispenser or 0.2 μℓ with the Hamilton Microlab, taking great care not to indent or puncture the nitrocellulose. As each row is completed, an end is labeled in order to keep track of the rows dotted. Since the dots dry rapidly, it is not possible to keep track visually of the number of rows dotted.

Blocking

After approximately 30 min at room temperature to complete the drying process, the nitrocellulose is briefly soaked in TBS until uniformly wetted and the excess binding capacity is then blocked with 1 hr gentle agitation at room temperature in TBS-1 mg/mℓ gelatin. This is followed by a rinse in TBS. The material can be stored dry at ambient temperature, in the dark, either before or after this blocking step.

Primary Incubation

Strips for individual samples are cut from the matrices and are immersed in TBS in the transfer trays. This preequilibration is conveniently done while the sample dilutions are being prepared. Preequilibration improves the reproducibility. Patient serum is diluted into TBS-10% (v/v) horse serum at 1:100 to 1:1000 dilution. Lower dilution factors result in some nonspecific inhibition effect, as shown by serum inhibition of the reaction with IgG dots. The effect is variable from serum to serum. The TBS is poured out of the trays, and 0.5 to 1 mℓ of the serum dilutions poured in. The pouring gives a uniform coverage and reaction start. The trays are gently agitated for 1 to 2 hr at ambient temperature. The degree of agitation should be just enough to give visible movement of the strips relative to the liquid. After this time, the serum dilutions are poured out and the strips washed five times with TBS applied with a wash bottle. The wash volumes need not be determined. The strips are agitated for 5 to 10 min in the last wash.

Secondary Incubation

Appropriate dilutions of the conjugate must also be determined by separate titration, but 1 ng/mℓ is reasonable. Care must be taken to obtain complete mixing of the stock with 50% (v/v) glycerol, as it is very difficult to get completely uniform mixing. It is advisable to first practice this with colored solution. Volumes of 0.5 to 1 mℓ of diluted conjugated antibodies are added rapidly to the wells, and incubation continued at ambient temperature for an additional 1 to 2 hr. The same washing procedure as in the previous step is used.

Development

The peroxidase substrate mixture is prepared and immediately used. Volumes of 1 mℓ are poured into the wells and color development continued for 15 min. At the end of this time, excess substrate is removed by washing three times with deionized water.

Mounting and Storage

The strips may be mounted for archiving and/or densitometry. Index cards are first thinly and uniformly coated with rubber cement. This is allowed to dry. The strips are taken from the water, placed on a paper towel, briefly blotted, and placed in the desired positions on the cards while still damp. When correctly aligned, the strips are affixed in position by pressure through the plastic sheet. This procedure permits accurate alignment and permanent adhesion.

COMMENTS

When the above procedure is followed, the color obtained is permanent for several years.

We have used the above protocol essentially in the form described above in several clinically oriented applications.[1,7,15,16] The only variation is the partial substitution of horse serum by gelatin in some diluents, confirming the usefulness of gelatin as a blocking agent as described by Saravis;[17] however, horse serum in the serum diluent still appears useful for maximization of the specificity. We presume this is due to a requirement for heterologous immunoglobulins.

This protocol is successful with a range of antigens from DNA[15,16] to viruses.[1] Others have adopted strategies to obtain higher binding to the nitrocellulose. For example, Polvino et al.[18] have used exposure to glutaraldehyde vapor to obtain better binding of carcinoembryonic antigen. Walsh et al.[19] used KOH in the diluent of allergens to increase the binding to the nitrocellulose.

REFERENCES

1. **Hawkes, R., Niday, E., and Gordon, J.,** A dot-immunobinding assay for monoclonal and other antibodies, *Anal. Biochem.,* 119, 142, 1982.
2. **Huet, J., Sentenac, A., and Fromageot, P.,** Spot-immunodetection of conserved determinants in eukaryotic RNA polymerases, *J. Biol. Chem.,* 257, 2613, 1982.
3. **Herbrink, P., Van Bussel, F. J., and Warnaar, S. O.,** The antigen spot test (AST): a highly sensitive assay for the detection of antibodies, *J. Immunol. Methods.,* 48, 293, 1982.
4. **Yen, T. S. B. and Webster, R. E.,** Translational control of bacteriophage f1 gene II and gene X proteins by gene V protein, *Cell,* 29, 337, 1982.
5. **Feinberg, J. G.,** A "microspot test" for antigens and antibodies, *Nature (London),* 192, 985, 1961.
6. **Feinberg, J. G.,** "Microspot test" applied to cellulose acetate membranes, *Nature (London),* 194, 307, 1962.
7. **Derer, M. M., Miescher, S., Johanssen, B., Frost, H., and Gordon, J.,** Application of the dot immunobinding assay to allergy diagnosis, *J. Allergy Clin. Immunol.,* 74, 85, 1984.
8. **Zollinger, W. D., Moran, E. E., Connelly, H., Mandrell, R. E., and Brandt, B.,** Monoclonal antibodies to serotype 2 and serotype 15 outer membrane proteins of *Neisseria meningitidis* and their use in serotyping, *Infect. Immun.,* 46, 260, 1984.
9. **Bode, L., Beutin, L., and Kohler, H.,** Nitrocellulose-enzyme-linked immunosorbent assay (NC-ELISA) — a sensitive technique for the rapid visual detection of both viral antigens and antibodies, *J. Virol. Methods,* 8, 111, 1984.
10. **Hibi, T. and Saito, Y.,** A dot immunobinding assay for the detection of tobacco mosaic virus in infected tissues, *J. Gen. Virol.,* 66, 1191, 1985.
11. **Brooks, R. G., Sharma, S. D., and Remington, J. S.,** Detection of *Toxoplasma gondii* antigens by a dot-immunobinding technique, *J. Clin. Microbiol.,* 21, 113, 1985.
12. **Parent, J. G., Belanger, F., Desjardins, S., and Brisson, J. D.,** Dot-immunobinding for the detection of tomato mosaic virus X infecting greenhouse tomatoes, *Phytoprotection,* 66, 53, 1985.
13. **Furuya, K., Noro, S., Yamagishi, T., and Sakurada, N.,** Adsorption of influenza viruses to nitrocellulose membrane filters by filtration and their quantitative densitometric determination, *J. Virol. Methods.,* 9, 193, 1984.
14. **Weckbach, L. A., Thompson, M. R., Staneck, J. L., and Bonventre, P. F.,** Rapid screening assay for toxic shock syndrome toxin production by *Staphylococcus aureus, J. Clin. Microbiol.,* 20, 18, 1984.
15. **Rordorf, C., Gambke, C., and Gordon, J.,** A multidot immunobinding assay for autoimmunity and the demonstration of novel antibodies against retroviral antigens in the sera of MRL mice, *J. Immunol. Methods,* 59, 103, 1983.
16. **Gordon, J. and Rosenthal, M.,** A multidot immunobinding assay for the diagnosis and management of connective tissue diseases, *J. Rheumatol.,* 12, 257, 1985.
17. **Saravis, C. A.,** Improved blocking of non-specific antibody binding sites on nitrocellulose membranes, *Electrophoresis,* 5, 54, 1984.
18. **Polvino, W. J., Saravis, C. A., Sampson, C. E., and Cook, R. B.,** Improved protein analysis on nitrocellulose membranes, *Electrophoresis,* 4, 368, 1985.
19. **Walsh, B. J., Sutton, R., Wrigley, C. W., and Baldo, B. A.,** Simultaneous detection of IgE binding to several allergens using a nitrocellulose "polydisc", *J. Immunol. Methods,* 73, 139, 1984.

Section 3

SEPARATION OF ANTIGENS BY ANALYTICAL GEL ELECTROPHORESIS

Karl-Erik Johansson

INTRODUCTION

This chapter is restricted to those separation methods that can be directly combined with immunoblotting, i.e., analytical gel electrophoresis. The choice of separation method is dependent on the number of proteins in the sample and the resolution required. Agarose gel electrophoresis,[1] which has a low resolution capacity of about 20 components, can be used for routine analysis of serum proteins. The resolution can be increased to approximately 50 components by combining agarose gel electrophoresis with immunoelectrophoresis, i.e., crossed immunoelectrophoresis.[2] Proteins are directly analyzed as antigens in crossed immunoelectrophoresis and the experiments have to be performed under native conditions. The immunoprecipitation pattern, which is obtained, can be further analyzed by electroblotting.[3] Isoelectric focusing (IEF) can be performed under native or denaturing conditions and it has a resolution capacity of about 100 components.[4] In this chapter it is discussed only in connection with two-dimensional polyacrylamide gel electrophoresis (2D PAGE). Electrophoresis in sieving polyacrylamide gels in the presence of sodium dodecyl sulfate (SDS-PAGE) with discontinuous buffer systems is an important method for protein analysis.[5] Proteins are separated mainly according to size and the resolution is 100 to 200 components. If whole-cell extracts are to be analyzed for total protein composition, the 2D PAGE is the method of choice[6] because the practical resolution capacity is 1000 to 2000 components.

SAFETY RULES

Toxic Chemicals

Acrylamide is neurotoxic and constitutes the most serious health hazard in connection with PAGE experiments. The acute oral LD_{50} in rodents is 170 mg/kg and the hazardous effects of acrylamide are cumulative. Work in a hood or a well-ventilated area, wear gloves, and use a surgical mask when weighing acrylamide. Remove waste immediately and never use mouth pipetting. A polyacrylamide gel is not regarded as toxic, provided it does not contain unpolymerized material as it might. TEMED (*N,N,N*^{1},*N*^{1}-tetramethylethylenediamine), bisacrylamide, 2-mercaptoethanol, iodoacetamide, and methanol are also poisonous and should be handled accordingly.

Electrical Hazard

Apparatuses not equipped with safety contacts or covered electrode vessels are the most dangerous. Keep the area around the electrophoresis apparatus neat and clean. Use a sign of warning for "High Voltage" when appropriate.

DISCONTINUOUS SDS-PAGE

Principles

The Polyacrylamide Gel

A common way to reduce convection in electrophoresis is to use an anticonvection or supporting medium. By using the synthetic gel polyacrylamide as supporting medium, the pore size can be decided. Thus, in the absence of SDS the proteins will be separated according to both charge and size. A polyacrylamide gel is produced by polymerization of acrylamide and bisacrylamide with TEMED and ammonium persulfate as "catalysts".[7] TEMED is activated by ammonium persulfate and initiates the chain reaction with acrylamide. Acrylamide polymerizes into long strands, which can be crosslinked to different degrees by changing the concentration of bisacrylamide. The separating properties of the gel can be conveniently expressed[8] by stating the total concentration (T) of acrylamide plus bisacrylamide and the degree of crosslinking (C), i.e., the fraction of bisacrylamide, as follows:

$$T = \frac{a + b}{v} \cdot 100\% \text{ (w/v)} \qquad C = \frac{b}{a + b} \cdot 100\% \text{ (w/w)}$$

Where a = amount of acrylamide in grams, b = amount of bisacrylamide in grams, and v = final volume of monomer solution in milliliters.

Molecular Weight Determinations

PAGE, in the presence of the anionic detergent SDS, is suitable for molecular weight (M_r) determinations of proteins,[9] as many proteins bind the same amount of SDS (1.4 g of SDS per gram of protein) and get the same negative charge per unit area.[10] Protein molecules are unfolded by SDS and the SDS-protein complex seems to be a rod-shaped particle where the length is proportional to the M_r of the protein.[10] Thus, the M_r of proteins is estimated by establishing a calibration curve for reference proteins of known M_r analyzed on the same gel. A linear relation within a certain M_r interval, which is dependent on the gel concentration, is obtained by plotting the migration distance against the logarithm of the M_r.[11] There are, however, exceptions to the rule that proteins bind the same amount of SDS. For instance, glycoproteins generally bind smaller amounts of SDS[12] and some membrane proteins bind larger amounts of SDS than should be expected from their size.[13] Very charged proteins might migrate anomalously in SDS-PAGE.[14] For M_r estimations, it is important to have the proteins completely unfolded and reduced, which is achieved with SDS and treatment with a reducing agent like 2-mercaptoethanol at high temperature (95 to 100°C). There are many examples of proteins which are difficult to reduce or denature at room temperature. For instance, the OmpC and OmpF proteins (porins) from the outer membrane of *Escherichia coli* appear as polymers if unheated, but as monomers after boiling with SDS.[15] The OmpA protein, has an apparent lower M_r in an unheated sample due to incomplete unfolding.[16]

Discontinuous Gel and Buffer Systems

Two different gels (stacking and separation gels) with different buffers in the electrode vessels and in the gels are used in the discontinuous systems. By this procedure, thin starting zones are created, because the proteins in the sample are stacked by displacement electrophoresis before they enter the separation gel.[17] The Neville[18] system is described in detail in this section.

Materials

The letters (L) and (O) in the list of materials indicate that it is also used for the pore gradient gel system (Laemmli) and/or the 2D PAGE system (O'Farrell), respectively.

Equipment

Various designs of electrophoresis apparatuses are shown in Figure 1.

1. Electrophoresis apparatus with electrodes, slot formers and gel cassettes (L,O)
2. Power supply (constant voltage, 250 V, 100 mA) (L)
3. Volt hour integrator (optional) (L, O)
4. Small water bath (100°C) (L)

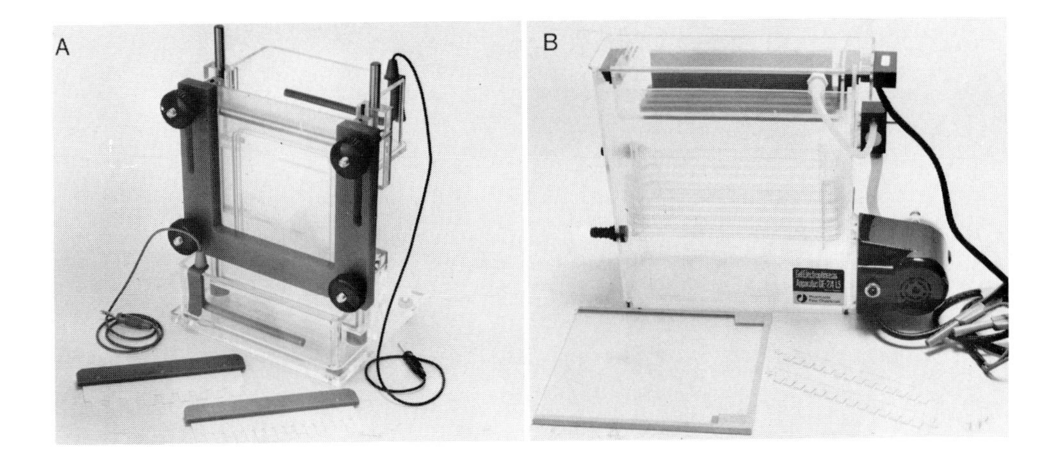

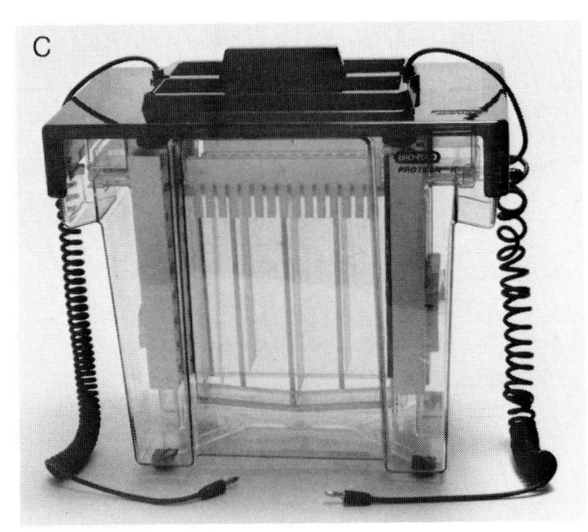

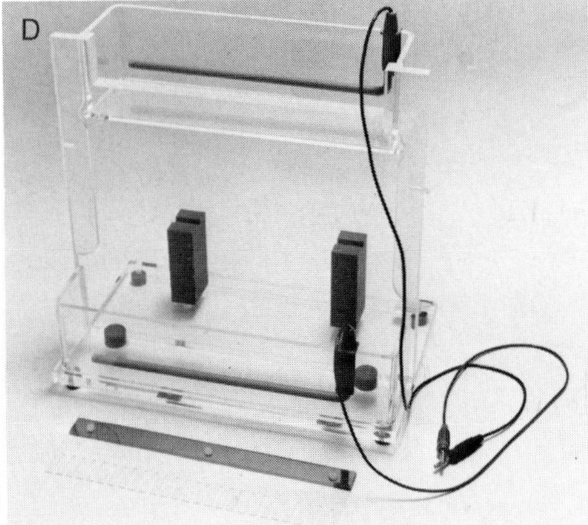

FIGURE 1. SDS-PAGE apparatuses. (A) Example of an apparatus, which is very useful for the Neville system and can be built in a workshop with reasonable resources (BMC-Workshop), Box 570, Uppsala, Sweden). (B) The GE-2/4 LS electrophoresis apparatus (Pharmacia), which is suitable for the Laemmli system. (C) The Protean II electrophoresis apparatus (Bio-Rad), suitable for the Neville or the Laemmli system. (D) An electrophoresis apparatus (BMC-Workshop) for large gels, which gives extremely good resolution with the Laemmli system. The gel cassette is shown in Figure 4.

5. Table centrifuge for microtubes (10,000 to 15,000 × *g*) (L, O)
6. Syringe (10 mℓ) with needles (L, O)
7. Pipetting device (L, O)
8. Automatic pipettes (L, O)
9. Repeating pipette (optional) (L)
10. Dispensers (optional) (L, O)
11. Sample applicator (e.g. Hamilton syringe or special tips for automatic pipettes (see Figure 2) (L, O)
12. Plastic vessels for staining and destaining (L,O)
13. Oscillating table capable of gentle agitation for staining and destaining (L,O)
14. Photographic equipment for documentation (L, O)
15. Gel scanner (optional) (L, O)
16. Drying frames and cellophane (see Figure 3) (optional) (L, O)

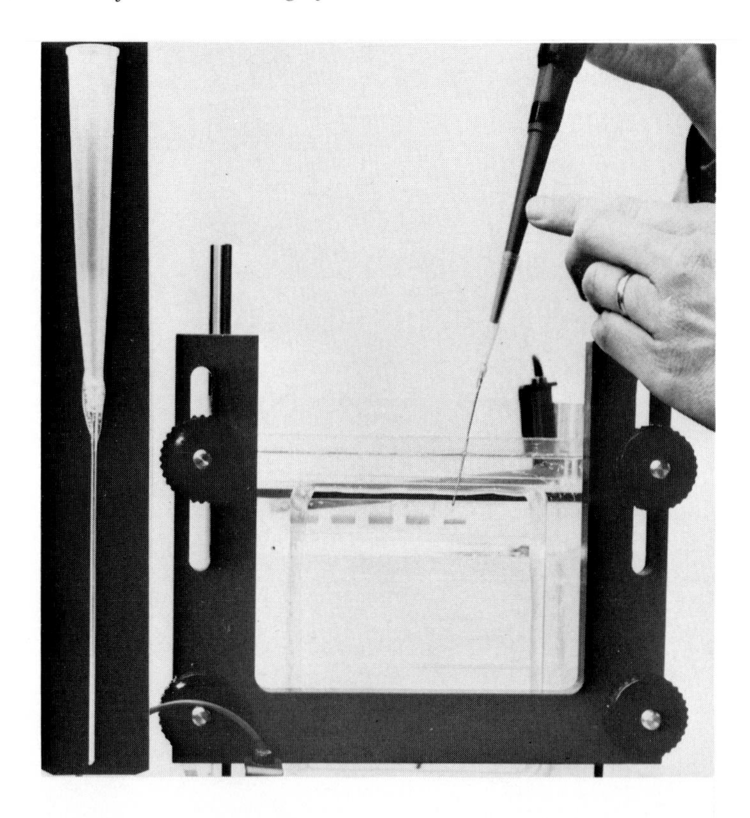

FIGURE 2. Sample application with an automatic pipette, equipped with a special tip (see insert), which has been made by gluing a piece of a cannula steel (see Figure 4) to a disposable plastic tip.

Chemicals

The quality of the chemicals will affect the reproducibility of the experiments; however, it is not necessary to select the most expensive quality of, e.g., acrylamide, bisacrylamide, and Tris for SDS-PAGE experiments. Chemicals and solutions can be stored at room temperature unless otherwise stated.

1. Acrylamide (store in the hood) (L, O)
2. Bisacrylamide (*N,N'*-methylenebisacrylamide) (L, O)
3. TEMED (*N,N,N',N'*-tetramethylethylenediamine) stable for several years if stored in a dark bottle at 4°C (L, O)
4. Ammonium persulfate (L, O)
5. SDS (sodium dodecyl sulfate) (L, O)
6. Boric acid (O)
7. Tris (L, O)
8. H_2SO_4 (O)
9. HCl (L, O)
10. NaOH (O)
11. 2-Butanol (isobutylalcohol) (L, O)
12. Sucrose (L, O)
13. EDTA
14. 2-Mercaptoethanol (O) stable for several years if stored at -20°C in tightly closed bottle

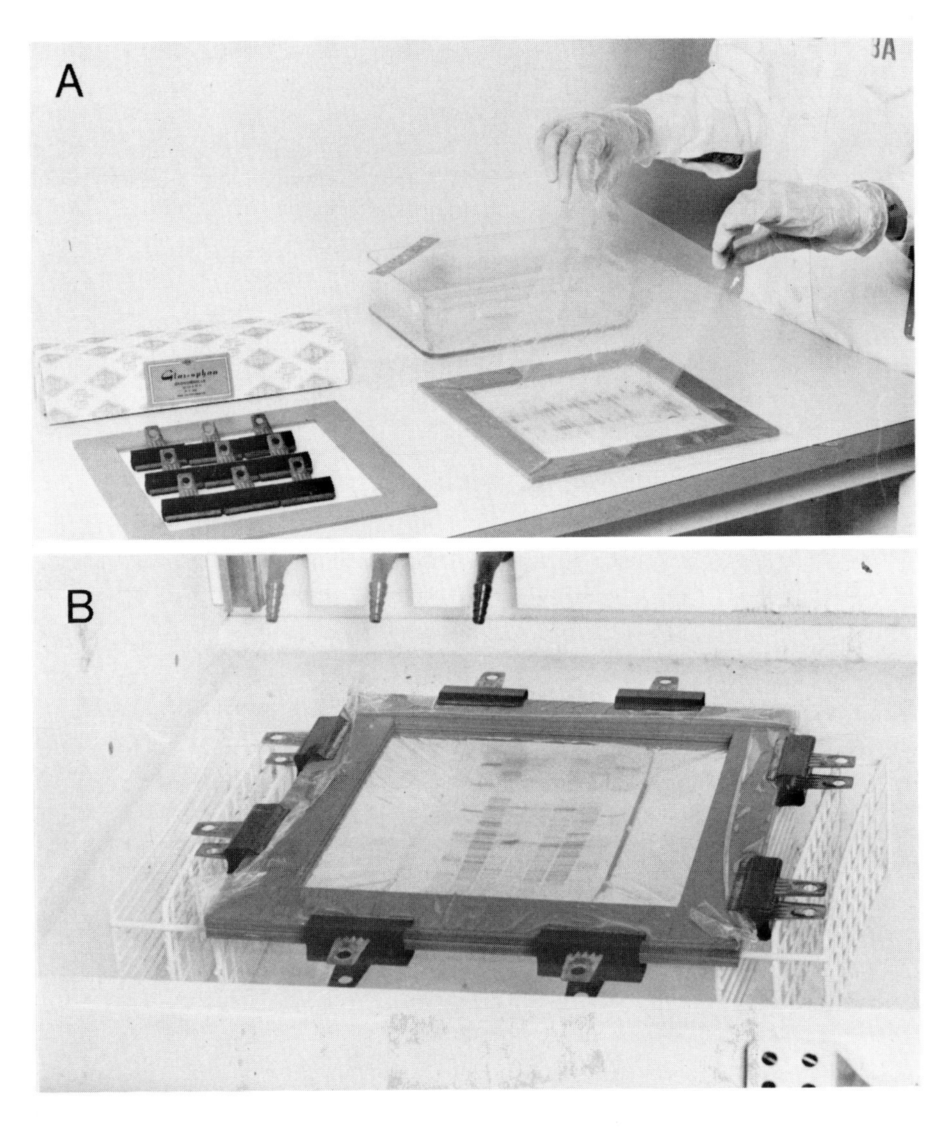

FIGURE 3. Drying of a polyacrylamide gel.[23] (A) Application of the second layer of cellophane on the gel. (B) Drying of the gel in the hood.

15. Bromophenol blue (L, O)
16. Methanol (L, O)
17. Acetic acid (L, O)
18. Coomassie brilliant blue R-250 (L, O)
19. Ethanol (L, O)
20. DNase I (optional) (L, O) from bovine pancreas, store desiccated at −20°C

Solutions

For optimal resolution, it is essential that the pH (at room temperature) of the buffers is correct after dilution. Use at least deionized water. The solutions used for gel casting should be filtered through a small pore size filter (Millipore, approximately 0.5 μm), if the gel is to be stained by a silver staining procedure.

1. Acrylamide solution (O): 33% (w/v) in water (stable for several months if stored in a dark bottle at 4°C). For safety reasons, it is convenient to store the solution in a dispenser.
2. Bisacrylamide solution (O): 1% (w/v) in water (stability as for the acrylamide solution)
3. Ammonium persulfate solution (L, O): 10% (w/v) in water (new solution should be prepared every week, store in a dark bottle).
4. Upper buffer (catholyte), pH 8.64 (O): 0.40 *M* boric acid (24.74 g/ℓ), 0.41 *M* Tris (49.65 g/ℓ), 1.0% (w/v) SDS (10.00 g/ℓ). Dissolve the components in 800 mℓ of H_2O and adjust the pH with NaOH (e.g., 5 *M*) to 8.64 and the volume to 1000 mℓ. Dilute the solution tenfold and check the pH before use.
5. Stacking buffer, pH 6.10 (O): 0.541 *M* Tris (6.55 g/100 mℓ). Dissolve in 80 mℓ of H_2O, adjust the pH with H_2SO_4 to 6.10, and the volume to 100 mℓ. Sterilize the solution by filtration and store at 4°C, inspect regularly for microbial growth. Refilter if necessary.
6. Separation buffer, pH 9.18 (O): 2.122 *M* Tris (256.97 g/ℓ). Dissolve in 600 mℓ of H_2O, adjust the pH with HCl to 9.18, and the volume to 1000 mℓ.
7. Anolyte (lower buffer), pH 9.18 (O): dilute the separation buffer five-fold, check the pH (should be 9.18), and adjust if necessary.
8. Overlay solution (L, O):
   a. 50 mℓ H_2O
      50 mℓ ethanol
      0.2 g SDS
   b. 2-Butanol saturated with H_2O
      Shake equal volumes of 2-butanol and H_2O and use the upper phase. This solution is easier to layer but more toxic than that of a. Solution a has to be degassed and 10 μℓ TEMED and 30 μℓ ammonium persulfate solution should be added to 10 mℓ before use.
9. Sample cocktail (Neville): 40 mℓ H_2O, 10 mℓ catholyte (undiluted), 5 mℓ 2-mercaptoethanol, 5 g sucrose, 185 mg EDTA, and 1.9 g SDS. Mix, dissolve, and adjust the pH to 8.64 and the volume to 100 mℓ. Check the pH. Stable for at least 6 months if stored at 4°C in a tightly closed bottle.
10. Bromophenol blue solution: 0.5 mg/mℓ in H_2O.
11. Molecular weight markers (L, O): available from companies dealing with electrophoresis equipment.
12. Staining solution (stock solution) (L, O): 0.5% (w/v) Coomassie blue dissolved in a fixative composed of 500 mℓ methanol, 430 mℓ H_2O, and 70 mℓ acetic acid. The stock solution should be filtered and diluted tenfold with fixative before use. Methanol and acetic acid should be of good quality to avoid precipitation of impurities on the gels.
13. Destainer (L, O): 800 mℓ H_2O, 100 mℓ methanol, and 100 mℓ acetic acid.
14. DNase I solution (optional) (L, O): dissolve in 0.15 *M* NaCl and 50% (v/v) glycerol to a final concentration of 1 mg/mℓ. Store at −20°C.

Procedures

Choice of Gel Concentration

T should be 10% or less if good resolution in the high M_r region (70,000 to 250,000) is required. Proteins smaller than 20,000 will migrate in the salt front in such a gel. Good resolution in the M_r-region 10,000 to 100,000 can be obtained with T = 15% and it is also possible to increase the degree of crosslinking. Higher T concentrations can be employed as well. For small peptides (down to 2500), special gel and buffer systems should be used.[19] The composition T = 14.8%; C = 1% (see Table 1) works within a rather broad range of M_r and can be recommended for general use.

Table 1
RECIPES FOR SEPARATION GELS OF DIFFERENT CONCENTRATIONS AND A SUITABLE STACKING GEL USED IN THE NEVILLE SYSTEM[18]

	Separation gels (C = 1%)				
Stock solutions	**T = 7.5%**	**T = 11.1%**	**T = 14.8%**	**T = 18.5%**	**Stacking gel (T = 4.4%, C = 5.7%)**
Acrylamide	10.0	15.0	20.0	25.0	2.5
Bisacrylamide	3.0	5.0	7.0	8.0	5.0
Separation buffer	10.0	10.0	10.0	10.0	—
Stacking buffer	—	—	—	—	2.5
H_2O	21.5	14.5	7.5	1.5	9.8
Ammonium persulfate	0.5	0.5	0.5	0.5	0.2
Final vol	45	45	45	45	20

Note: The volumes are in mℓ and the amount of stacking gel is, in general, enough for two gels (18 × 15 × 0.15 cm).

Preparation of the Gels

The Separation Gel

It is recommended to use a gel thickness of 0.07 to 0.15 cm. Thinner gels are more rapidly stained and destained, give lower background in silver staining, and a more efficient transfer in electroblotting. They are, however, more fragile. A stiff plastic foil can be used as a support for transfering. Determine the inner volume of the gel cassette and recalculate the volumes given in Table 1, if necessary. Clean the gel cassette carefully with detergent, water, ethanol, and water again. Let it dry in the air and assemble the apparatus. Select a suitable gel concentration and prepare the monomer solution according to Table 1. Degas the monomer solution by using a water suction pump and add 20 to 25 μℓ of TEMED. Transfer the monomer solution with a pipette to the gel cassette and remove any air bubbles. Fill the cassette to 3 to 4 cm from the top, depending on the shape of the slot former. The length of the stacking gel should be about 2 cm. Apply a layer (2 to 3 mm) of the overlay solution on top of the monomer solution with a syringe and a needle. Polymerization should proceed within 10 to 20 min, but can vary with the quality of the catalysts. If polymerization is too slow or too fast, try first to change the amount of TEMED. Use less catalysts if several gels are to be cast at the same time from the same batch of monomer solution, because polymerization proceeds faster in a large volume. When the polymerization is complete, one sharp boundary is formed if the ethanol overlay is used and two boundaries are formed with the butanol overlay. Remove the overlay solution and rinse the gel surface carefully with water. Remove drops of water with a piece of filter paper without touching the gel surface and cast the stacking gel.

The Stacking Gel

Prepare the monomer solution for the stacking gel according to Table 1. Degas the monomer solution and add 20 μℓ of TEMED. The polymerization proceeds rather fast. Transfer the monomer solution to the gel cassette with a pipette. Introduce a suitable slot former into the cassette and wait at least 20 min before removing. Rinse the slots carefully with catholyte to remove any traces of unpolymerized material. Apply stacking buffer (diluted eightfold), if the gel is not to be used the same day. It is preferable to use the gel as soon as possible, but it can be stored for a couple of days at 4°C without affecting the resolution much.

Preparation of Samples

Prepare the samples in conical microcentrifuge tubes. Dilute all samples to the same volume (e.g., 20 to 50 μℓ) which facilitates the sample application. Extract containing DNA can be very difficult to apply due to the high viscosity. In such a case, add DNase I to a final concentration of 10 to 100 μg/mℓ and it might be necessary to add trace amounts of Mg^{2+}. Add 5 μℓ of the bromophenol blue solution to all samples and add 1 vol (25 to 55 μℓ) of the sample cocktail. Leave the tubes in an ice bath for 5 to 10 min, or until the viscosity has been reduced, if DNase has been added. Puncture the covers of the tubes and insert them in a boiling water bath for 3 min. Clarify the samples by centrifugation at room temperature for 5 min at 10,000 × *g* prior to analysis. Apply the samples with, for example, a Hamilton syringe or an automatic pipette and special tips as shown in Figure 2. Rinse the tip or syringe very carefully between every sample. Any empty slot should be filled with diluted sample cocktail to get a uniform migration over the gel. The two sample lanes closest to the edges of the gel do not normally give a good resolution and should not be used for important samples. In case DNase has been used, it will appear as a band on the gel (M_r = 35,000).

Protein Amount

If the gel is to be stained with Coomassie blue, 2 to 3 μg of protein per band is recommended, but it should be possible to detect 0.5 μg. The silver staining procedure[20] can be used to detect from 1 ng of protein per band.[21] A gel stained with Coomassie blue can be silver stained afterward. For immunoblotting it is often enough to apply 1/10 to 1/2 of the amount needed for Coomassie blue staining. It is possible to detect proteins, remaining on the gel after electroblotting, by silver staining or even Coomassie blue staining.

Electrophoresis

Perform the electrophoresis at 75 V, constant voltage (corresponding to about 30 mA with T = 15% and a gel size of 15 × 15 × 0.15 cm) for 15 to 20 min or until the samples have entered the stacking gel. Increase the voltage, but do not exceed 150 V (should give about 45 mA). Continue the electrophoresis for 600 to 700 V · hr or until the dye front has reached the bottom of the gel. During the electrophoresis, the catholyte of lower conductivity will enter the gel and the current will drop to about 20 mA (at 150 V).

Interruption of the Electrophoresis

A warm gel is more fragile than a cold gel, so before picking the cassette apart, allow it to cool. Use gloves and transfer the gel to a suitable plastic vessel. Rinse the gel quickly with water and add 200 mℓ (for the above size) of transfer buffer, if it is to be electroblotted (see Chapter 4), otherwise add 150 mℓ of the staining solution.

Staining and Drying of SDS-Polyacrylamide Gels

Leave the gel in the staining solution for 2 to 3 hr under slow agitation, for example, on an oscillating table. Elevated temperature (e.g., 37°C) will speed up the procedure. Change destainer every 3 hr until the background has been almost removed. Formaldehyde fixation[22] can be employed for proteins, which are basic or of low M_r. Documentation is obtained by drying the gel, or scanning it in a gel scanner or by taking a photo (with a yellow filter, e.g., Kodak® Wratten No. 12). The gel can be dried between sheets of cellophane and sandwiched between two plastic frames[23] (see Figure 3). It is recommended to keep the gel in 20% ethanol at 4°C overnight, whereby the gel will shrink about 10%, but never crack during the drying procedure.

Concluding Remarks

Discontinuous SDS-PAGE according to Neville is useful. An immunoblotting experiment performed with such a gel is shown in Chapter 6.3.2 (see Figure 1). The resolution is in general very good and the reproducibility is high, provided that the pH of the buffers is correct. However, if none of the gel concentrations gives satisfactory resolution, it could be worthwhile to try one of the pore gradient gel systems described in the next section.

SDS-POLYACRYLAMIDE GEL ELECTROPHORESIS IN PORE GRADIENT GELS

Principles

The Pore Gradient Gel and the Buffer System

The system described in this section is also discontinuous, but the same buffer is used as catholyte and anolyte. The pH of the buffers is not as critical as for the Neville system. A gel gradient is used to sharpen up the protein bands and improve resolution. Thus, the buffers are easier to prepare but the gel a little more tricky. Without this gradient, the system will not give satisfactory resolution, particularly for proteins with M_r below 30,000. A polyacrylamide gel gradient can be made shallow (e.g., T = 10 to 15%) whereby the proteins become more and more retarded during migration. With T = 6 to 30%, the pores will become so small that the proteins finally get stuck in the gel matrix. These gels are not useful for blotting purposes, since it will be difficult to electroelute the proteins. When a shallow gradient gel is used for blotting purpose, the normally higher electroelution speed for smaller proteins will be partly compensated for by the smaller pore size of the gel matrix. Thus, an electroblotting pattern from such a gel will give a better reflection of the protein composition of the original sample.

The gel and buffer system described in this section has been used by Laemmli[24] and by Blobel and Dobberstein,[25] and some of the buffers were also used by Maizel.[26] Here, it is referred to as the Laemmli system. Sucrose is included in the heavy gradient solution to increase the density difference, which facilitates the gradient formation. Since it takes a long time to introduce a gradient in the gel cassette, the polymerization has to be delayed, for instance, by using photopolymerization with riboflavin[27] or as described here by using small amounts of catalysts. This system is sensitive to the quality of these chemicals and the temperature. It might, therefore, be necessary to modify the amounts suggested and use completely fresh ammonium persulfate. Furthermore, plastic material such as perspex in the spacers sometimes inhibits the polymerization. Gradient systems are particularly sensitive due to the minimal amount of catalysts used. Incomplete polymerization close to the edges will result in nonuniform migration or even leakage of buffer. If it takes about 10 min to introduce the gradient, it should take 30 to 45 min before the first sign of polymerization appears (one or two sharp boundaries) between the gel and the overlay solution.

Denaturation of Sample Proteins

Unfolding is achieved by using SDS in the sample and gel and reduction of disulfides with dithiothreitol (DTT) in the sample.[28] The chemical structure is $HS–CH_2$ $(CHOH)_2$ $CH_2–SH$.

Disulfide bonds may reform during electrophoresis when the reducing agent is left behind,[26] which can be prevented by treating the samples with iodoacetic acid or iodoacetamide. The alkylation with iodoacetic acid introduces new charges of the protein which might affect the electrophoretic migration.

Determination of Molecular Weights

When the migration distances for reference proteins in a gradient gel containing SDS are plotted against the logarithm of the M_r, a sigmoid curve is obtained. However, a linear relation (within a certain M_r range) can be obtained if $\log M_r$ is plotted against logR, where R is the relative mobility.[29]

Materials

Only materials not previously mentioned are listed here.

Equipment

1. Gradient mixing system composed of gradient mixer, magnetic stirrer, peristaltic pump, and hollow needle (25 × 0.1 cm)

Chemicals

1. Dithiothreitol or dithioerythritol: store desiccated at 4°C and dissolve immediately before use
2. Iodoacetamide or iodoacetic acid: store desiccated at −20°C and dissolve immediately before use
3. Glycine

Solutions

1. Acrylamide-bisacrylamide (30:0.8): 30 g acrylamide and 0.8 g bisacrylamide are dissolved in H_2O to a final volume of 100 mℓ, which can be expressed as T = 30.8%; C = 2.6%; stable for several months if stored at 4°C in a dark bottle
2. 2.0 *M* Tris-HCl (pH 8.80): store at 4°C
3. 0.5 *M* Tris-HCl (pH 6.80): store at 4°C; keep sterile by aseptic working conditions and inspect regularly for microbial growth; refilter if necessary
4. 20% (w/v) SDS: store at room temperature
5. 60% (w/v) sucrose: store at +4°C
6. 0.5 *M* iodoacetamide solution: add 250 μℓ of H_2O to 23 mg of iodoacetamide immediately before use; the solution is not stable
7. Sample buffer: 1 mg bromophenol blue, 0.5 mℓ 2.0 *M* Tris-HCl (pH 8.80), 6.0 mℓ 60% sucrose, and 3.5 mℓ H_2O; mix and store at 4°C; check for microbial growth
8. Sample cocktail (Laemmli): 3.1 mg dithiothreitol, 800 μℓ sample buffer, 180 μℓ SDS (20%), and 20 μℓ H_2O; dissolve dithiothreitol in sample cocktail immediately before use; the solution is not stable
9. Stock solution (5×) of electrophoresis buffer: 360 g glycine and 150 g Tris; dissolve the components in H_2O to a final volume of 2.5 ℓ; the pH should be 8.7 and does, in general, not have to be adjusted; store at room temperature
10. Electrophoresis buffer: 400 mℓ stock solution (5x) and 10 mℓ SDS (20%): dilute with H_2O to a final volume of 2.0 ℓ; this buffer is used both as catholyte and anolyte

Procedures

Preparation of Pore Gradient Polyacrylamide Gels

The volume to be occupied by the separation gel (3 to 4 cm from top) in the assembled gel cassette should be determined. If half of that volume is much different from the final volume of 30 mℓ in Table 2, the figures in the table should be recalculated. Assume for simplicity that all 50 mℓ of monomer solution must be introduced. Select the range of the gradient and mix the corresponding light and heavy monomer solutions according to Table 2 in two suction flasks. Degas the solutions and add to each flask 75 μℓ SDS (20%), 10 μℓ TEMED, and 50 μℓ ammonium persulfate (10%). Close the valves of the gradient mixer

Table 2
RECIPES FOR MONOMER SOLUTIONS FOR GEL CASTING IN THE LAEMMLI SYSTEM[24]

	Light monomer solution			Heavy monomer solution		Stacking gel solution	
Stock solutions	**5%**	**7%**	**10%**	**15%**	**18%**	**2.4%**	**4.8%**
Acrylamide-bis	5.0	7.0	10.0	15.0	18.0	2.0	4.0
Tris-HCl, pH 8.8	6.0	6.0	6.0	6.0	6.0	—	—
Tris-HCl, pH 6.8	—	—	—	—	—	3.5	3.5
60% sucrose	—	—	—	6.0	6.0	5.0	5.0
H_2O	18.8	16.8	13.8	2.8	—	14.4	12.4
Final vol	30	30	30	30	30	25	25

Note: The volumes are in mℓ and the concentrations of monomer solutions refers to the concentration of acrylamide only. The amounts of catalysts are included in the final volumes.

and pour 25 mℓ of the heavy monomer solution in the chamber closest to the outlet and 25 mℓ of the light monomer solution in the other chamber (feeding chamber). Introduce the hollow needle between the glass plates of the gel cassette, as shown in Figure 4, start the stirrer and the pump, and open the valve between the gradient chambers. Use a pump speed of approximately 5 mℓ/min. Move the needle back and forth and keep the orifice of the needle about 1 cm above the surface of the monomer solution. Apply 2 to 3 mℓ of overlay solution on top of the gradient with a syringe and a needle. It is important to allow the polymerization to proceed completely and it usually takes at least 4 hr before the gel can be used for electrophoresis. The stacking gel can in general be prepared after about 2 hr.

Prepare a suitable monomer solution for a stacking gel according to Table 2. This volume is usually enough for two to three gels. Remove the overlay solution and rinse the gel surface carefully with water. Degas the monomer solution and add 62 μℓ SDS (20%), 50 μℓ TEMED, and 100 μℓ ammonium persulfate (10%) This monomer solution polymerizes fast, so transer it immediately to the gel cassette, introduce the slot former, and wait at least 1 hr. Do not remove the slot former until the samples are ready. If the slot former is removed too early, the sucrose will diffuse out from the gel and interfere with sample application. A gel can usually be stored for 2 to 3 days at 4°C in the cassette with the slot former, without greatly affecting the resolution.

Selection of Gradient

Linear Gradients

A linear gradient will often give a high resolution, but it is difficult to know the concentration range (of acrylamide) that gives optimal resolution for a certain protein mixture. As a general rule, first try a gradient of 10 to 15%. For proteins of relatively low M_r (10,000 to 50,000), a 10 to 18% gradient may be better. Larger proteins (100,000 to 200,000) are separated in gradients of 5 to 15% or 7 to 15%. The stacking gels of 2.4 and 4.8% should be chosen for gradients starting at 5 or 7 and 10%, respectively.

Exponential Gradients

An exponential gradient can easily be prepared with an equipment similar to the one shown in Figure 4;[30] however, the outlet chamber should be tightly covered with a rubber stopper, because it works only as a mixing chamber. Assume again that the volume of the

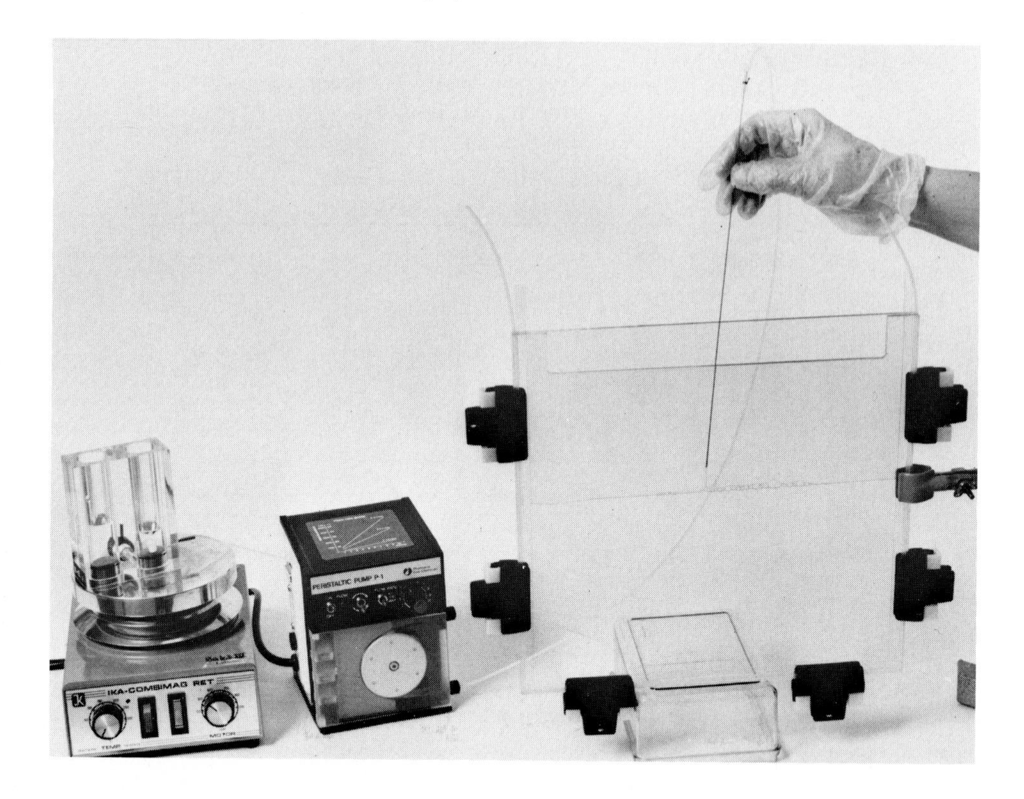

FIGURE 4. Preparation of the gel gradient for the Laemmli system. The monomer solution is introduced in the cassette (BMC-Workshop) with a peristaltic pump equipped with a long hollow needle (cannula steel). Move the needle back and forth during pumping in order to distribute the solution evenly. This cassette is to be used with the apparatus shown in Figure 1D.

monomer solution to be introduced in the gel cassette is 50 mℓ. Then add 50 mℓ of (e.g., 7%) light monomer solution to the feeding chamber and 25 mℓ of (e.g., 15%) monomer solution to the outlet chamber. Start the gradient formation as described earlier.

Preparation of Samples

The preparation follows the same rules as in "Discontinuous SDS-PAGE" with the following exception. Add 5 μℓ of the freshly prepared iodoactamide solution, if there is a risk of reoxidation of –SH groups.[26] Leave the samples for 15 min at room temperature and clarify them by centrifugation as described earlier.

Electrophoresis

Rinse the sample slots with electrophoresis buffer and mount the gel cassette in the electrophoresis apparatus. Apply the samples and start at approximately 100 V. Increase the voltage but do not exceed 160 V and continue for 2500 V · hr (for a gel size of 18 × 15 × 0.15 cm). Dissassemble the cassette and stain the gel as described earlier or use it for a blotting experiment.

Concluding Remarks

Both the discontinuous and pore gradient systems give very good resolution, but certain samples may be more suited for analysis in one of these systems. The gradient gels work very well in electroblotting experiments and that is illustrated in Chapter 6.3.2 (see Figure 2).

TWO-DIMENSIONAL POLYACRYLAMIDE GEL ELECTROPHORESIS (2D PAGE)

Principles

General Considerations

Two-dimensional (2D) electrophoresis was introduced to improve the analysis of complex biological samples.[31] The method was converted to a high-resolution technique by O'Farrell,[32] who used IEF in the first direction and SDS-PAGE for the second direction.[32] He was able to separate and detect 1100 proteins in cell extracts of *Escherichia coli,* and presently, 157 of these have been identified.[33] The resolution capacity for each of the two employed techniques is at least 100 components, so the theoretical resolution capacity should be 10,000 components. However, in practice, it is about 2000 components, since proteins in biological samples are seldom evenly distributed over the actual separation ranges and often present at very different concentrations. If the amount of components in a particular sample is expected to be limited to 100 to 200, 2D PAGE should not be chosen because it is a rather complicated and time-consuming technique unless it is performed on a microscale;[34] however, for whole cell extracts, or certain subcellular fractions, it is the method of choice.

Three important technical developments should be mentioned. First, computer-aided pattern analysis, which has made it possible to store and handle the enormous amount of data contained in a 2D gel.[35,36] Second, the introduction of blotting techniques, which makes it possible to identify protein spots and is said to have introduced the third dimension of 2D PAGE.[37] Third, silver staining techniques for detection of proteins in polyacrylamide gels, which makes it possible to develop the protein pattern of 2D gels without radiolabeling.[38]

The literature on 2D PAGE is continuously growing and several reviews,[39,40] manuals,[6,41-43] and application publications[6,44-46] have been published.

Sample Solubilization

The aim of a 2D PAGE is often to analyze as many proteins as possible from a whole cell extract, for instance, membrane proteins,[47] ribosomal proteins,[48] and histones. Therefore, the conditions for solubilization have to be rather harsh but still compatible with IEF. SDS cannot easily be used, since its charges affect the pH gradient. Neutral detergents such as Triton® X-100 or Nonident P-40 are, therefore, often used in combination with high concentration of urea (8 to 9 *M*). Dithiothreitol or 2-mercaptoethanol is used to secure complete dissaggregation of the proteins into their constituent polypeptide chains. Some authors[43,49] are successfully using small amounts of SDS in the solubilization step. DNase and RNase can be added.[43]

Isoelectric Focusing (IEF)

In IEF, the proteins will migrate to their isoelectric points where they focus into very sharp bands.[4] The pH gradient is created by introducing carrier ampholytes in the system. Such ampholytes are substances which have buffer capacity in a physiological pH range. By introducing a number of ampholytes with different isoelectric points in an electric field, a pH gradient can be created.

The pH gradient can be manipulated by changing the proportions of the different ampholytes.[50] Ampholytes of different pH ranges are sold by many companies: Pharmalyte® (Pharmacia, Sweden). Ampholine® (LKB, Sweden), Servalyt® (Serva, West Germany) and Biolyte® (Bio-Rad, U.S.). Some authors[39,51] prefer to combine ampholytes of different pH ranges from different companies. One should always try to get the proteins evenly distributed in the isoelectric focusing direction. This means that the gradient often should be less steep in the acid region as compared to the alkaline region.[32,52]

Nonequilibrium pH Gradient Electrophoresis (NEPHGE)

It is often difficult to get a good pH gradient in the alkaline region. This effect, termed the cathodic drift, seems to be more pronounced in rod-shaped gels than in slab gels and results in streaking of the basic proteins. The problem can be partly overcome by interrupting the experiment after about 25% of the number of V · hr necessary to reach equilibrium. This NEPHGE system is very useful for basic proteins such as ribosomal proteins.[53]

SDS-PAGE

After interruption of the IEF experiment, the gel should be "equilibrated" with a solution containing SDS and 2-mercaptoethanol and a suitable buffer. The IEF gel is then fixed to a precast polyacrylamide gel by using a hot agarose solution as glue. The SDS-PAGE system can be of any type. Slots can be formed in the agarose on each side of the IEF gel and used for molecular weight markers or the original sample.

Materials

Equipment

Standard equipment can often be used; however, for comparative work it might be necessary to use specially designed apparatuses[43,54,55] which are also commercially available from, for example, Bio-Rad. The lists below include new equipment, chemicals, and solutions not mentioned earlier.

1. High voltage (1000 V) power supply, constant voltage, or power
2. Electrophoresis apparatus for IEF in rod gels (see Figure 5)
3. Transfer tool for IEF gels (see Figure 6)
4. A syringe with a long and thin needle (see Figure 7)
5. Rack of IEF gel tubes
6. Test tubes with screw caps
7. Gel cassette with a ditch (groove) where the IEF gel can be applied

Chemicals

1. Acrylamide and bisacrylamide; high quality, presence of charged substances such as acrylic acid might ruin the pH gradient in IEF
2. Triton® X-100, (Serva) stable for several years at 4°C
3. Urea: high quality since presence of cyanate can cause charge modification (carbamylation) of proteins; the quality ultrapure from BRL (U.S.), is sufficient
4. Carrier ampholytes (see an earlier section); stable for 2 to 5 years at 4°C, check on the bottle
5. Phosphoric acid
6. Glycerol
7. Agarose: should have low electroendosmosis, as for instance, type C ($m_r = -0.02$) from Pharmacia

Solutions

The solutions used for gel casting should be filtered through a 0.5 μm filter (Millipore).

1. Monomer solution (T = 23%; C = 5.3%; dissolve 5.68 g acrylamide and 0.32 g bisacrylamide in 20 mℓ of H_2O and filtrate the solution; store the solution in a dark bottle at 4°C and prepare new each second week; acrylic acid is formed during storage due to hydrolysis of acrylamide
2. Triton® X-100 solution: dissolve 10 g Triton® X-100 in 75 mℓ of H_2O under magnetic stirring; adjust the volume to 100 mℓ and filter the solution; store at 4°C and inspect regularly for microbial growth

FIGURE 5. Apparatus for isoelectric focusing (IEF) in rod-shaped gels (BMC-Workshop).

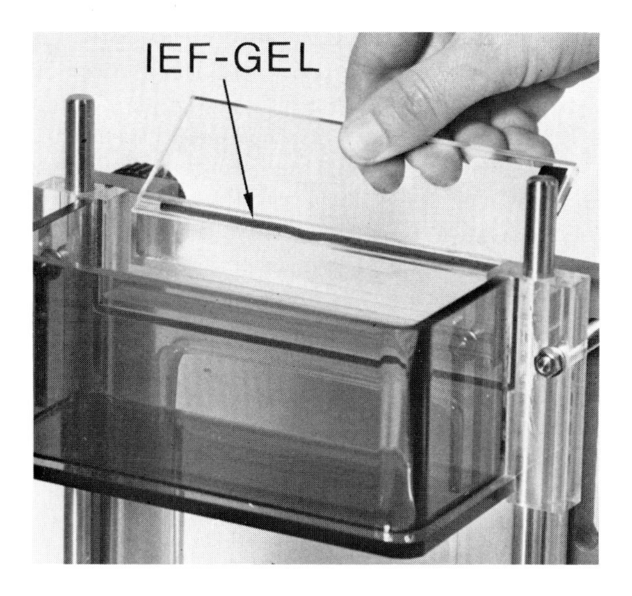

FIGURE 6. Transfer of the IEF gel to the ditch of the gel cassette for the second dimension in 2D PAGE.

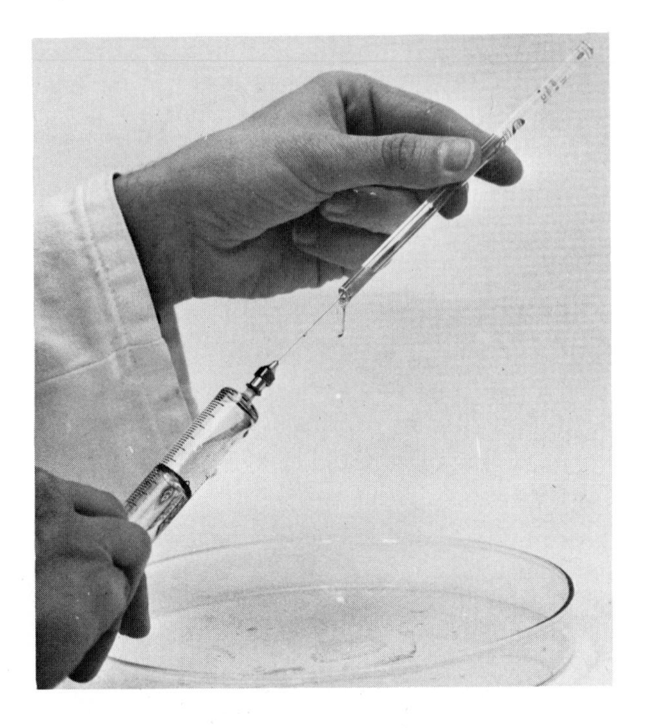

FIGURE 7. Removal of the IEF gel from the glass tube with a syringe and a needle.

3. Anolyte: 10 m*M* H_3PO_4; a 100 m*M* stock solution can be used
4. Catholyte: 20 m*M* NaOH; should be prepared in degassed water and used fresh to avoid CO_2 dissolving in the solution
5. Sample cocktail (O'Farrell): add to 5.71 g of urea in a graded test tube, 2.0 mℓ Triton® X-100 solution, 0.5 mℓ 2-mercaptoethanol, 0.2 mℓ ampholytes, (pH range 3 to 10), and H_2O to a final volume of 10 mℓ; freeze the solution in aliquots at −20°C and prepare new each second week
6. Overlay cocktail: is prepared as the sample cocktail, but with 2.85 g of urea
7. Equilibration buffer: dissolve the following chemicals in 175 mℓ of H_2O 5.75 g SDS, 1.90 g Tris, 30.0 mℓ glycerol, and 12.5 mℓ 2-mercaptoethanol; adjust the pH to 6.8 with HCl and the volume to 250 mℓ; stable for several months if stored at 4°C in a tightly closed bottle
8. Agarose solution: dissolve 0.3 g of agarose in boiling stacking buffer diluted to the concentration used in the stacking gel; stacking buffer must be chosen according to the SDS-PAGE system which is to be used

Procedures

Casting of IEF Gels

The gels should be polymerized in glass tubes with an inner diameter of 2 to 3 mm. The length of the tubes depends upon the size of the final 2D gel (here 15 cm). The tubes should be carefully cleaned with a detergent (any strong laboratory detergent will do) and ethanol prior to use. Seal one end of the tubes with two layers of Parafilm (3M Company, U.S.) and keep them in a rack.

Prepare the IEF gel solution by adding the following to a 25 mℓ suction flask: 5.5 g urea, 2.0 mℓ H_2O, 2.0 mℓ Triton® X-100 solution 1.33 mℓ Monomer solution, and 0.50 mℓ Ampholytes (a suitable cocktail). Degas the solution for a short time (to avoid crystallization of urea due to evaporation of water) and add 40 μℓ ammonium persulfate solution followed by 20 μℓ TEMED. Transfer the IEF gel solution to the glass tubes by using a peristaltic pump equipped with tubing and a needle (see Figure 4). Start with the needle close to the bottom of the tube and check that air bubbles are replaced by solution. Fill the tubes up to 1 to 2 cm from the top. A pasteur pipette can also be used. Apply a layer of H_2O on top of the IEF gel solution with a syringe and a needle. Let the polymerization proceed for at least 1 hr, eventually at 30 to 37°C to secure complete polymerization.

IEF

Remove H_2O from the IEF gel tubes and add 50 to 100 μℓ of sample cocktail. Change the sample cocktail after 10 min and leave the gels for another 10 min period. Remove the parafilm carefully and dip the tubes in H_2O to get a drop at the bottom. Fix a dialysis membrane without air bubbles at the bottom with an elastic band cut from a piece of silicon or latex tubing. Fill the lower electrode vessel with catholyte, remove the sample cocktail from the tubes, and mount them in the apparatus. Add overlay cocktail in the tubes and fill the upper electrode vessel with anolyte. Remove any remaining air bubbles and apply samples as described previously for SDS-PAGE. The needle of the applicator should be introduced into the overlay cocktail to avoid the sample coming into contact with the anolyte. Start the IEF experiment at 300 V, do not exceed 1 mA per gel, and check that the current decreases. Run the IEF for a total of 8000 to 10,000 V · hr, the final 3 hr at 1000 V.

In O'Farrell's[32] original method, the samples were applied on the basic side of the gel and the gels were prerun; however, it does not always seem to be necessary to prerun the gels. Apply the sample on the acid side of the gel to avoid interference of 2-mercaptoethanol with the pH gradient.[56]

Determination of the pH Gradient

Two strategies can be used for determination of the pH profile, either by use of markers or by measurement of the pH in a parallel gel. One problem is that the pH is not well defined in a solution containing a high concentration of urea. If pH markers are used, they should preferably be mixed with the sample, assuming they can be distinguished from spots originating from sample proteins in the final 2D map. Suitable pH markers can be prepared by carbamylation of proteins,[57] but are also commercially available from Pharmacia as Carbamylyte™. A carbamylation train can be used in combination with M_r markers to introduce a coordinate system, which will facilitate positioning of protein spots.[57,58] When the pH profile is determined on a parallel gel, it is withdrawn from the electrophoresis tube with the aid of a syringe and a long needle as shown in Figure 7. The gel is then cut into 5- to 10-mm pieces and each piece is eluted with 0.2 to 0.5 mℓ of degassed water overnight and the pH measured. The drawback is that the pH gradient of the gel used for the second dimension can never be determined.

4. Electrophoresis in the Second Dimension

SDS-PAGE with discontinous buffer systems are recommended for the second dimension in 2D PAGE and both of the two earlier described systems are suitable.

Prepare the SDS gels in advance to make sure they are ready when it is time to interrupt the IEF. The gels should be prepared without using slot formers. Instead, the stacking gel solution is filled up to the bottom of the ditch and an overlay solution is then applied on top of the stacking gel solution.

After completion of the IEF, the gels are withdrawn from the tubes directly into test tubes with screw caps. Add 5 mℓ of equilibration buffer and place the tubes in the horizontal position on an oscillating table for 15 to 30 min. It is important to standardize the procedure, since the loss of proteins from the IEF gel is pronounced and dependent on the "equilibration time''; however, the equilibration is necessary to avoid streaking in the second dimension. Prepare the agarose solution, which should be used hot, during equilibration. Transfer the gel from the tube to a plastic mesh and place it on the transfer tool for IEF gels. Rinse the gel with a few milliliters of diluted stacking buffer, add 2 to 3 mℓ of hot agarose solution in the ditch of the gel cassette, and introduce the IEF gel as shown in Figure 6. If molecular weight markers are also to be run, insert the slot formers before the agarose congeals.

After a few minutes the apparatus can be loaded with buffer, carefully to avoid making the IEF gel loose. Start the electrophoresis at 50 V for 30 min, then increase the voltage to 50 to 100 V, and run for the number of V · hr recommended for the system used. Interrupt the electrophoresis as described earlier and use the gel for electroblotting or staining.

Concluding Remarks

The difference in resolution between 1- and 2D electrophoresis is more than one order of magnitude, but unfortunately the difference in investment of work per sample is also at least one order of magnitude. Therefore, try first to solve the analytical problem with 1D gels (SDS-PAGE or IEF).

ACKNOWLEDGMENTS

I thank Prof. Bengt Hurvell and Dr. Göran Bölske for constant support during preparation of the manuscript. I am thankful to Bengt Ekberg for photographic assistance and to Carina Bohlin for typing the manuscript. This work was supported by grants from the foundations of Magn. Bergvall, O. E. and Edla Johansson, and Helge Ax:son Johnson.

REFERENCES

1. **Johansson, B. G.,** Agarose gel electrophoresis. *Scand. J. Clin. Lab. Invest.*, 29 (Suppl. 124), 7, 1972.
2. **Axelsen, N. H., Ed.,** *Handbook of Immunoprecipitation-in-Gel Techniques,* Blackwell Scientific, Oxford 1983.
3. **Bhakdi, S., Jenne, D., and Hugo, F.,** Electroimmunoassay-immuno blotting (EIA-IB) for the utilization of monoclonal antibodies in quantitative immunoelectrophoresis: the method and its applications, *J. Immunol. Methods,* 80, 25, 1985.
4. **Righetti, P. G.,** *Isoelectric Focusing: Theory, Methodology and Applications,* Elsevier, Amsterdam, 1983.
5. **Hames, B. D.,** An introduction to polyacrylamide gel electrophoresis, in *Gel Electrophoresis of Proteins: A Practical Approach,* Hames, B. D. and Rickwood, D., Eds., IRL Press, Oxford, 1981, 1.
6. **Celis, J. E. and Bravo, R., Ed.** *Two-Dimensional Gel Electrophoresis of Proteins: Methods and Applications,* Academic Press, New York, 1984.
7. **Tanaka, T.,** Gels, *Sci. Am.*, 244, 110, 1981.
8. **Hjertén, S.,** ''Molecular sieve'' chromatography on polyacrylamide gels, prepared according to a simplified method, *Arch. Biochem. Biophys.*, Suppl. 1, 147, 1962.
9. **Shapiro, A. L., Vinuela, E., and Maizel, J. V., Jr.,** Molecular weight estimation of polypeptide chains by electrophoresis in SDS-polyacrylamide gels, *Biochem. Biophys. Res. Commun.*, 28, 815, 1967.
10. **Reynolds, J. A. and Tanford, C.,** The gross conformation of protein-sodium dodecyl sulfate complexes, *J. Biol. Chem.*, 245, 5161, 1970.
11. **Weber, K. and Osborn, M.,** The reliability of molecular weight determinations by dodecyl sulfate-polyacrylamide gel electrophoresis, *J. Biol. Chem.*, 244, 4406, 1969.
12. **Glossman, H. and Neville, D. M., Jr.,** Glycoproteins of cell surfaces. A comparative study of three different cell surfaces of the rat, *J. Biol. Chem.*, 246, 6339, 1971.
13. **Banker, G. A. and Cotman, C. W.,** Measurement of free electrophoretic mobility and retardation coefficient of protein-sodium dodecyl sulfate complexes by gel electrophoresis. A method to validate molecular weight estimates, *J. Biol. Chem.*, 247, 5856, 1972.

14. **Miyake, J., Ochiai-Yanagi, S., Kasumi, T., and Takagi, T.,** Isolation of a membrane protein from *R. rubrum* chromatophores and its abnormal behaviour in SDS-polyacrylamide gel electrophoresis due to a high binding capacity for SDS, *J. Biochem.*, 83, 1679, 1978.
15. **Osborn, M. J. and Wu, H. C. P.,** Proteins of the outer membrane of gram-negative bacteria, *Ann. Rev. Microbiol.*, 34, 369, 1980.
16. **DiRienzo, J. M., Nakamura, K., and Inouye, M.,** The outer membrane proteins of gram-negative bacteria: biosynthesis, assembly, and functions, *Ann. Rev. Biochem.*, 47, 481, 1978.
17. **Hjelmeland, L. M. and Chrambach, A.,** The impact of L. G. Langsworth (1905—1981) on the theory of electrophoresis, *Electrophoresis*, 3, 9, 1982.
18. **Neville, D. M., Jr.,** Molecular weight determination of protein-dodecyl sulfate complexes by gel electrophoresis in a discontinuous buffer system, *J. Biol. Chem.*, 246, 6328, 1971.
19. **Anderson, B. L., Berry, R. W., and Telser, A.,** A sodium dodecyl sulfate-polyacrylamide gel electrophoresis system that separates peptides and proteins in the molecular weight range of 2,500 to 90,000, *Anal. Biochem.*, 132, 365, 1983.
20. **Tunón, P. and Johansson, K.-E.,** Yet another improved silver staining method for the detection of proteins in polyacrylamide gels, *J. Biochem. Biophys. Methods*, 9, 171, 1984.
21. **Eschenbruch, M. and Bürk, R. R.,** Experimentally improved reliability of ultrasensitive silver staining of protein in polyacrylamide gels, *Anal. Biochem.*, 125, 96, 1982.
22. **Steck, G., Leuthard, P., and Bürk, R. R.,** Detection of basic proteins and low molecular weight peptides in polyacrylamide gels by formaldehyde fixation, *Anal. Biochem.*, 107, 21, 1980.
23. **Wallevik, K., Jensenius, J. C., Andersen, I., and Poulsen, A. M.,** A simple and reliable method for the drying of polyacrylamide slab gels, *J. Biochem. Biophys. Methods*, 6, 17, 1982.
24. **Laemmli, U. K.,** Cleavage of structural proteins during the assembly of the head of bacteriophage T4, *Nature (London)*, 227, 680, 1970.
25. **Blobel, G. and Dobberstein, B.,** Transfer of proteins across membranes. I. Presence of proteolytically processed and unprocessed nascent immunoglobulin light chains on membrane-bound ribosomes of murine myeloma, *J. Cell. Biol.*, 67, 835, 1975.
26. **Maizel, J. V.,** Polyacrylamide gel electrophoresis of viral proteins, in *Methods in Virology*, Vol. 5, Maramorosch, K. and Koprowski, H., Eds., Academic Press, New York, 1971.
27. **Elson, E. and Jovin, T. M.,** Fractionation of oligodeoxynucleotides by polyacrylamide gel electrophoresis, *Anal. Biochem.*, 27, 193, 1969.
28. **Cleland, W. W.,** Dithiothreitol, a new protective reagent for SH-groups, *Biochemistry*, 3, 480, 1964.
29. **Poduslo, J. F. and Rodbard, D.,** Molecular weight estimation using sodium dodecyl sulfate-pore gradient electrophoresis, *Anal. Biochem.*, 101, 394, 1980.
30. **Mouches, C. and Bové, J. M.,** Electrophoretic characterization of mycoplasma proteins, in *Methods in Mycoplasmalogy*, Vol. 1 Razin, S. and Tully, J. G., Eds., Academic Press, New York, 1983, 241.
31. **Smithies, O. and Poulik, M. D.,** Two-dimensional electrophoresis of serum proteins, *Nature (London)*, 177, 1033, 1956.
32. **O'Farrell, P. H.,** High resolution two-dimensional electrophoresis of proteins, *J. Biol. Chem.*, 250, 4007, 1975.
33. **Neidhardt, F. C., Vaughn, V., Phillips, T. A., and Bloch, P. L.,** Gene-protein index of *Escherichia coli* K-12, *Microbiol. Rev.*, 47, 231, 1983.
34. **Poehling, H.-M. and Neuhoff, V.,** One- and two-dimensional electrophoresis in micro-slab gels, *Electrophoresis*, 1, 90, 1980.
35. **Anderson, N. L., Taylor, J., Scandora, A. E., Coulter, B. P., and Anderson, N. G.,** The TYCHO system for computer analysis of two-dimensional gel electrophoresis patterns, *Clin. Chem.*, 27, 1807, 1981.
36. **Garrels, J. I., Farrar, J. T. and Burwell, C. B.,** IV, The QUEST system for computer-analyzed two-dimensional electrophoresis of proteins, in *Two-Dimensional Gel Electrophoresis of Proteins: Methods and Applications*, Celis, J. E. and Bravo, R., Eds., Academic Press, New York, 1984, 37.
37. **Towbin, H., Stahelin, T., and Gordon, J.,** Electrophoretic transfer of proteins from polyacrylamide gels to nitrocellulose sheets: procedure and some applications, *Proc. Natl. Acad. Sci. U.S.A.*, 76, 4350, 1979.
38. **Merril, C. and Goldman, D.,** Detection of polypeptides in two-dimensional gels using silver staining, in *Two-Dimensional Gel Electrophoresis of Proteins: Methods and Applications*, Celis, J. E. and Bravo, R., Eds., Academic Press, New York, 1984, 93.
39. **Dunn, M. J. and Burghes, A. H. M.,** High resolution two-dimensional polyacrylamide gel electrophoresis. I. Methodological procedures, *Electrophoresis*, 4, 97, 1983.
40. **Dunn, M. J. and Burghes, A. H. M.,** High resolution two-dimensional polyacrylamide gel electrophoresis. II. Analysis and applications, *Electrophoresis*, 4, 173, 1983.
41. **Sidman, C.,** Two-dimensional gel electrophoresis, In *Immunological Methods*, Vol. 2, Lefkovits, I. and Pernis, B., Ed., Academic Press, New York, 1981, 57.
42. **Sinclair, J. and Rickwood, D.,** Two-dimensional gel electrophoresis, in *Gel Electrophoresis of Proteins: A Practical Approach*, Hames, B. D. and Rickwood, D., Eds., IRL Press, Oxford, 1981, 189.

43. **Garrels, J. I.,** Quantitative two-dimensional gel electrophoresis of proteins, *Methods Enzymol.*, 100, 411, 1983.
44. **King, J. S., Ed.,** Special issue on "two-dimensional gel electrophoresis", *Clin. Chem.*, 28, 737, 1982.
45. **King, J. S., Ed.,** Special issue on "two-dimensional electrophoresis", *Clin. Chem.*, 30, 1897, 1984.
46. **Neuhoff, V., Ed.,** *Electrophoresis 84,* Verlag Chemie, Weinheim, 1984.
47. **Rubin, R. W. and Leonardi, C. L.,** Two-dimensional polyacrylamide gel electrophoresis of membrane proteins, *Methods Enzymol.*, 96, 184, 1983.
48. **Böck, A.,** Analysis of ribosomal proteins by two-dimensional gel electrophoresis, in *Methods in Microbiology,* Vol. 18, Gottschalk, G., Ed., Academic Press, New York, 1985, 109.
49. **Manning, P. A., Beutin, L., and Achtman, M.,** Outer membrane of *Escherichia coli:* properties of the F sex factor traT protein which is involved in surface exclusion, *J. Bacteriol.*, 142, 285, 1980.
50. **Johansson, K.-E.,** Two-dimensional polyacrylamide gel electrophoresis of *Escherichia coli* proteins with Pharmalytes, *Protides. Biol. Fluids,* 33, 475, 1985.
51. **Thompson, B. J., Dunn, M. J., Burghes, A. H. M., and Dubowitz, V.,** Improvements of isoelectric focusing in agarose for direct tissue isoelectric focusing, *Electrophoresis,* 3, 307, 1982.
52. **O'Farrell, P. H. and O'Farrell, P. Z.,** Two-dimensional polyacrylamide gel electrophoretic fractionation, *Methods Cell Biol.*, 16, 407, 1977.
53. **O'Farrell, P. Z., Goodman, M. M., and O'Farrell, P. H.,** High resolution two-dimensional electrophoresis of basic as well as acidic proteins, *Cell,* 12, 1133, 1977.
54. **Anderson, N. G. and Anderson, N. L.,** Analytical techniques for cell fractionation. XXI. Two-dimensional analysis of serum and tissue proteins: multiple isoelectric focusing, *Anal. Biochem.*, 85, 331, 1978.
55. **Anderson, N. L. and Anderson, N. G.,** Analytical techniques for cell fractionation. XXII. Two-dimensional analysis of serum and tissue proteins: multiple gradient-slab gel electophoresis, *Anal. Biochem.*, 85, 341, 1978.
56. **Righetti, P. G., Tudor, G., and Gianazza, E.,** Effect of 2-mercaptoethanol on pH gradients in isoelectric focusing, *J. Biochem. Biophys. Methods,* 6, 219, 1982.
57. **Anderson, N. L. and Hickman, B. J.,** Analytical techniques for cell fractionations. XXIV. Isoelectric point standards for two-dimensional electrophoresis, *Anal. Biochem.*, 93, 312, 1979.
58. **Johansson, K.-E.,** Two-dimensional polyacrylamide gel electrophoresis of whole cell extracts from *Escherichia coli* with Pharmalytes, optimization of the conditions and standardization, in *Recent Progresses in Two-Dimensional Electrophoresis,* Galteau, M.-M. and Siest, G., Eds., Presses Universitaires de Nancy, Nancy, France, 1986, 37.

Section 4

TRANSFER OF PROTEINS

4.1 CAPILLARY BLOTTING AND CONTACT DIFFUSION

Keith B. Elkon

INTRODUCTION

The major value of immunoblotting has been the ability to detect and analyze molecules of interest in a heterogeneous protein mixture. The properties of protein binding and water permeability as well as easy handling of nitrocellulose have made it an excellent matrix for diffusion transfer of proteins from large pore gels. Although the methods and applications cited in this chapter refer exclusively to nitrocellulose, the same principles may be applied to other matrices when specific needs arise (see Section 5). Optimal transfer will depend upon the nature of the resolved protein, the fractionation method used, and the equipment available. The simplest method for transfer of gel-fractionated proteins is diffusion blotting of which three varieties are illustrated in Figure 1.

The principle of capillary blotting is that protein migrates from gel to nitrocellulose facilitated by a capillary flow of buffer through the gel[1] (similar to DNA migration in the original technique of Southern[2]) (Figure 1a) or by minimal flow of fluid from the gel itself[3,4] (Figure 1c). In both cases, capillary flow of buffer is produced by overlaying the gel with a moist nitrocellulose sheet followed by dry absorbent filter paper or towels. With contact diffusion, on the other hand (Figure 1b), the whole blot sandwich is immersed in buffer so that transfer occurs mainly by diffusion of proteins from both surfaces of the gel to the adjacent nitrocellulose.[5]

The efficiency of diffusion blotting relates directly to the porosity of the gel and, to a lesser extent, to the size of the protein. Thus, diffusion transfer is the blotting method of choice for agarose gels and is a useful alternative to electrotransfer for large-pore (acrylamide concentration 6% or less) polyacrylamide gels and composite polyacrylamide-agarose gels. A comparison between electrophoretic and capillary transfer is shown in Table 1.

General techniques for agarose gels,[6] agarose isoelectric focusing,[4,7] and composite gel electrophoresis[8] are described in detail elsewhere. A method for transfer of proteins from polyacrylamide gels directly to antibody impregnated agarose (thus, eliminating the need for nitrocellulose) has also been described.[9] Only diffusion transfer and probing of the nitrocellulose replicas are described in detail here.

EQUIPMENT AND REAGENTS

Since size is not limiting for diffusion transfer (Table 1), almost any size or volume of equipment and reagents could be used. For convenience, we usually work with agarose gels 11 × 11 cm and 0.5- to 1-mm thick and composite and polyacrylamide gels 16 cm (height) × 14 cm (width) × 1.5 mm (thickness).

Equipment

For continuous flow transfers (Figures 1a and b), a buffer tank is required. In the case of the Southern type transfer (a) a shallow glass baking dish (larger and wider than the gel) is adequate, whereas for contact diffusion (b) a Perspex or plexiglass tank capable of holding the gel and at least 2 ℓ of buffer are required. In addition, a gel support (plastic or stack of

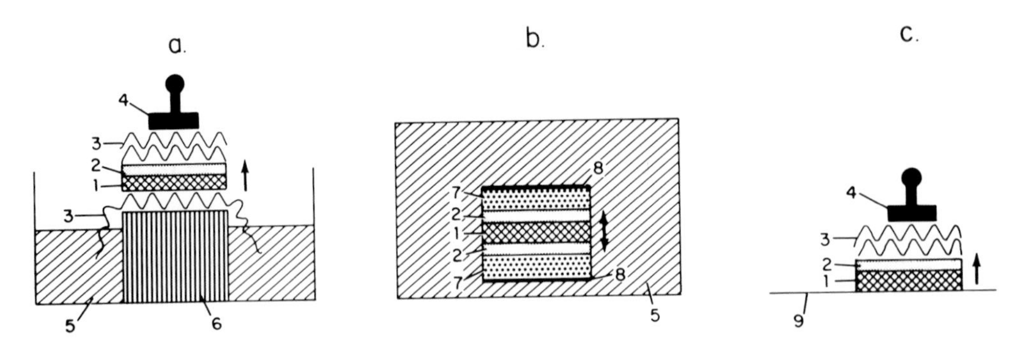

FIGURE 1. Diagram of three types of diffusion blotting. In (a) and (c) proteins are transferred from gel (1) to nitrocellulose (2) by capillary action exerted by dry filter paper (3) or towels on top of the gel. In (b) transfer of proteins to nitrocellulose on both sides of the gel occurs by diffusion. Key: 1, gel; 2, nitrocellulose paper; 3, filter paper; 4, weight; 5, buffer, 6, gel support, 7, gel pads, 8, sandwich holder; 9, plastic film. See text for details.

Table 1
COMPARISON BETWEEN PROTEIN BLOTTING TO NITROCELLULOSE BY DIFFUSION AND ELECTROTRANSFER

	Diffusion	Electrotransfer
Special apparatus	No	Yes
Buffer necessary	Some types (see Figure 1)	Yes
Agarose gels[a]	Most useful (1—2 hr)	Difficult because of fragility
Composite gels	Useful (2—3 hr)	Useful (2 hr)
Polyacrylamide gels	Slow, often incomplete (1—2 days)	Most efficient (2—14 hr)
Charge-dependent transfer	No	Yes[b]
Protein denatured	No	Yes (methanol and SDS)
Sample no. and gel size limitations	No	Yes

[a] Time for usual transfer indicated in parentheses.
[b] Uniform charge usually provided by protein-SDS complex.

glass plates larger than the gel), gel cassette (stainless steel or plastic obtainable from Bio-Rad, U.S., Hoefer, U.S., or LKB, Sweden) and foam pads (e.g., Scotch Brite®, 8-mm thick) may be required (see Figure 1). In the Southern type transfer, a 1 to 2 cm wad of thick absorbent filter paper may be used on top of the gel support (item 6 in Figure 1a).

Most of the above materials are easily obtained without special orders. For further details of tank construction, the original paper by Southern[2] and a molecular cloning manual[10] should be consulted.

Reagents

The following reagents are used: nitrocellulose, 0.2 μm (e.g., Schleicher & Schuell, BA83); thick, absorbent filter paper (e.g., Whatman 3MM, Schleicher & Schuell blot block #2 or #4); paper towels (any firm, absorbent towel with a smooth surface can be used); and gel-bond film (FMC, Marine Colloids); 500 g weight. All of these should be slightly larger than the gel.

Blocking Solution: 0.01 *M* phosphate buffer, pH 7.4 with 0.14 *M* NaCl, 0.05% (w/v) NaN_3, and 3% (w/v) bovine serum albumin (PBS-BSA).

Wash Solution: PBS containing 0.05% (v/v) Tween® 20 (PBS-Tw).

Probe Solution: PBS-BSA-Tw containing antibody or lectin. Where necessary, 1 to 5% (v/v) of normal animal (the same species as used for second antibody) serum is used to reduce background staining.

PROCEDURES

Large-Pore Gels (Agarose and Composite; (Figure 1c)

Nitrocellulose and two filter paper sheets both slightly larger than the gel are presoaked in PBS for at least 10 min for uniform wetting. After completion of electrophoresis, the gel is placed onto gel-bond film to facilitate handling and subsequent staining. Agarose isoelectric focusing gels are usually cast directly onto Gel-Bond[4,7] so that they are immediately ready for blotting.

The wet nitrocellulose is allowed to drain excess buffer for a few seconds, and is then carefully placed on top of the gel without trapping air bubbles. Two moist filter papers and an approximately 4-cm thick wad of paper towels complete the sandwich (Figure 1c). A 500 g weight ensures good contact during transfer. Either the weight can be placed on top of a glass plate to allow even pressure over the whole gel surface or, more conveniently, a large, heavy book (e.g., 1kg) is placed on top of the towels. Transfer is usually performed at room temperature, but can be performed at 4°C. For 1% agarose gels 0.5 to 1-mm thick, transfer is complete in 60 to 90 min, even for molecules with a M_r as large as 900,000.[4] For composite gels (1.5-mm thick, 2% acrylamide, 0.8% agarose), high molecular weight (900,000) proteins are transferred within 3 hr.[8]

After blotting, the sandwich is disassembled. Before removing the nitrocellulose sheet, however, it can be marked for subsequent orientation or cut into strips for individual probing. Usually, the paper remains moist, but occasionally overlaying with PBS facilitates removal of the strips.

The methods for blocking protein-free sites and probing with labeled antigens, antibodies, or lectins are similar to those described in Section 6 and are only briefly outlined here (see ''Reagents'' for details of solutions):

1. Block protein-free sites for 90 min at room temperature
2. Add first antibody antigen or lectin in probe solution for 2 hr at room temperature
3. Wash for 10 min × 3 in PBS-Tw
4. Where appropriate, add second (radiolabeled) antibody in probe solution for 90 min at room temperature
5. Wash for 10 min × 4 in PBS-Tw
6. Air dry nitrocellulose and expose to X-ray film

Steps 1 to 5 are performed on a rocker or rotary shaker.

Small-Pore Gels (Polyacrylamide; Figure 1a and b).

SDS-polyacrylamide gels are prepared according to standard methods using either Tris-HCl[11] or phosphate buffers.[12] To facilitate protein transfer, reversible crosslinkers may be included in the resolving gel as follows[1]: stock (30%) acrylamide solutions contain either 1.5 g of *N,N*'-dialyltartardiamide (DADT, Bio-Rad, Calif.) or 1.1 g ethylene diacrylate (EDA, Pfalz and Bauer, Conn.) per 100 mℓ solution. The resolving gels (30 mℓ) are polymerized with either 300 μℓ ammonium persulfate (10%, w/v) and 30 μℓ TEMED (*N,N,N*', *N*'-tetramethylethylenediamine) for DADT gels or 100 μℓ ammonium persulfate and 10 μℓ TEMED for EDA gels. To preserve the integrity of the gel, 1% agarose is included in the resolving, but not in the stacking, gel. The 3% stacking gel is polymerized with DADT.

After completion of electrophoresis, the crosslinker is cleaved by two 30-min incubations in 250 mℓ 2% (88 m*M*) periodic acid (DADT gels) or two 30-min incubations in 25 m*M* ammonium hydroxide (EDA gels). The gels are finally equilibrated in transfer buffer (10 m*M* Tris-50 m*M* saline or 50 m*M* phosphate buffer near neutral pH are suitable) for 30 to 60 min and nitrocellulose is presoaked in the same buffer.

Blotting sandwiches are prepared as illustrated in Figures 1a or b. (Figure 1a), a sheet of thick filter paper is moistened in the desired buffer and placed on top of the gel support with the edges submerged in the buffer. The gel is then placed on top of the filter paper-covered support. The gel is overlaid successively with presoaked nitrocellulose, two sheets of moist filter paper, a 4-cm wad of absorbent paper towels, and a weight. At all stages, care is taken to avoid trapping air bubbles. If all the paper towels above the gel become moist, they are replaced. The time necessary for transfer should be determined by pilot experiments but it is usually greater than 12 hr. After removal the nitrocellulose is blocked and probed as described earlier.

Assembly of the contact diffusion sandwich (Figure 1b) is more similar to that used for electrophoretic blotting (see Chapter 4.3.1), except that nitrocellulose is placed on both sides of the gel. The sandwich is compressed by clips or rubber bands and placed in a 2-ℓ buffer tank for 1 to 2 days. The buffer (Tris-saline, phosphate-saline as above) is changed once during transfer. Blocking and probing of nitrocellulose is performed as described earlier.

COMMENTS ON THE PROCEDURES

A comparison between diffusion and electrophoretic blotting methods is presented in Table 1.

Methods a and b (from Figure 1) are slow and relatively inefficient. Reported percentages of transfer of radiolabeled proteins from 10 to 12% polyacrylamide gels varied from 16 to 75.[1,5] Efficiency of transfer decreases with increasing protein molecular weight and increasing concentration of acrylamide. Although cleavage of reversible crosslinkers facilitates protein transfer, both sodium periodate and strong alkali may irreversibly denature proteins leading to loss of antigenic sites and failure of detection. Transfer from thin gels (0.8-mm thick) may also facilitate transfer, but restricts volume applied to each lane.

For the transfer by capillary flow (Figure 1a), care should be taken not to allow the buffer to pass from lower to upper filter papers, thus bypassing the gel. In the original description of the contact diffusion method (Figure 1b), presoaking gel strips in buffer containing 4 *M* urea for 3 hr was recommended in order to remove SDS and possibly renature proteins. Since renaturation may be more important for retention of protein function than exposure of antigenic determinants, this step will not be necessary for all studies.

The direct gel contact diffusion method (Figure 1c) has a restricted number of applications, but is simple, fast, and efficient, even for proteins of M_r of 10^6. Efficient transfer is obtained after isoelectric focusing of proteins in 1% agarose (usual concentration) as well as size fractionation of high molecular weight proteins on composite gels (acrylamide concentrations from 1.8 to 4.5%). As for all blotting methods, however, the rate of transfer is inversely related to the protein size. Ampholines, urea, and small amounts (0.1%, w/v) of SDS do not appreciably interfere with capillary transfer so that washing of gels is unnecessary and may, especially in the case of focusing gels, cause loss of band resolution. Prolonged washing will cause the proteins to be lost from the gel entirely. The only disadvantage of the method are artefacts caused either by sample application to the agarose gel or cracking of the nitrocellulose (see Figure 3). These artifacts can usually be avoided by applying the protein sample in several dots on an agarose focusing gel and blotting for the minimum period of time necessary to achieve transfer. Minimizing blotting time prevents drying and subsequent cracking of the nitrocellulose. The relative ease of transfer of proteins from 7.5% polyacrylamide isoelectric focusing (IEF) gels but not 7.5% SDS polyacrylamide slab gels probably relates to the presence of proteins either on the surface (focusing) or interior (SDS-PAGE) of the gel.[3]

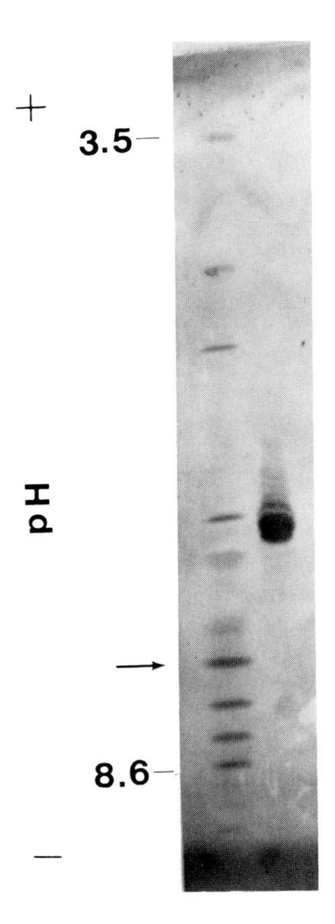

FIGURE 2. Amido Black staining of broad range pH markers (Pharmacia Fine Chemicals, Sweden) (left lane) and a purified Waldenstrom's monoclonal IgM (right lane) after focusing and transfer to nitrocellulose. Method C was employed and blotting time was 60 min. The position of myoglobin (M_r 17,500) is indicated by the arrow. Anode is indicated (+). (From Elkon, K. B., *J. Immunol. Methods*, 66, 313, 1984. With permission.)

APPLICATIONS

The continuous flow transfer techniques (Figures 1a and b) are used as an alternative to electrophoretic transfer of proteins resolved on small pore polyacrylamide gels when electrophoretic blotting equipment is not available. The capillary flow method has also been used to transfer proteins from 5% IEF polyacrylamide gels,[3] although electrophoretic transfer is again more efficient.

The direct gel contact diffusion method has been used to blot proteins from polyacrylamide[3] and agarose[4,13,14] IEF gels, from agarose zone electrophoresis gels,[15,16] and from composite gels.[8] The ability to focus all immunoglobulin classes in agarose gels makes the agarose focusing-diffusion nitrocellulose blot an excellent combination of methods for analysis of antibody heterogeneity. Applications include analysis of oligoclonal bands in urine or cerebrospinal fluid and analysis of polymeric rheumatoid factors.[17]

Examples of results obtained in this laboratory are illustrated in Figures 2 to 5. Figure 2 shows that proteins of vastly different size and charge are rapidly transferred from an agarose IEF gel. In Figure 3, acidic antigens present in a soluble human spleen extract are detected on nitrocellulose by probing with serum from a patient with the autoimmune disease systemic lupus erythematosus.[13] Conversely, Figure 4 illustrates how the heterogeneity of antibodies can be analyzed by focusing the antibody (not necessarily purified), transfer to nitrocellulose

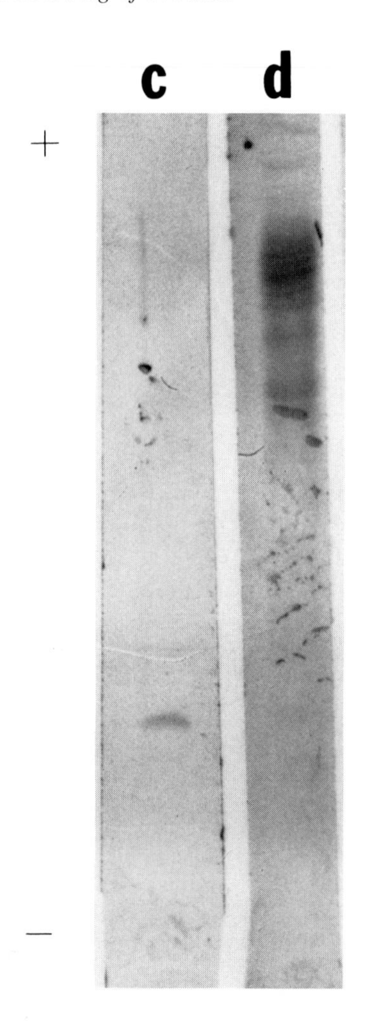

FIGURE 3. Autoradiographic detection of the acidic antigens Ro and La from a complex mixture of soluble human spleen proteins (100 μg) using agarose isoelectric focusing, diffusion blotting, and probing with a 1/100 dilution of a serum from a patient with systemic lupus erythematosus (lane d) or a normal serum (lane c). The second antibody was ^{125}I-labeled goat antihuman IgG (Tago, Calif.) at 500,000 cpm/mℓ of probe solution. Note artifactual markings due to drying of nitrocellulose in the center of lane d. The positions of the anode (+) and cathode (−) are shown on the left. (From Elkon, K. B. and Culhane, L., *J. Immunol.*, 132, 2350, 1984. With permission.)

and probing with the radiolabeled antigen. Rapid transfer of immunoglobulins from composite gels is also possible using either electrophoretic or diffusion blotting (Figure 5) where antibodies of more than one immunoglobulin class occur in the same serum. In this study, polymeric IgA antibodies were also detected by antigen overlay.[8]

ACKNOWLEDGMENT

The excellent assistance of Jia Li Chu during these studies is acknowledged.

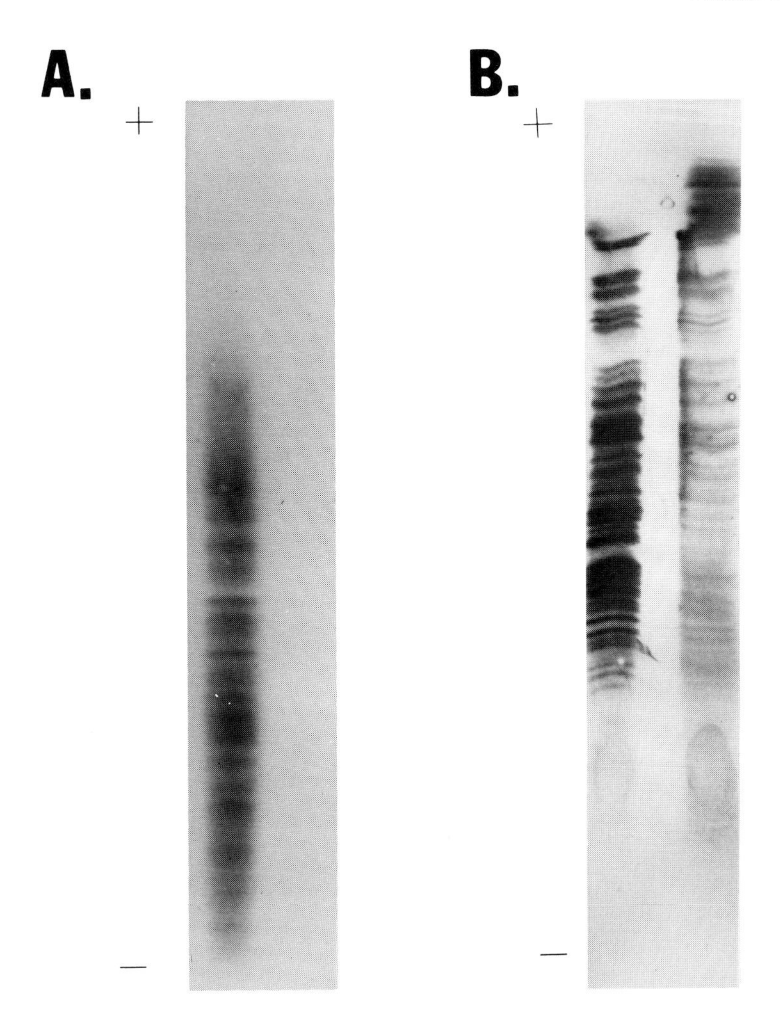

FIGURE 4. Clonotypic analysis of antibodies. (A) Isoelectric focusing of a polyclonal rabbit anti-BSA (bovine serum albumin) antiserum (left lane) and normal rabbit serum (right lane) followed by capillary transfer and 90 min overlay with ^{125}I-labeled BSA. (500,000 cpm/mℓ in probe solution with ovalbumin substituted for BSA. (B) IEF of the serum IgA fraction (isolated by sucrose density ultracentrifugation) from two patients with IgA rheumatoid factors. After capillary transfer to nitrocellulose, the lanes were probed for 90 min with ^{125}-labeled aggregated IgG (500,000 cpm/mℓ).

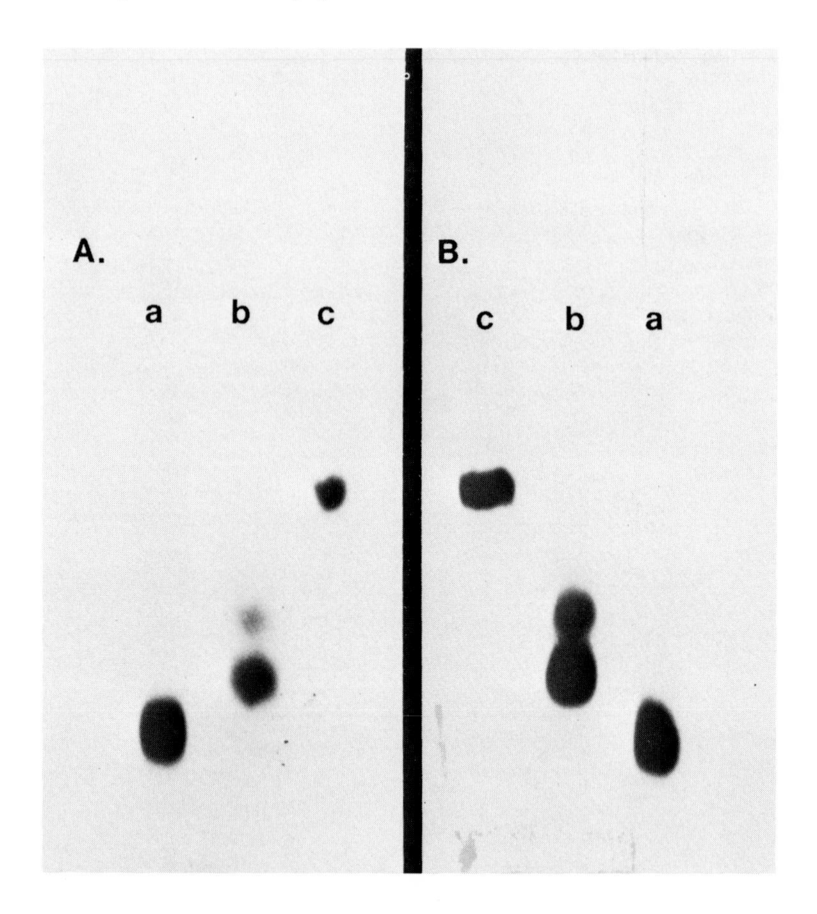

FIGURE 5. Electrophoresis of 4×10^{-4} cpm of ^{125}I-labeled IgG (lane a), IgA monomer and dimer (lane b), and IgM (lane c) on a composite gel (0.8% agarose, 2.1% polyacrylamide in 100 m*M* Na phosphate buffer, pH 7.2 containing 0.1%, w/v SDS) followed by transfer to nitrocellulose by capillary diffusion (A) or electrophoresis (B). Diffusion transfer time was 3 hr.[8] Autoradiograph exposure times were 16 hr for A and B (Kodak® X-Omat film). (From Elkon, K. B., Jankowski, P. W., and Chu, J. L., *Anal. Biochem.*, 140, 208, 1984. With permission.)

REFERENCES

1. **Renart, J., Reiser, J., and Stark, G. R.,** Transfer of proteins from gels to diazobenzyloxymethyl-paper and detection with antisera: a method for studying antibody specificity and antigen structure, *Proc. Natl. Acad. Sci. U.S.A.*, 76, 3116, 1979.
2. **Southern, E. M.,** Detection of specific sequences among DNA fragments separated by gel electrophoresis, *J. Mol. Biol.*, 98, 503, 1975.
3. **Reinhart, M. P. and Malamud, D.,** Protein transfer from isoelectric focusing gels: the native blot, *Anal. Biochem.*, 123, 229, 1982.
4. **Elkon, K. B.,** Isoelectric focusing of human IgA and secretory proteins using thin layer agarose gels and nitrocellulose capillary blotting, *J. Immunol. Methods*, 66, 313, 1984.
5. **Bowen, B., Steinberg, J., Laemmli, U. K., and Weintraub, H.,** The detection of DNA-binding proteins by protein blotting, *Nucl. Acids Res.*, 8, 1, 1980.
6. **Axelsen, N. H., Krøll, J., and Weeke, B., Eds.,** A manual of quantitative immunoelectrophoresis. Methods and applications, *Scand. J. Immunol.*, 2(Suppl. 1), 1973.
7. **Rosen, A., Ek, K., and Aman, P.,** Agarose isoelectric focusing of native human immunoglobulin M and macroglobulin, *J. Immunol. Methods*, 28, 1, 1979.

8. **Elkon, K. B., Jankowski, P. W., and Chu, J. L.,** Blotting intact immunoglobulins and other high molecular weight proteins after composite agarose-polyacrylamide gel electrophoresis, *Anal. Biochem.*, 140, 208, 1984.
9. **Elkon, K. B. and Chu, J. L.,** Counterimmunoblotting: detection of non-denatured or denatured antigens in antibody-containing agarose gels following polyacrylamide gel electrophoresis, *J. Immunol. Methods*, 70, 211, 1984.
10. **Maniatis, T., Fritsch, E. F., and Sambrook, J.,** Molecular Cloning, A Laboratory Manual, Cold Spring Harbor Laboratory, Cold Spring Harbor, N.Y., 1982, chap. 11.
11. **Laemmli, U. K.,** Cleavage of structural proteins during the assembly of the head of bacteriophage T_4, *Nature (London)*, 227, 680, 1970.
12. **Weber, K. and Osborn, M.,** The reliability of molecular weight determination by dodecyl sulfate polyacrylamide gel electrophoresis, *J. Biol. Chem.*, 244, 4406, 1969.
13. **Elkon, K. B. and Culhane, L.,** Partial immunochemical characterization of the Ro and La proteins using antibodies from patients with the sicca syndrome and lupus erythematosus, *J. Immunol.*, 132, 2350, 1984.
14. **Elkon, K. B.,** Charge heterogeneity of human secretory component: immunoglobulin and lectin binding studies, *Immunology*, 53, 131, 1984.
15. **McMichael, J. C., Greisiger, L. M., and Millman, I.,** The use of nitrocellulose blotting for the study of hepatitis B surface antigen electrophoresed in agarose gels, *J. Immunol. Methods.*, 45, 79, 1981.
16. **Keir, G., Walker, R. W. H., Johnson, M. H., and Johnson, E. J.,** Nitrocellulose immunofixation following agarose electrophoresis in the study of immunoglobulin G subgroups in unconcentrated cerebrospinal fluid, *Clin. Chim. Acta*, 121, 231, 1982.
17. **Chu, J. L., Gharavi, A. E., and Elkon, K. B.,** Spectrotypic analysis of IgM and IgA rheumatoid factors, *Clin. Exp. Immunol.*, 63, 601, 1986.

4.2 VACUUM BLOTTING — TRANSFER OF PROTEINS FROM POLYACRYLAMIDE GELS ONTO NITROCELLULOSE BY VACUUM SUCTION

Marnix Peferoen

INTRODUCTION

At the beginning of 1982 we used a capillary blotting system to transfer proteins from polyacrylamide gels onto nitrocellulose.[1] The gel was blotted overnight, the amount of protein transferred was low, but the system worked. We were quite satisfied with the high resolution, especially compared to immunoelectrophoresis in agarose, but still it was time-consuming and the transfer efficiency was rather low. Electrophoretic blotting[2,3] was another possibility, but there were also financial restrictions for the department.

We had noticed before that drying of Coomassie blue stained polyacrylamide gels on a slab gel dryer removed some staining from the gel, leaving a nice "protein pattern" on the underlying filter papers. We wondered whether it would be possible to remove unfixed proteins from polyacrylamide gels by using the same set-up, so, a flow of buffer was forced through the gel by vacuum suction; this was in contrast to the capillary force used by Renart et al.[1] Vacuum blotting seemed to be a straightforward name, referring to the force responsible for the transfer.[4]

EQUIPMENT AND REAGENTS

Apparatus

Proteins were separated in a vertical electrophoresis apparatus (Bio-Rad, Protean Cell) and were transferred from the gel to a nitrocellulose membrane by the use of a slab gel dryer (Bio-Rad, Model 1125), connected to a suction pump (Heraeus, 250 W, 1390 ℓ/min). Staining was performed on a rocking table and radioactivity was counted in a liquid scintillation counter (Beckman, LS 9000).

Materials

Nitrocellulose membrane (pore size 0.22 μm, GSWPO2500, Millipore). The vacuum transfer also required a flexible plastic sheet and about 20 filter papers (Whatman No. 2, 16 × 16 cm). The silica gel column was made with a transparent plastic tube (50 × 5 cm) filled with silica gel. Radiolabeled proteins were visualized on Kodak X-Omat film.

Chemicals

Acrylamide and bisacrylamide were purchased from Serva. Glycine, Tris, sodium dodecyl sulfate (SDS), methanol, Tween® 20, $Na_2HPO_4 \cdot 2H_2O$, KH_2PO_4, and urea were obtained from Merck. High M_r calibration proteins (M_r 67,000 to 669,000), and a ^{14}C-methylated protein mixture (M_r 14,300 to 200,000) were supplied by Pharmacia Fine Chemicals and Amersham, respectively. Female *Locusta migratoria* hemolymph was collected from our own locust culture. Reagents used for silver staining were similar to those used by Heukeshoven and Dernick.[7] AuroDye is a Janssen Life Sciences Product.

Solutions

1. Transfer buffer pH 8.3	
25 m*M* Tris	3.03 g
150 m*M* glycine	11.26 g
20% (v/v) methanol	200 mℓ
Distilled water	ad 1000 mℓ
2. Rinsing buffer pH 7.2	
14 m*M* $Na_2HPO_4 \cdot 2H_2O$	2.49 g
6 m*M* KH_2PO_4	0.82 g
0.3% (v/v) Tween® 20	0.90 mℓ
Distilled water	ad 300 mℓ

PROCEDURES

Gel Electrophoresis

Polypeptides were separated in SDS-polyacrylamide gradient (5 to 15%) gels, with the discontinuous buffer system of Laemmli.[5] Proteins were separated by discontinuous electrophoresis applied to polyacrylamide gradient (5 to 15%) gels.[5] Female *L. migratoria* hemolymph was separated for 2000 V · hr in polyacrylamide-urea tubes with a pH gradient from 3.5 to 10 (nonequilibrium pH gradient gel electrophoresis, NEpHGE[6]). The gels were then electrophoresed at right angles on SDS-polyacrylamide gradient (5 to 15%) gels with the discontinuous buffer system. All polyacrylamide gels were 1.5-mm thick. Proteins and polypeptides were silver stained with a procedure based on the method of Heukeshoven and Dernick.[7] Radiolabeled polypeptides were visualized by autoradiography.

Blotting Set-Up

After electrophoresis, the stacking gel was removed and the separating gel was soaked in transfer buffer for 20 to 30 min. The gel was frozen for 1 to 15 days, preferably at −70°C, in a plastic bag containing 30 mℓ of transfer buffer, or it was immediately processed. In the slab gel dryer, a sheet of dry filter paper was laid on a porous polyethylene sheet, which created an evenly distributed suction force. A nitrocellulose membrane, wetted in 10 mℓ transfer buffer, was placed on the dry filter paper. A window 5 mm smaller than the gel was cut in a plastic sheet, leaving outer margins of at least 6 cm width. This plastic sheet was put on the dry filter paper, with the window on the nitrocellulose membrane. The gel was placed on the nitrocellulose so that the edges of the gel rested on the plastic sheet (Figure 1A). Air bubbles disturbing the contact between the gel and the nitrocellulose membrane were removed. Filter paper, 15 sheets, trimmed to the same size as the gel, were wetted in transfer buffer and placed on the gel.

Trapping of air bubbles between the layers should be avoided. The four outer margins of the plastic sheet were then folded on top of the wet filter papers (Figure 1B, C). The slab gel dryer was closed with its rubber sheet and it was connected to the suction pump. The suction pump was protected against moisture by interconnecting a 1000-mℓ side arm flask and a column with dried silica gel between the slab gel dryer and the suction pump (Figure 1D).

Protein Transfer

The transfer buffer contained in the layer of filter papers on top of the gel was driven through the gel by the vacuum. The plastic sheet prevented leakage of the buffer around the gel. Normally, the gel was blotted for 1 hr. However, if a quantitative transfer was required, the blotting time was extended for several hours. In that case it was necessary to pour 100 mℓ of transfer buffer on the filter papers every hour, after removing the rubber sheet and unfolding the edges of the plastic sheet. This prevented the gel from drying and sticking to the filter paper.

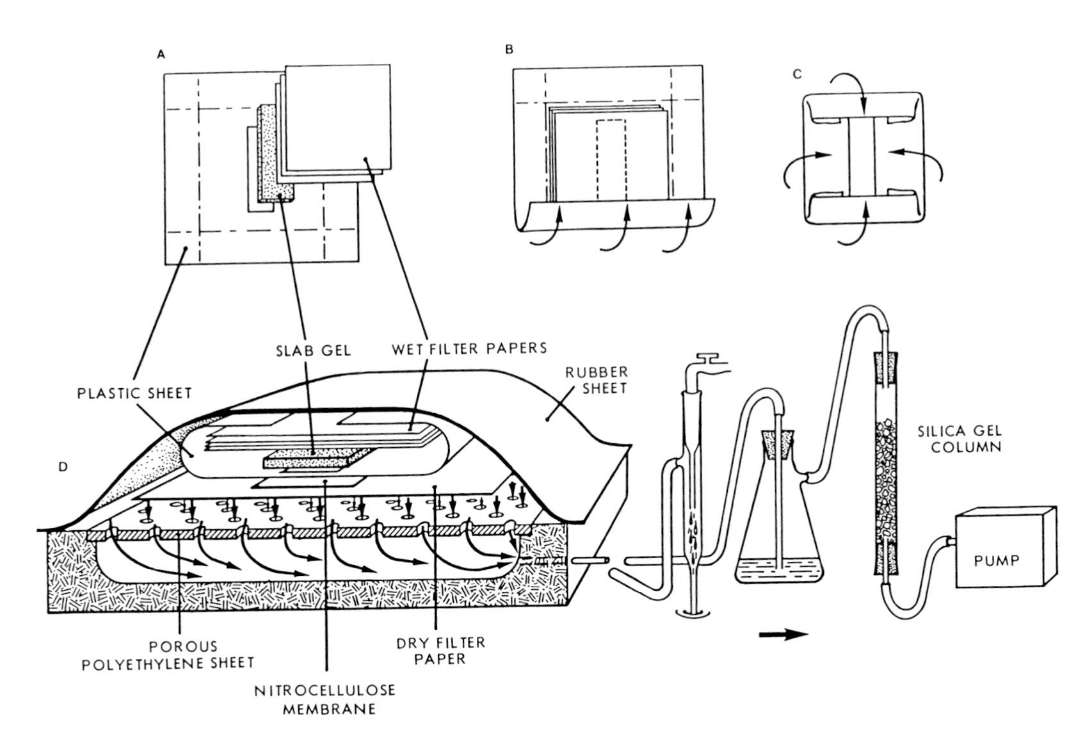

FIGURE 1. Vacuum blotting set-up. (A) A nitrocellulose membrane was wetted in blotting buffer and placed on a dry filter paper in the slab gel dryer. The membrane was covered with a plastic sheet with a window 5 mm smaller than the gel. The gel was placed on the nitrocellulose membrane with its edges resting on the plastic sheet. Some 15 sheets of filter paper, wetted in 500 mℓ blotting buffer, were laid on the gel. (B, C) The outer margins of the plastic sheet were folded on top of the filter papers. (D) The slab gel dryer was closed with its rubber sheet and was connected to the suction pump or a water jet pump.

Visualization of Nitrocellulose-Bound Proteins

After the transfer, the nitrocellulose membrane was washed in rinsing buffer for 30 min at 37°C and three times for 15 min at room temperature. The membrane was then incubated for 4 to 12 hr in AuroDye.[8] More details about this staining technique are presented in Chapter 6.1.3. Radiolabeled polypeptides were visualized by autoradiography.

Quantitation of the Transfer

After electrophoresis of ^{14}C-methylated polypeptides, gels were cut into strips containing two separation lanes, which were blotted onto nitrocellulose by vacuum suction. Polypeptides in the gel and on the nitrocellulose membrane were visualized by autoradiography and were excised and counted in a liquid scintillation counter.

COMMENTS ON THE PROCEDURE

After electrophoresis, gels may be kept frozen for weeks without any loss of resolution. Gels were preferably stored at −70°C since we noticed that gels stored at −20°C occasionally cracked on thawing.

Gels with less than 5% acrylamide were not suitable for vacuum blotting since these gels stuck to the filter paper and to the nitrocellulose. In order to adsorb impurities from the transfer buffer or from the filter papers, a sheet of nitrocellulose was inserted between the gel and the stack of wet filter papers. Thus, the gel was sandwiched between two nitrocellulose membranes. The AuroDye staining especially benefitted from the filtering activity of

Table 1
TRANSFER EFFICIENCY OF VACUUM BLOTTING

	% of ^{14}C-methylated proteins transferred onto nitrocellulose after SDS-polyacrylamide gradient (5 to 15%) electrophoresis		
M_r of proteins	**1 hr**	**2 hr**	**4 hr**
14,300	28	45	64
30,000	23	38	54
46,000	18	31	45
69,000	23	32	44
92,500	18	26	36
200,000	17	21	32

Note: After electrophoresis in SDS-polyacrylamide gradient (5 to 15%) gels (14 × 14 cm), ^{14}C-methylated polypeptides (M_r 14,300 to 200,000) were blotted by vacuum blotting for 1, 2, and 4 hr. About 100 mℓ of transfer buffer was added to the stack of filter papers every 30 min. Radiolabeled polypeptides were visualized by autoradiography and were cut from the gel and the nitrocellulose membrane for liquid scintillation counting. Counts in separated polypeptides which were not blotted were referred to as 100%.

the nitrocellulose membrane lying on top of the gel. If the blotting time was extended for 4 hr without adding buffer to the filter papers, the gel dried and stuck to the nitrocellulose paper. Rehydrating the gel made it swell and the nitrocellulose could be easily removed. Polypeptides of low M_r (lower than 20,000) were adsorbed to a lesser extent by the nitrocellulose membrane with a pore size of 0.45 μm.[3,9] This resulted in an insufficient immobilization, causing leakage or diffusion of the blotted proteins.[4] We therefore decided to use a nitrocellulose membrane with a smaller pore size (0.22 μm).

The amount of transfer buffer driven through the gel was of more importance for the transfer than the blotting time. For a standard polyacrylamide gradient gel (5 to 15%, 14 × 14 cm), 500 mℓ of transfer buffer was necessary to obtain a satisfactory transfer. A semiquantitative transfer, however, required at least 1500 mℓ of transfer buffer. The transfer efficiency was, of course, very much dependent on the characteristics of the proteins and the gel (e.g., M_r, solubility in the blotting buffer, pore size).

A suction pump proved not to be indispensable since a good water jet pump did work as well (results not shown). As a rule-of-thumb, if your set-up functions for drying polyacrylamide gels, it will also be suitable for vacuum blotting.

EVALUATION

Transfer Efficiency

A mixture of ^{14}C-methylated polypeptides with M_r from 14,300 to 200,000 was separated on SDS-polyacrylamide gradient (5 to 15%) gels (14 × 14 cm). After electrophoresis, polypeptides were blotted by vacuum suction for 1, 2, and 4 hr. Every 30 min, about 100 mℓ of transfer buffer was added. The transfer efficiency for different proteins was determined by liquid scintillation counting after cutting the ^{14}C-methylated protein bands from the gel and from the nitrocellulose. Counts in polypeptide bands in gels, which were not used for blotting, were referred to as 100%. The percentage of counts in different polypeptides which were transferred onto nitrocellulose membranes is shown in Table 1. The transfer efficiency was dependent on the M_r of the protein. After 4 hr of vacuum blotting, the amount of lysozyme eluted (M_r 14,300) was twice as high as the amount of myosin (200,000): 64 and 32%, respectively.

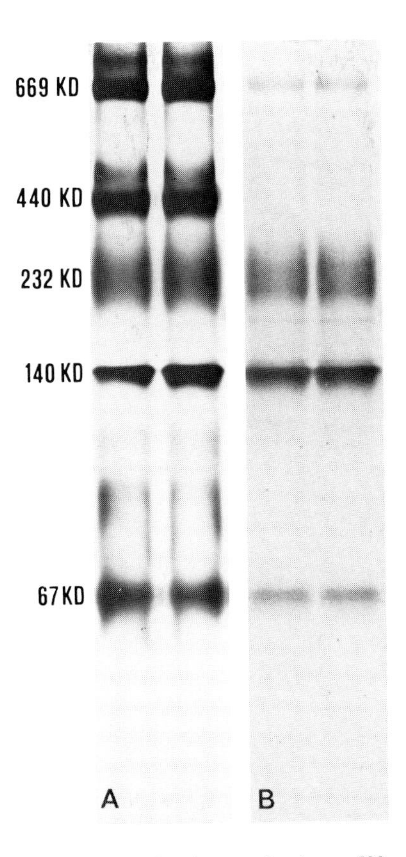

FIGURE 2. Silver staining of high M_r calibration proteins (some 500 ng each), separated in a polyacrylamide gradient (5 to 15%) gel (A). A duplicate gel was blotted for 1 hr with 500 mℓ blotting buffer onto nitrocellulose, which was then stained with colloidal gold, i.e., AuroDye (B).

The sum of the counts of lysozyme (M_r 14,300) in the gel and on the corresponding sheet on nitrocellulose compared to counts obtained from the corresponding control gels, not subjected to blotting, showed that even after 4 hr of blotting, there was no leakage through a nitrocellulose membrane with 0.22 μm pore size.

Applications of Vacuum Blotting

Transfer of Native Proteins

Native proteins separated in polyacrylamide (gradient) gels were easily transferred onto nitrocellulose by vacuum blotting. This was demonstrated by the blotting of a mixture of proteins with M_r varying between 67,000 and 669,000. The gel (14 × 14 cm) in Figure 2 contained about 500 ng of each protein. After 1 hr of blotting with 500 mℓ of blotting buffer, the proteins were detectable on the nitrocellulose membrane with no loss of electrophoretic resolution (Figure 2B). As it was demonstrated for SDS-denatured polypeptides, the transfer efficiency of large proteins was lower than the transfer efficiency of small proteins. The fact that the elution rate of ferritin (M_r 400,000) was lower than the elution rate of thyroglobulin (M_r 669,000), strongly suggested that other factors, such as solubility, affected the elution as well. Increasing the amount of buffer and the blotting time should enhance the transfer.

Transfer from Two-Dimensional Gels

Hemolymph of female *L. migratoria* was separated by two-dimensional (2D) electrophoresis according to the method of O'Farrell et al.[6] Figure 3A shows a part of the gel containing

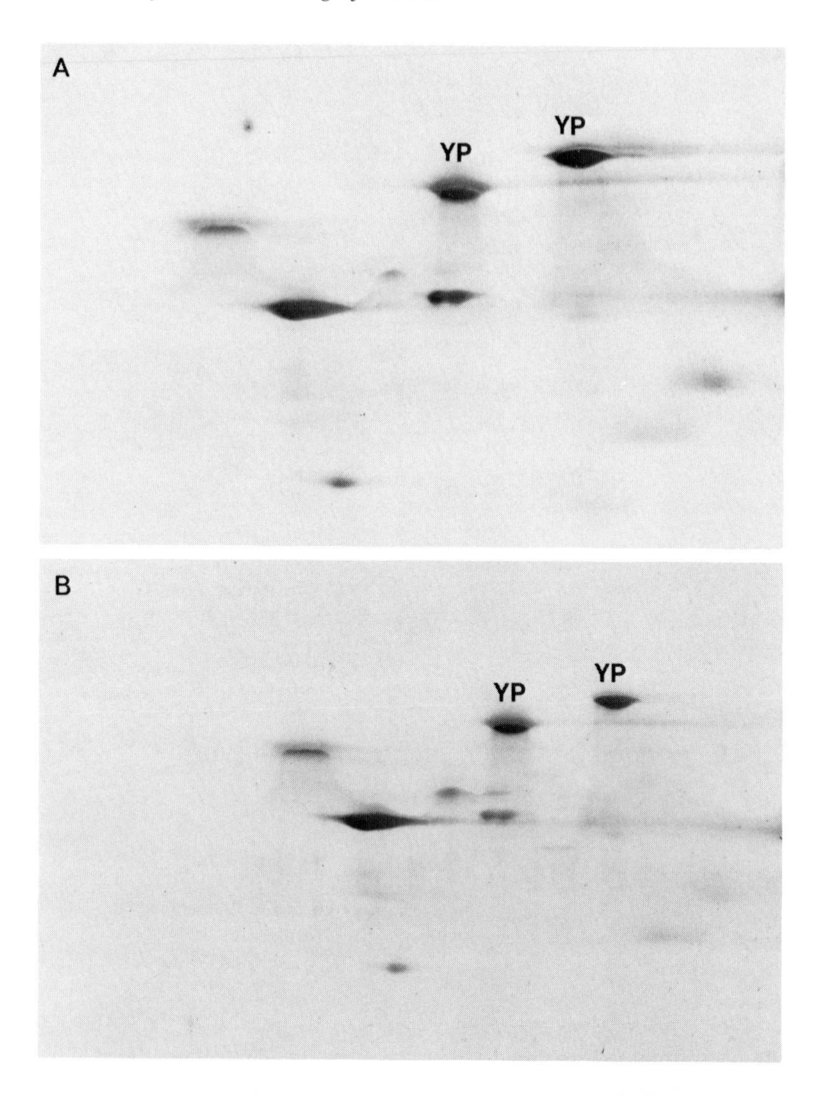

FIGURE 3. 2D electrophoresis (NEpHGE-SDSPAGE) of *Locusta migratoria* female hemolymph. The gel was stained with silver (A) or blotted for 1 hr with 500 mℓ blotting buffer onto nitrocellulose, which was stained with AuroDye (B). Only a part of the gel (10 × 7 cm) containing 2 female specific yolk polypeptides (YP), is shown.

two female specific yolk polypeptides (YP). These polypeptides have M_r in the range of 120,000.[10] The gel was blotted for 1 hr with 500 mℓ of blotting buffer. The high resolution of the 2D electrophoresis was preserved (Figure 3B).

USE AND LIMITATIONS

The transfer characteristics of vacuum blotting are comparable to the transfer profile of low powered electroblotting,[3,9] meaning that the elution is dependent on the M_r of the proteins. Erickson et al.[11] have reported a quantitative transfer of proteins with M_r larger than 100,000 by adding 0.1% SDS to the transfer buffer and by performing the transfer with ten times more energy. Kyhse-Andersen[12] has described an electrophoretic transfer based on isotachophoretic principles which usually resulted in a complete blotting in 1 hr (see Chapter 4.3.2.). Thus, if a quantitative transfer of proteins from a polyacrylamide gel is required, high powered or isotachophoretic electroblotting should be used.

Still, vacuum blotting may offer some advantages. It is a very cheap system since laboratories doing electrophoresis routinely can transfer proteins onto immobilizing paper by using their slab gel dryer system. In vacuum blotting there are, of course, no problems with increasing conductivity and temperature during transfer or with proteins which do not migrate at their isoelectric point.[13] Finally, there is no restriction concerning the choice of buffer so that the buffer chosen should provide the best conditions for transferring and binding proteins to the adsorbing membrane. If necessary, by adjusting the buffer, proteins can even be selectively removed from the gel.

Vacuum blotting has been used in many immunological identifications of proteins[14-17] and it has also been adapted to transfer DNA.[18]

REFERENCES

1. **Renart, J., Reiser, J., and Stark, G. R.,** Transfer of proteins from gels to diazobenzyloxymethyl-paper and detection with antisera: a method for studying antibody specificity and antigen structure, *Proc. Natl. Acad. Sci. U.S.A.*, 76(7), 3116, 1979.
2. **Towbin, H., Staehelin, T., and Gordon, J.,** Electrophoretic transfer of proteins from polyacrylamide gels to nitrocellulose sheets: procedure and some applications, *Proc. Natl. Acad. Sci. U.S.A.*, 76(9), 4350, 1979.
3. **Burnette, W. N.,** "Westernblotting": electrophoretic transfer of proteins from sodium dodecyl sulfate-polyacrylamide gels to unmodified nitrocellulose and radiographic detection with antibody and radioiodinated protein A, *Anal. Biochem.*, 112, 195, 1981.
4. **Peferoen, M., Huybrechts, R., and De Loof, A.,** Vacuum-blotting: a new simple and efficient transfer of proteins from sodium dodecyl sulfate-polyacrylamide gels to nitrocellulose, *FEBS Lett.*, 145, 369, 1982.
5. **Laemmli, U. K.,** Cleavage of structural proteins during assembly of the head of bacteriophage T4, *Nature (London)*, 227, 680, 1970.
6. **O'Farrell, P. Z., Goodman, H. M., and O'Farrell, P. H.,** High resolution two-dimensional electrophoresis of basic as well as acidic proteins, *Cell*, 12, 1133, 1977.
7. **Heukeshoven, J. and Dernick, R.,** Simplified method for silver staining of proteins in polyacrylamide gels and the mechanism of silver staining, *Electrophoresis*, 6, 103, 1985.
8. **Moeremans, W., Daneels, G., and De Mey, J.,** Sensitive colloidal metal (gold or silver) staining of protein blots on nitrocellulose membranes, *Anal. Biochem.*, 145, 315, 1985.
9. **Lin, W. and Kasamatsu, H.,** On the electrotransfer of polypeptides from gels to nitrocellulose membranes, *Anal. Biochem.*, 128, 302, 1983.
10. **Gellissen, G., Wajc, E., Cohen, E., Emmerich, H., Applebaum, S. W., and Flossdorf, J.,** Purification and properties of oocyte vitellin from the migratory locust, *J. Comp. Physiol.*, 108, 287, 1976.
11. **Erickson, P. F., Minier, L. N., and Lasher, R. S.,** Quantitative electrophoretic transfer of polypeptides from SDS polyacrylamide gels to nitrocellulose sheets: a method for their re-use in immunoautoradiographic detection of antigens, *J. Immunol. Methods*, 51, 241, 1982.
12. **Kyhse-Andersen, J.,** A simple horizontal apparatus without buffer tank for electrophoretic transfer of proteins from polyacrylamide to nitrocellulose, *J. Biochem. Biophys. Methods.*, 10, 203, 1984.
13. **Legocki, R. P. and Verma, D. P. S.,** Multiple immunoreplica technique: screening of specific proteins with a series of different antibodies using one polyacrylamide gel, *Anal. Biochem.*, 111, 385, 1981.
14. **Briers, T. and Huybrechts, R.,** Control of vitellogenin synthesis by ecdysteroids in *Sarcophaga bullata, Insect Biochem.*, 14, 121, 1984.
15. **Geysen, J., De Loof, A., and Vandesande, F.,** How to perform subsequent or "double" immunostaining of two different antigens on a single nitrocellulose blot within one day with an immunoperoxidase technique, *Electrophoresis*, 5, 129, 1984.
16. **Peferoen, M. and De Loof, A.,** Intraglandular and extraglandular synthesis of proteins secreted by the accessory reproductive glands of the Colorado potato beetle, *Leptinotarsa decemlineata, Insect Biochem.*, 14, 407, 1984.
17. **Ishida, I., Ichikawa, T., and Deguchi, T.,** Immunochemical and immunohistochemical studies on the specificity of a monoclonal antibody to choline acetyltransferase of rat brain, *Neurosci. Lett.*, 42, 267, 1983.
18. **Zaitsev, I. Z. and Yakovlev, A. G.,** Vacuum transfer of DNA to filters for detecting interindividual polymorphism by Southern blotting hybridization method, *B. Exp. B. Med.*, 96, 1442, 1983.

4.3 ELECTROPHORESIS

4.3.1 ELECTROTRANSFER IN EQUIPMENT CONTAINING BUFFER

Michael Bittner and Edwin Rowold, Jr.

INTRODUCTION

Blotting combines powerful ways of resolving molecular mixtures with precise ways of recognizing specific molecules. Blotting followed by immunodetection can be used in conjunction with discontinuous gel electrophoresis,[1] denaturing discontinuous electrophoresis,[2] isoelectric focusing (IEF),[3] or two-dimensional (2D) electrophoresis[4] to assign a physical characteristic (mass, isoelectric point, etc.) to a particular antigen. This form of analysis can provide information useful in determining whether the antigen under examination is a single, discrete species, how abundant the antigen is in the mixture, and what characteristics may prove useful in further purification of the antigen. The great discrimination[5] and sensitivity[6] of the antigen-antibody interaction allow one to apply immunoblotting analyses to protein mixtures in which the antigen is present as a very minor component.

In order to make proteins resolved on a gel readily accessible to antibodies, methods which allowed proteins to be eluted from gels and deposited onto freely accessible surfaces were required. Solutions to a similar problem involving nucleic acids had been devised.[7,8] In these techniques, nucleic acids were eluted from a gel by either liquid flow[7] or an electric field,[8] and deposited onto a nitrocellulose filter in contact with one side of the gel. This type of methodology was dubbed blotting, and the filter replica produced by blotting became known as a blot. These same approaches were found to be applicable to the elution of proteins from gels. Both fluid flow[9] and electric fields[10,11] could be used to transfer proteins out of gels and onto a closely apposed filter. Simpler approaches in which the protein is allowed to diffuse out of a gel were also found to be suitable for preparing blots.[12] The advantages of capillary blotting and diffusion contact blotting have been reviewed in the preceding chapters. The particular advantage which derives from the use of an electrical field to transfer protein from a gel to a filter is the speed with which extensive transfer can be accomplished. Numerous improvements on the original methods for electroblotting have been reported in the past few years, and one can now expect to achieve nearly quantitative transfer of most proteins within a few hours.

This chapter deals with the practical aspects of protein blotting in an electrical field. We discuss the apparatus and reagents used, two protocols which are useful for a variety of applications, and most importantly, methods of evaluating the performance of any given protocol. Evaluation is extremely important, since a particular protein within a heterogeneous mixture of proteins will exhibit idiosyncratic behavior with regard both to conditions appropriate for elution[13,14] and conditions appropriate for binding to a given matrix.[15] Finally, we present a cursory listing of reports which detail conditions which have been successfully used with a variety of gel systems and immobilizing matrices.

EQUIPMENT

Blotting Devices

Several designs are available in the literature of the apparatus required for electrically driven protein blotting[10,11,16] and many are commercially available from suppliers such as Hoefer Scientific Instruments, San Francisco, Bio-Rad, Richmond, Calif., E-C Apparatus Corp., St. Petersburg, Fla., and Idea Scientific, Corvallis, Ore. A sketch of such an apparatus

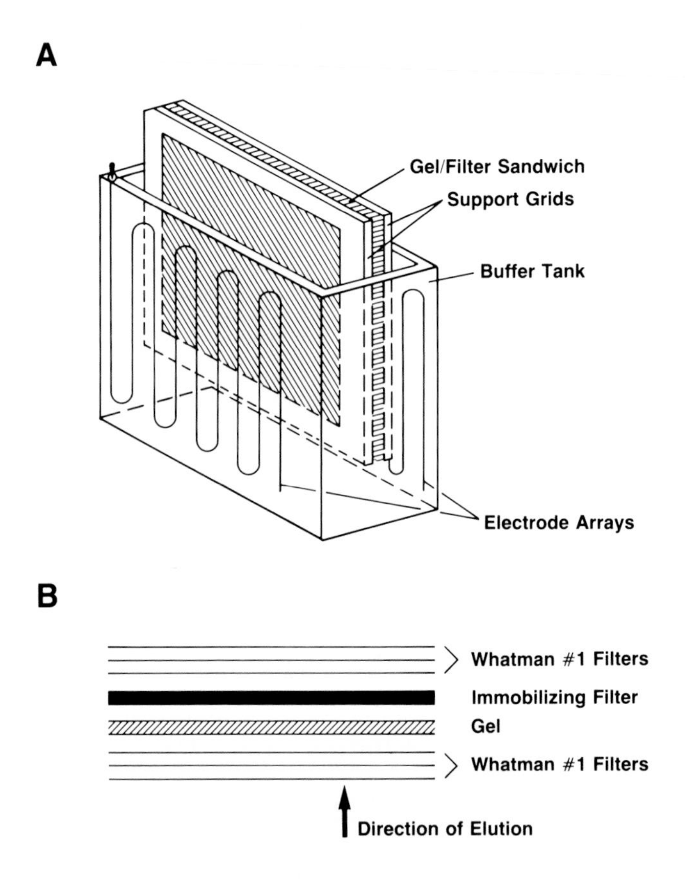

FIGURE 1. An electrophoretic transfer apparatus. Panel A depicts the arrangement of the electrodes and gel/matrix sandwich holder in the buffer tank. Panel B shows the arrangement of the gel and immobilizing matrix within the supporting cassette.

depicting the essential elements is given in Figure 1. The electrical field is generated between a pair of electrodes in a tank filled with buffer. The electrodes may be either a continuous slab of conductor or an array of conductors which generate a homogeneous electrical field. The gel and the filter are held in intimate contact and supported perpendicular to the electrical field by two supporting grids which can be fastened together and held in place within the buffer tank. The electrical field is produced by a DC power supply, typically one capable of producing 40 to 60 V at 1 to 2 A. Two other features often incorporated are a circulation system to exchange buffer between the anodic and cathodic sections of the buffer tank and a heat exchanger which may be connected to an external cooling source. Ordinary laboratory refrigerated, circulating water baths such as those manufactured by Precision Scientific, Fisher Scientific, Lauda, and Haake-Buchler provide sufficient cooling capacity. The high current flows often employed during electroelution lead to considerable electrolytic generation of acid and base at the electrodes and to considerable joule heating of the buffer, either of which may be undesirable.

Immobilizing Matrices

Matrices used in immunoblotting are reviewed in Section 5. Here, it is sufficient to note that the matrices fall into two broad classes. Filters such as nitrocellulose,[10] Zeta-bind (Zeta-probe),[14] Biodyne A, Gene Screen, and Nytran retain electroeluted proteins by noncovalent attachment. Filters such as diazobenzyloxymethyl (DBM)[9] cellulose paper, diazophenylthioether (DPT)[16] cellulose paper, and cyanogen bromide activated paper (CBP)[17] retain proteins by the formation of covalent linkages between the paper and the protein.

It is often useful to compare a set of the above filters for their abilities to bind a particular antigen and their compatibility with the chosen detection system. A simple and economic way to do such a comparison is to spot a small amount of antigen containing fluid directly onto the set of filters to be compared, wash each filter in a buffer solution appropriate for transfer to that particular filter, and then block the filters and subject them to the detection regimen.[18] The results from this dot immunobinding approach can generally be interpreted as the best result one may expect from the filter type tested. The signal strength to background noise ratio on a blot resulting from electrophoretic separation and transfer is unlikely to be higher than that seen on a dot binding assay.

Gel Systems

While the type of resolving gel system required for a particular line of investigation is often fixed, there are important gel variables which are usually discretionary. The thickness and the porosity of the resolving gel makes a considerable difference in transfer results.[19] Poor mobility of a protein during electroelution can require the use of very long periods of electroelution.[13] If activated filters are being used to immobilize the eluted proteins, side reactions of the active groups can lead to severe decreases in binding capacity with time.[15] Long periods of electroelution increase the chance of bubbles forming between the gel and the filter, which can reduce or eliminate transfer from a region of the gel. For these reasons, excellent rules of thumb are limit gel thickness to the minimum required to accomodate the required protein loads and to achieve reasonable mechanical stability, and use a gel porosity which gives the desired separation but allows maximum mobility of the antigen being investigated.

RECIPES

	Transfer Buffers	**(mℓ)**
1.	**Phosphate transfer buffer**	
	1 *M* phosphate buffer (0.19 *M* $NaHPO_4$, 0.81 *M* Na_2PO_4)	15
	20% (w/v) sodium dodecyl sulfate (SDS) in water	5
	Deionized water	980
2.	**Tris transfer buffer**	
	1.92 *M* glycine	100
	2 *M* Tris (base)	12.5
	20% (w/v) SDS in water	5
	Deionized water	880
3.	**Tris-methanol transfer buffer**	
	Tris transfer buffer	800
	Methanol	200

PROCEDURES

The prime requirement for any electroblotting protocol is that it provide an efficient transfer of the antigen or antigens under study from a gel onto a filter. A quantitative transfer of all proteins, while desirable, is seldom achievable with a convenient protocol. The protocols which are described in this section were derived by amalgamating many published procedures.[9-11,21] The treatments indicated in these protocols provide very effective elution for most proteins, depending primarily on the characteristics of the matrix itself, and the amount of protein loaded.[10,14]

Covalent Immobilization

Step 1 — Pretreatment of Gels

Rinse the gel by gently rocking in phosphate transfer buffer at room temperature, changing the solution every 20 min over a period of 1 hr. The volume of the rinse solution should be at least five times the volume of the gel. If freshly activated paper (for example, Transabind aminophenolthioether (APT) paper treated per Schleicher & Schuell's instructions) is to be used, the final steps in activation should be timed to finish at the time pretreatment of the gel is complete. The activated paper should be briefly rinsed in the transfer buffer before assembling the electroblotter.

Step 2 — Electroblotting

Place the gel on the activated filter and sandwich the gel and filter between two three-sheet sets of Whatman No 1 filter paper which have been thoroughly soaked in the transfer buffer. Make sure that no air bubbles are trapped between layers of the sandwich. Place the gel sandwich between the support grids of the electroblotting device and clamp the grids together, again making sure that no air bubbles are trapped at the junctures of the grid and the sandwich. Place the assembled sandwich carrier into the buffer tank of the electrophoresis unit. The buffer tank should contain sufficient precooled (4°C) buffer to cover the filter sandwich. Arrange the carrier so that the activated filter is between the gel and the electrode toward which the antigen will migrate. SDS-complexed proteins are anions, and so the activated filter is placed between the gel and the anode.

Potentials of 5 to 30 V/cm are used to elute the gel. A primary limit to the field strength which can be employed is the heat produced by current flux. A tank containing 4 ℓ of buffer cooled to 4°C, placed in a 4°C room, is heated to approximately 25°C by a current of 1.5 A applied for 2 hr. The use of a supplemental cooling system allows adequate control of heating at even higher current densities. A 2-hr blot at 5 V/cm is sufficient for elution of most proteins of M_r <100,000 from gels <0.8 mm in thickness and <18% (w/v) polyacrylamide (acrylamide:bisacrylamide = 40:1).

Step 3 — Gel Staining

After blotting, the gel and filter are removed from the apparatus, and the filter is treated to quench unreacted and nonspecific binding sites, and then treated with the desired detection system (see Chapter 6). The gel is then stained with either Coomassie brilliant blue R, or silver stain[21] to determine whether elution occurred in an even fashion throughout the gel. This step is of considerable importance, since poor elution from a specific area of a gel might easily produce very misleading data.

Noncovalent Filters

The physical steps used for gel pretreatment and electroelution are the same as those described above for transfers to filters which retain proteins by means of covalent attachment. The differences in procedure are the use of nitrocellulose or derivatized nylon filters and sometimes different transfer buffers. For nitrocellulose filters, either Tris transfer buffer or a mixture of 80% Tris transfer buffer and 20% methanol are often used. The addition of methanol both increases the binding capacity (micrograms of protein per square centimeter) of the nitrocellulose[10] and slows down the rate of elution of protein from the gel.[14] For transfers to derivatized nylon filters, Tris transfer buffer is recommended as the buffer system. The protein binding capcity of derivatized nylon membranes is typically high compared to nitrocellulose,[14] and the capacity is not significantly increased by the addition of methanol.[14]

For transfers in the absence of methanol, a 2-hr blot at 5 V/cm is generally sufficient for elution of most proteins of M_r <100,000 from gels <0.8 mm in thickness and <18% (w/v) polyacrylamide (acrylamide:bisacrylamide = 40:1).

COMMENTS ON THE PROCEDURES

SDS Transfer Buffers

The use of SDS in transfer buffers[20] produces several effects which are useful in immunoblotting. Pretreatment of the gel in a buffer containing SDS converts all of the proteins in the gel to anions of high charge density by virtue of the formation of SDS-protein complexes. This effect has been demonstrated to considerably enhance the rate at which proteins can be eluted from a gel at a given field strength.[20] We find that including SDS in transfer buffers increases the reliability of transfer, especially from gel systems which themselves contain SDS. We suspect that the variability noted during transfers in the absence of SDS is due to variable loss of SDS from the SDS-protein complexes in the gel during pretreatment.

When eluting native gels, SDS pretreatment allows one to avoid the problem of determining the charge of the protein of interest under the conditions chosen for transfer. Treatment of the native gel with SDS causes all of the proteins to be negatively charged, and thus mobilizable in a predictable fashion. The possible drawback of using a buffer containing SDS is that the antigenicity of the protein of interest will be either diminished or lost due to exposure to SDS. Loss of antigenicity due to transfer in SDS-containing buffers is not a frequently reported event. The possibility may be checked by using the dot-immunobinding procedure described in Section II.B on a protein sample which has been exposed to the denaturing conditions.

A Broad-Spectrum System

The procedure described above as covalent immobilization (Section IV.A) has proven to be easily adaptable for a very wide range of applications. Using this protocol and diazothiophenolether paper, we have been able to transfer and retain polypeptides ranging in M_r from 2400 to 140,000. These include some peptides with properties similar to insulin, which were either very poorly initially adsorbed or poorly retained by nitrocellulose.[2,3] Some of the problems of adsorption and retention can be overcome with nylon membranes which have a high binding capacity for protein; however, these membranes require very extensive quenching to eradicate nonspecific antibody binding.[14] Other approaches to overcome the problem of poor protein retention by nitrocellulose involve treatments with agents such as glutaraldehyde which crosslink the transferred proteins either directly to the matrix, or to other proteins stably adsorbed to the matrix.[23]

While diazopapers avoid many of the problems of protein adsorbtion and retention, and are very easily quenched, they have the defects of being intensely colored (red-orange), and very coarse. The color of reacted diazopaper makes it harder to see the deposition of many of the insoluble chromophores produced by current immunochemical enzyme-linked detection systems. The coarseness of the papers used as substrate for the attachment of reactive diazo groups leads to some loss of sharpness of resolved protein bands after transfer. Recent work with cyanogen bromide-activated paper indicates that it has the desirable property of covalent retention of protein without the disadvantage of coloration of the matrix.[17] Like diazopapers, cyanogen bromide papers can be quenched by treatments as simple as a 1-hr incubation in 0.25% (w/v) gelatin, 100 m*M* Tris/HCl (pH 9.0).

TESTING TRANSFER EFFICIENCY

Transfer of Prestained Gels

A simple way to estimate the effectiveness of protein transfer is to subject a gel which has been stained with Coomassie brilliant blue R to transfer. Proteins from SDS-discontinuous gels retain much of their stain upon transfer to aminophenylthioether paper using the protocol described in the Procedures section.[24] This method allows a quick, if somewhat qualitative, estimation of both the efficiency and evenness of transfer. The method also allows very precise determination of which stained bands give rise to immunologically detected signals.[24]

Use of Isotopic Methods

The most quantitative methods for analyzing the efficiency of a transfer procedure utilize radiolabeled proteins. Efficient labeling of proteins in vivo or in vitro allow one to determine how much protein was applied to a gel, how much was eluted from the gel, and how much was bound by the filter toward which transfer was directed.[14-16] Radiography or staining of the gel and filter allow one to determine whether transfer was efficient for all of the labeled species, or preferential for subclasses of proteins.[13-16] Scintillation counting of gel and filter samples allows more exact determinations of quantity, however, the differential counting efficiencies of isotopes in gels and on filters must be taken into account.[16]

A Sample Evaluation of Transfer

An example of an analysis of transfer efficiency is presented in Figure 2. Bacterial cells producing atriopeptigen, a precursor form of atrial natriuretic factor with M_r of about 19,000[25] were labeled with ^{35}S methionine and separated on a 1D denaturing gel.[2] The proteins in the gel were then electrophoretically transferred using either the phosphate or Tris methanol transfer buffer protocols described earlier. Efficiency of removal of the labeled protein from the gel during transfer was followed by autoradiography of gel slices pre- and post-transfer, and a variety of matrices were crudely analyzed for their ability to bind the labeled proteins.

The capture of labeled proteins by the nitrocellulose and derivatized nylon membranes (Figure 2A, lanes 4 to 6), appears very efficient compared to the matrices which form covalent attachments (Panel A, lanes 2 and 3). Much of this apparent difference is due to the localization of labeled molecules near the surface of the noncovalent matrices compared to immobilization throughout the coarser covalent matrices. Careful treatment to quantitatively liberate and quantify the label (especially when it is as weak an emitter as ^{35}S) would be required to make statements as to the exact quantity of label retained by the various matrices. In the case of deposition of protein onto the various membranes, the radiographic method employed above is only sufficient to give a qualitative indication that protein is being retained. Radiography of the gel to determine extent of elution is quite quantitative, since the quenching due to the acrylamide is relatively constant before and after elution.

In Figure 2, the low rate of transfer of precursor out of the gel relative to all of the other labeled proteins present in the gel is also observed. This was true for both the phosphate and Tris methanol transfer buffer systems. In the latter system, atriopeptigen was quantitatively removed from the gel only with a very long (18 hr) period of elution (Figure 2B). Atriopeptigen was the only protein of low M_r which was significantly retained in the gel (Figure 2A, lane 7). Proteins which exhibit anomalously low rates of transfer have been previously reported.[14] In this case it is interesting to note that the atriopeptin precursor has a much lower migration in SDS gels than expected. Atriopeptigen migrates with proteins of M_r approximately 19,000, but has M_r of approximately 12,000. The reason for the resistance of atriopeptigen to electroelution is not clear, but it would be reasonable to speculate that the properties which retard its migration in the separating gel also slow its rate of electroelution.

Clearly, it is unwise to base a strategy for electrophoretic blotting of a particular protein on the average behavior of even a wide variety of other proteins. Where it is possible to determine the behavior of the protein under study, it is worth the small amount of extra effort required to tailor the blotting system to produce the optimal elution and recovery of the protein in question.

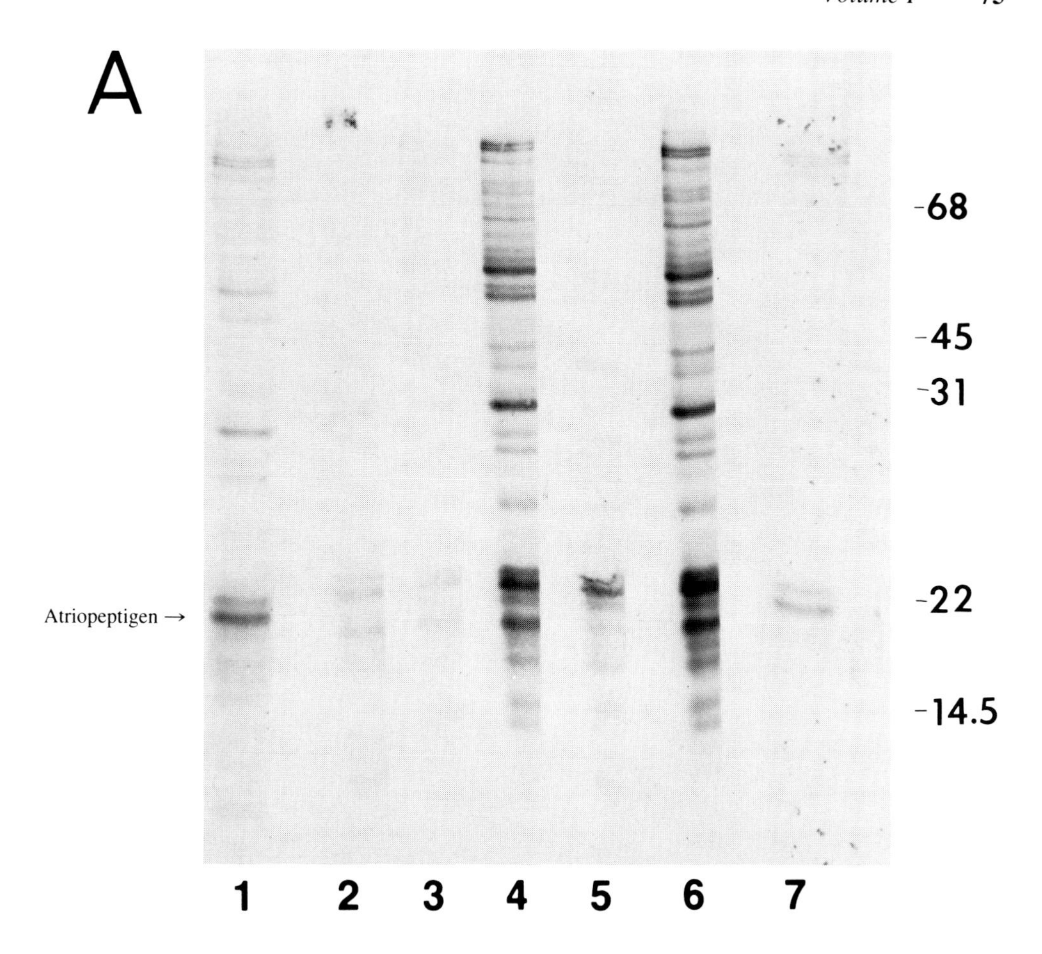

FIGURE 2. Autoradiograph of transfer of bacterially synthesized labeled proteins from a discontinuous-SDS gel (15%) to a variety of immobilizing matrices. (Panel A) Transfer in a phosphate buffer system. Lanes 1 and 7 are dried gel strips before and after a transfer for 2 hr at 5 V/cm. Lanes 2 through 6 are the result of transfer to (2) azobenzyloxy methyl paper (Transa-Bind, Schleicher & Schuell, Keene, N.H.), (3) CNBr activated paper,[17] (4) derivatized nylon (Pall Biodyne A, Pall Ultrafine Filtration Corp., Glen Cove, N.Y.), (5) Pall Biodyne Immunoaffinity Membrane, and (6) nitrocellulose (BA85, Schleicher & Schuell). (Panel B) Transfer in a Tris-methanol buffer system. Lanes 1 and 4 are dried gel strips before and after an 18 hr transfer. Lanes 2 and 3 are the strips resulting from transfer for 6 or 18 hr at 5 V/cm to nitrocellulose. The distance migrated by protein markers of known $M_r \times 10^{-3}$ is indicated. The products of protein synthesis in an *E. coli* cell producing a precursor form of atriopeptin[25] were labeled by the incorporation of ^{35}S methionine. Total bacterial extracts (approximately 80,000 cpm of incorporated ^{35}S) were prepared by boiling labeled cells in SDS sample buffer. The position of a precursor to rat atriopeptin is marked. The difference in spacing of proteins on the dried gel and on the matrices is due to swelling of the gel. Dried strips were autoradiographed for 8 hr on Kodak® XAR-5 film.

EXAMPLES OF ELECTROBLOTTING METHODOLOGY

In order to provide the reader with easy access to a variety of protocols which have been succesfully utilized to produce electrophoretic blots, Table 1 has been compiled.

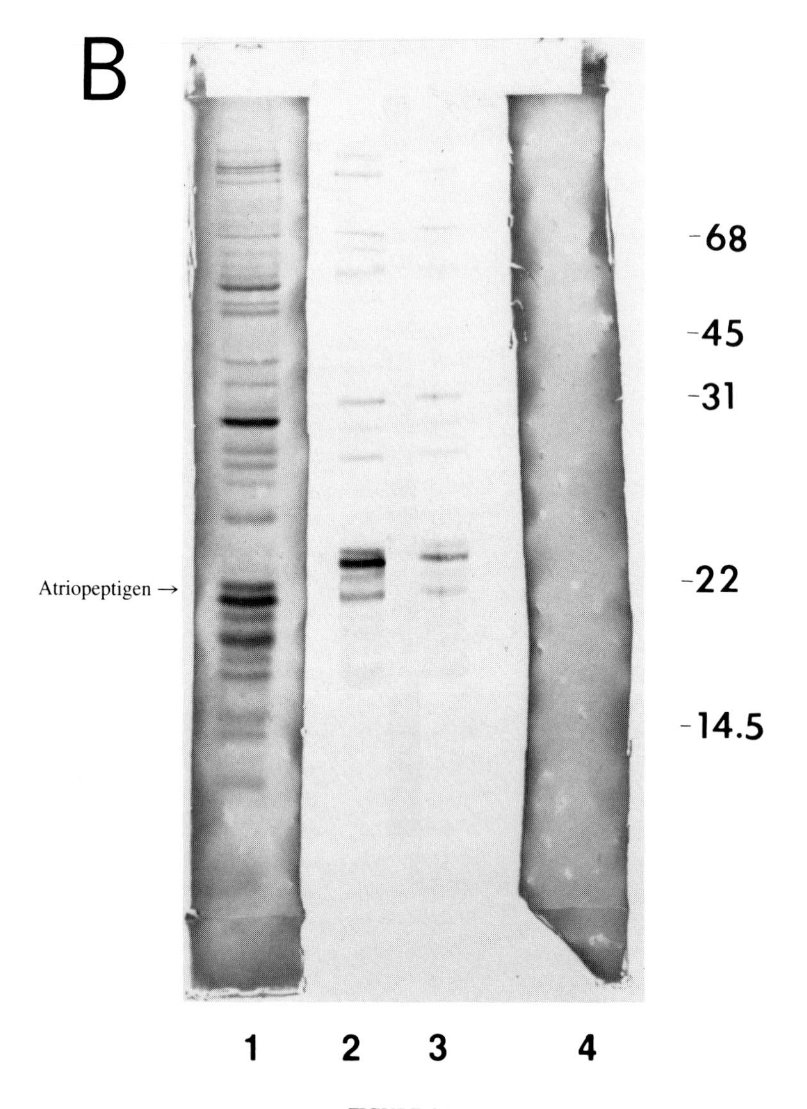

FIGURE 2B.

Table 1
A LISTING OF SOME SUCCESSFUL USES OF ELECTROPHORETIC TRANSFER PROCEDURES

Gel	Matrix	Ref.
Acid/urea	Nitrocellulose	10
Discontinuous/SDS	Nitrocellulose	10
Discontinuous/SDS	DBM paper	11, 15
Discontinuous/SDS	APT paper	16
Nonequilibrium pH gradient	APT paper	16
Discontinuous/SDS	Derivatized nylon	14
IEF	Nitrocellulose	12
Discontinuous/SDS	CNBr-activated paper	17
2D gel	DBM-paper	15, 26
2D gel	APT-paper	16

Note: APT: aminophenylthioether; DBM: diazobenzyloxymethyl.

REFERENCES

1. **Davis, B. J.,** Disc electrophoresis. II. Method and application to human serum proteins, *Ann. N.Y. Acad. Sci.*, 121, 404, 1964.
2. **Laemmli, U. K.,** Cleavage of structural proteins during the assembly of the head of bacteriophage T4, *Nature (London)*, 227, 680, 1970.
3. **Righetti, P. and Drysdale, J. W.,** Isoelectric focusing in polyacrylamide gels, *Biochem. Biophys. Acta*, 236, 17, 1971.
4. **O'Farrell, P. H.,** High resolution two-dimensional electrophoresis of proteins, *J. Biol. Chem.*, 250, 4007, 1974.
5. **Ausgustin, R.,** Fundamental aspects of single versus double diffusion methods for immunological assays, *Int. Arch. Allergy Appl. Immunol.*, 11, 153, 1957.
6. **Berson, S. A. and Yalow, R. S.,** Iodoinsulin used to determine the specific activity of Iodine-131, *Science*, 152, 205, 1966.
7. **Southern, E. M.,** Detection of specific sequences among DNA fragments separated by gel electrophoresis, *J. Mol. Biol.*, 98, 503, 1975.
8. **Arnheim, N. and Southern, E. M.,** Heterogeneity of the ribosomal genes in mice and men, *Cell*, 11, 363, 1977.
9. **Renart, J., Reiser, J., and Stark, G. R.,** Transfer of proteins from gels to diazobenzyloxymethyl-paper and detection with antisera: a method for studying antibody specificity and antigen structure, *Proc. Natl. Acad. Sci. U.S.A.*, 76, 3116, 1979.
10. **Towbin, H., Staehelin, T., and Gordon, J.,** Electrophoretic transfer of proteins from polyacrylamide gels to nitrocellulose sheets: procedure and some applications, *Proc. Natl. Acad. Sci. U.S.A.*, 76, 4350, 1979.
11. **Bittner, M., Kupferer, P., and Morris, C. F.,** Electrophoretic transfer of proteins and nucleic acids from slab gels to diazobenzyloxymethyl cellulose or nitrocellulose sheets, *Anal. Biochem.*, 102, 459, 1980.
12. **Reinhart, M. P. and Malamud, D.,** Protein transfer from isoelectric focusing gels: the native blot, *Anal. Biochem.*, 123, 229, 1982.
13. **Howe, J. G. and Hershey, J. W. B.,** A sensitive immunoblotting method for measuring protein synthesis initiation factor levels in lysates of *Escherichia coli*, *J. Biol. Chem.*, 256, 12836, 1981.
14. **Gershoni, J. M. and Palade, G. E.,** Electrophoretic transfer of proteins from sodium dodecyl sulfate polyacrylamide gels to a positively charged membrane filter, *Anal. Biochem.*, 124, 396, 1982.
15. **Stellwag, E. J. and Dahlberg, A. E.,** Electrophoretic transfer of DNA, RNA and protein onto diazobenzyloxymethyl (DBM) paper, *Nucl. Acids Res.*, 8, 299, 1980.
16. **Reiser, J. and Wardale, J.,** Immunological detection of specific proteins in total cell extracts by fractionation in gels and transfer to diazophenylthioether paper, *Eur. J. Biochem.*, 114, 569, 1981.
17. **Bhullar, B. S., Hewitt, J., and Candido, E. P. M.,** The large high mobility group proteins of rainbow trout are localized predominantly in the nucleus and nucleoli of a cultured trout cell line, *J. Biol. Chem.*, 256, 8801, 1981.
18. **Towbin, H. and Gordon, J.,** Immunoblotting and dot immunobinding — current status and outlook, *J. Immunol. Methods.*, 72, 313, 1984.
19. **Gershoni, J. M. and Palade, G. E.,** Protein blotting: principles and applications, *Anal. Biochem.*, 131, 1, 1983.
20. **Erickson, P. F., Minier, L. N., and Lasher, R. S.,** Quantitative electrophoretic transfer of polypeptides from SDS polyacrylamide gels to nitrocellulose sheets: a method for their re-use in immunoautoradiographic detection of antigens, *J. Immunol. Methods.*, 51, 241, 1982.
21. **Wray, W., Boulikas, T., Wray, V. P., and Hancock, R.,** Silver staining of proteins in polyacrylamide gels, *Anal. Biochem.*, 118, 197, 1981.
22. **Legocki, R. P. and Verma, D. P. S.,** Multiple immunoreplica technique: screening for specific proteins with a series of different antibodies using one polyacrylamide gel, *Anal. Biochem.*, 111, 385, 1981.
23. **Kakita, K., O'Connell, K., and Permutt, M. A.,** Immunodetection of insulin after transfer from gels to nitrocellulose filters. A method of analysis in tissue extracts, *Diabetes*, 31, 648, 1982.
24. **Gierse, J. K.,** personal communication, 1985.
25. **Seidman, C. E., Duby, A. D., Choi, E., Graham, R. M., Haber, E., Homcy, C., Smith, J. A., and Seidman, J. G.,** The structure of rat preproatrial natriuretic factor as defined by a complementary DNA clone, *Science*, 225, 324, 1984.
26. **Symington, J., Green, M., and Brackmann, K.,** Immunoautoradiographic detection of proteins after electrophoretic transfer from gels to diazo-paper: analysis of adenovirus encoded proteins, *Proc. Natl. Acad. Sci. U.S.A.*, 78, 177, 1981.

4.3.2 SEMIDRY ELECTROBLOTTING — TRANSFER USING EQUIPMENT WITHOUT BUFFER VESSEL

Jan Kyhse-Andersen

INTRODUCTION

One of the crucial steps in the blotting technique is the transfer of proteins from the gel matrix to the immobilizing membrane. Electrophoretic transfer is rapid and can be performed with high efficiency.[1] The semidry electroblotting is carried out in a horizontal apparatus without buffer vessel using wetted filter paper as the only buffer reservoir.[2] Several articles have been published using this method[3-5] (see also Chapters 1, 8.5, 8.6, and 9.2). Other horizontal electroblotting apparatus have been described.[6-8]

APPARATUS AND SET-UP

The semidry electroblotting apparatus (Figure 1) consists of two plastic parts in which graphite electrode plates are fixed. The bottom is the anode, and the lid is the cathode. The plates ensure a uniform voltage gradient over the narrow gab. The connection to the apparatus ends in two female banana plugs of which one is especially designed as a safety cut-out, making it impossible to disassemble the apparatus without disconnecting the current. The apparatus is obtainable from JKA-Biotech (Vaerlose, Denmark).

The assembly of the filter papers, nitrocellulose membrane, slab gel, and dialysis membrane (a Trans-unit) is shown in Figure 2. By stacking Trans-units in the apparatus it is possible to transfer simultaneously from several gels. The dialysis membrane prevents penetration of polypeptides from one gel to the next.

The speed and efficiency of the transfer in this apparatus depend on the buffer system, which must be of low ionic strength in order to increase the transfer efficiency of the proteins and to minimize the joule heat development. Thus, the conditions have to be chosen so that the proteins "carry" a considerable part of the current. Either a continuous or a discontinuous buffer system can be used in the wetted filter paper stack. Here, a discontinuous buffer system based on the isotachophoretic theory is employed.[2,9]

MATERIALS AND REAGENTS

Whatman No. 1 filter paper cut for 14 × 14 cm was employed. Dialysis membrane was obtained from Dansk Emballage Industri, Søborg, Denmark. Nitrocellulose membrane (code SM 11306, 0.45 μm) was purchased from Sartorius, Göttingen, West Germany. 6-Amino-*n*-hexanoic acid was from BDH, Poole, England. Power supply: any apparatus which delivers a constant current up to about 200 mA.

BUFFER SOLUTIONS

Anode solution no. 1:	
0.3 *M* Tris	36.3g
20% (v/v) methanol	200 mℓ
Distilled water	ad 1000 mℓ
Anode solution no. 2:	
0.025 *M* Tris	3.03 g
20% (v/v) methanol	200 mℓ
Distilled water	ad 1000 mℓ

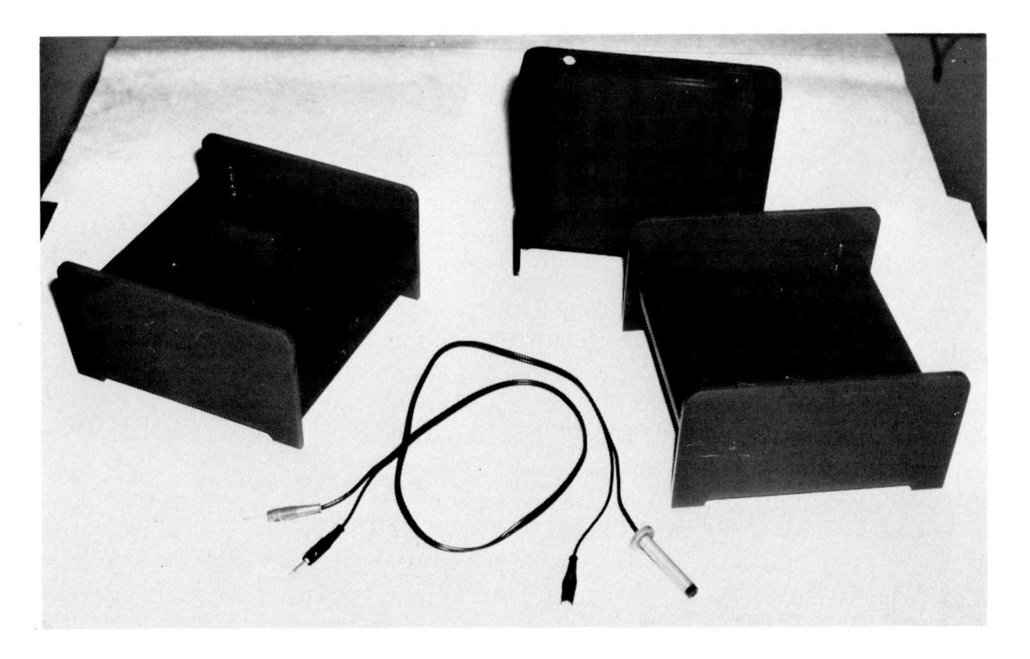

FIGURE 1. The semidry electroblotting apparatus. At left the apparatus is shown in a closed position and at the right in an open position.

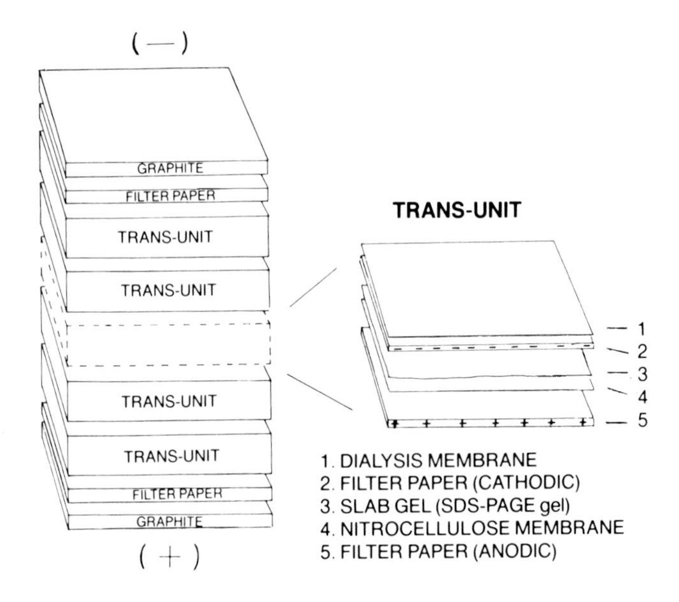

FIGURE 2. Build-up of the Trans-units for semidry electroblotting. In order to perform simultaneous protein transfer from several gels in one operation, the Trans-units can be stacked. Filter papers soaked in either anodic or cathodic buffers are placed on the electrodes as buffer reservoirs. The Trans-unit is assembled by placing the gel on the nitrocellulose membrane between filter papers soaked in buffer with a dialysis membrane on top of the sandwich.

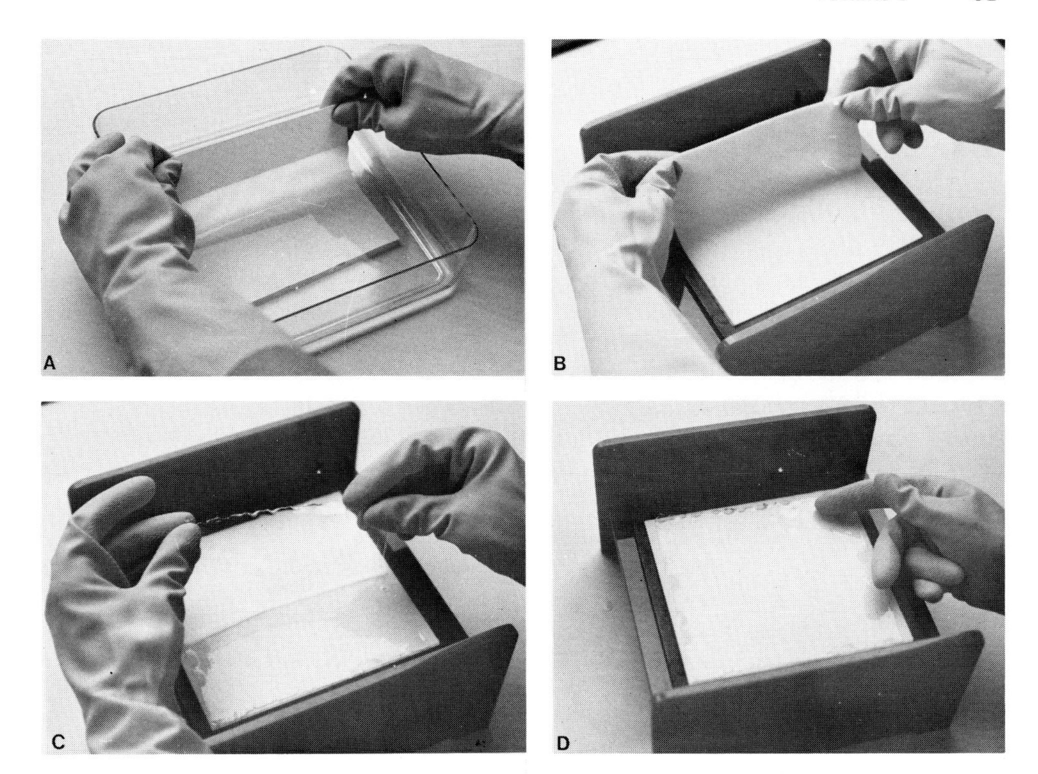

FIGURE 3. Assembly of the Trans-unit: (A) Filter papers are soaked by slow immersion in buffer. (B) The soaked filter papers are placed on the prerinsed graphite plate. The filter paper stack is positioned on the electrode plate without entrapments of air bubbles. (C) the SDS-PAGE gel is "rolled" into position. (D) If air bubbles appear between the nitrocellulose membrane and the gel, the surface is wetted with a few drops of cathodic buffer which allows the air bubbles to be gently squeezed out.

Cathode solution:

0.040 *M* 6-amino-*n*-hexanoic acid	5.2g
20% (v/v) methanol	200 mℓ
Distilled water	ad 1000 mℓ

PROCEDURE

1. Rinse the graphite plates with distilled water.
2. Place six layers of filter paper which have been soaked in anode solution no. 1 (Figure 3A) on the anodic graphite plate (Figure 3B).

3a. Assemble a Trans-unit as shown in Figure 2. Apply on top of the six layers: three layers of filter paper soaked in anode solution no. 2 followed by the nitrocellulose membrane prerinsed in distilled water (14 × 14 cm). The SDS-PAGE gel (13.5 × 13.5 cm) is then placed on top of the nitrocellulose membrane (Figure 3C). If air bubbles appear, the gel should be wetted with a few drops of cathode solution and the bubbles gently squeezed out (Figure 3D). Another three layers of filter paper soaked in cathode solution are placed on top of the SDS-PAGE gel. Finally, a dialysis membrane (14 × 14 cm) softened in water is placed on top of the Trans-unit.

3b. Several of these Trans-units can be placed on top of each other in a stack in the apparatus.

4. Cover the stack with six layers of filter paper soaked in cathode solution.
5. Finally, place the lid containing the cathode plate. Connect the apparatus to the power supply with a constant current of 0.8 mA/cm^2 of gel for 1 hr at room temperature.

COMMENTS ON THE PROCEDURE

The filter papers, nitrocellulose, and dialysis membranes should be trimmed to the size of the slab gel so that the current has to pass through the gel. Gloves should be worn when handling polyacrylamide gels, nitrocellulose, and filter papers.

The filter papers should be soaked by slow immersion in buffer solution (Figure 3A). Wetting the filter papers by capillary force eliminates entrapment of air bubbles.

At a constant current of 0.8 mA/cm^2 of gel, the described buffer system develops and maintains a voltage gradient of about 10 V/cm. This causes a sufficient migration velocity of the proteins at a pH range between 9 and 10. The SDS, glycine, and SDS-complexes move toward the anode in the stated order according to their net mobilities. From the surrounding buffer (cathodic side), 6-amino-*n*-hexanoic acid is the terminating ion in the isotachophoretic train, and "pushes" the proteins out of the polyacrylamide gel.

EVALUATION

The semidry electroblotting procedure has the following advantages: (1) highly reduced transfer time — 1 hr compared to 4 to 22 hr for other systems;[10,11] (2) low joule heat development; (3) electrophoretic transfer from several (at least six) gels simultaneous with equally efficient transfer from all gels in the stack; (4) simple assembly of the horizontal electroblotting cell.

The efficiency of the electrophoretic transfer is shown in Figure 4, where human serum proteins are analyzed by simultaneous transfer from six gels. The nitrocellulose membranes stained after transfer (Figure 4B and 4C) show that the pattern of the separation in the SDS-PAGE is retained. The SDS-PAGE gel stained with Coomassie blue after transfer shows only traces of a few high M_r proteins (Figure 4D, arrowheads).

Electroblotting of immunoprecipitates from agarose gels can be performed in different ways.[14,15] With the procedure described in the legend to Figure 5, all precipitates are transferred to the nitrocellulose membrane with about 90% yield. Subsequent immunodetection of antigens with monoclonal antibodies can be performed.[14,15]

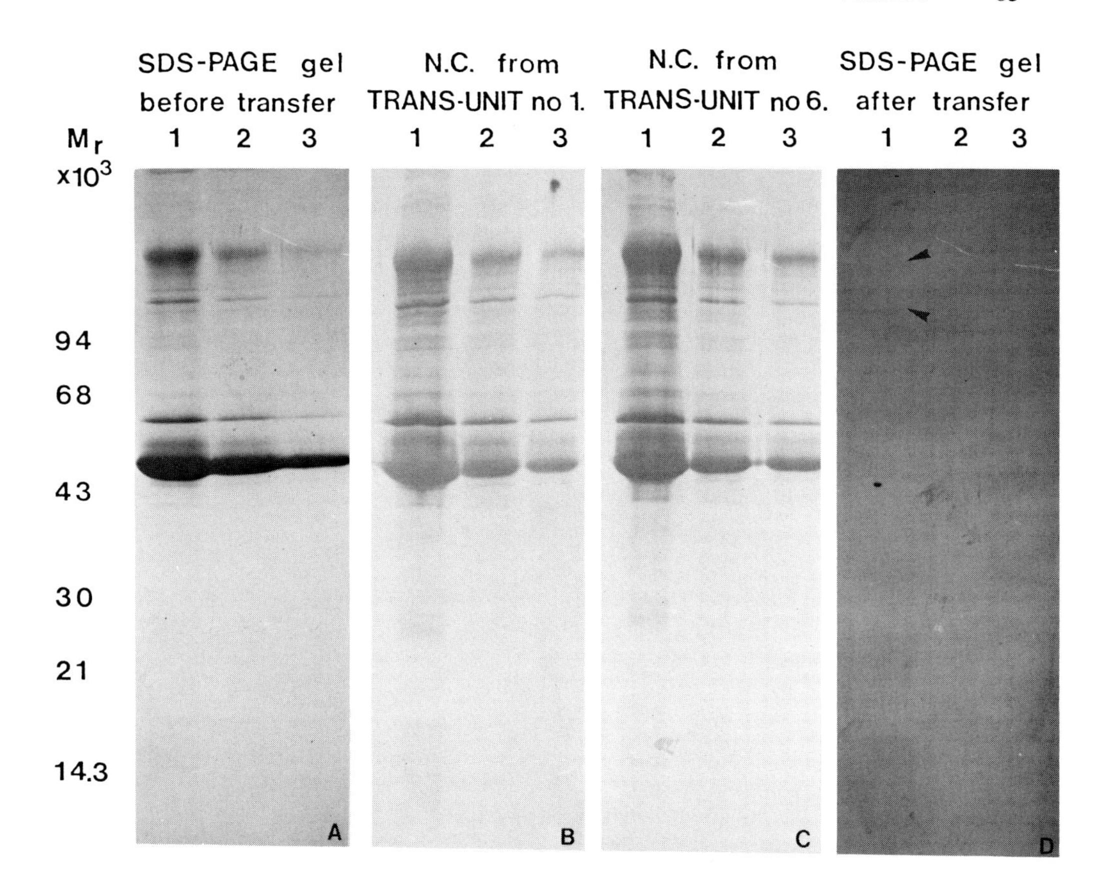

FIGURE 4. Efficiency of the electroblotting in a stack of six Trans-units. SDS-PAGE was carried out in a modified Laemmli-system.[12] Only the separation gel, prepared to give a linear gradient from 12% T-acrylamide/0.6% C-bisacrylamide to 16% T-acrylamide/0.8% C-bisacrylamide, is used for electroblotting. One SDS-PAGE gel is stained with Coomassie brilliant blue G 250 before (A) and another after (D) electroblotting. The nitrocellulose membranes (N.C.) from Trans-unit no. 1 (B) and no. 6 (C) (top and bottom) in the stack are stained with Amidoblack after electroblotting. Only traces of a few high M_r proteins are left after the transfer (D, arrowheads). Human serum proteins were analyzed, diluted as follows: lane 1, 140 μg; lane 2, 70 μg; and lane 3, 14μg. (From Kyhse-Andersen, J., *J. Biochem. Biophys. Methods,* 10, 203, 1984. With permission.)

FIGURE 5. Electroblotting from agarose gels with immunoprecipitates. Crossed immunoelectrophoresis of 2 μℓ human serum proteins was performed with 10 μℓ/cm² of rabbit-antihuman serum proteins in the second dimension electrophoresis (code: A206, Dakopatts Glostrup, Denmark).[13] After immunoelectrophoresis the agarose gel was pressed with six layers of filter papers for 10 min, followed by solubilization of the immunoprecipitates by incubation for 20 min in 8 m*M* dithioerytritol (Sigma) and 0.5% (w/v) SDS at room temperature. Electroblotting was performed for 20 min with 0.04 *M* 6-amino-*n*-hexanoic acid + 0.1% (w/v) SDS (pH = 6.9) and 0.025 *M* Tris (pH = 10.3) as buffers. The transferred proteins were detected by incubation overnight with rabbit antibodies against human serum proteins (1 + 100) and stained with peroxidase conjugated swine antirabbit IgG antibodies (1 + 2000) Dakopatts and 3-amino-9-ethyl carbazole as described in the appendix to Chapter 1.

REFERENCES

1. **Towbin, H., Staehelin, T., and Gordon, J.,** Electrophoretic transfer of proteins from polyacrylamide gels to nitrocellulose sheets: procedure and some applications, *Proc. Natl. Acad. Sci. U.S.A.*, 76, 4350, 1979.
2. **Kyhse-Andersen, J.,** Electroblotting of multiple gels; a simple apparatus without buffer tank for rapid transfer of proteins from polyacrylamide to nitrocellulose, *J. Biochem. Biophys. Methods*, 10, 203, 1984.
3. **Naaby-Hansen, S., Lihme, A. O. F., Bøg-Hansen, T. C., and Bjerrum, O. J.,** Lectin blotting of normal and derivatised membrane proteins, in *Lectins — Biology, Biochemistry, Clinical Biochemistry*, Bøg-Hansen, T. C. and Breborowicz, J., Eds., Walter de Gruyter, Berlin, 1985, 241.
4. **Friis, S. U., Norén, O., Sjöström, H., and Gudman-Høyer, E.,** Patient with coeliac disease a characteristic gliadin antibody pattern, *Clin. Chim. Acta*, 155, 133, 1986.
5. **Heegaard, N. H. H., Juhl, A., and Bjerrum, O. J.,** Immunaftrykning-immunoblotting, *Ugeskr. Lag.*, 3798, 1985.
6. **Gibson, W.,** Protease-facilitated transfer of high-molecular-weight proteins during electrotransfer to nitrocellulose, *Anal. Biochem.*, 118, 1, 1981.
7. **Manabe, T., Takahashi, Y., and Okuyama, T.,** An electroblotting apparatus for multiple replica technique and identification of human serum proteins on micro two-dimensional gels, *Anal. Biochem.*, 143, 39, 1984.
8. **Vaessen, R. T. M. J., Kreike, J., and Groot, G. S. P.,** A simple method for quantitation of single proteins in complex mixtures, *FEBS Lett.*, 124, 193, 1981.
9. **Schäfer-Nielsen, C., Svendsen, P. J., and Rose, C.,** Separation of macromolecules in isotachophoresis systems involving single or multiple counterions, *J. Biochem. Biophys. Methods*, 3, 97, 1980.
10. **Burnette, W. N.,** "Western blotting": electrophoretic transfer of proteins from sodium dodecyl sulphate-polyacrylamide gels to unmodified nitrocellulose and radiographic detection with radioiodinated protein A, *Anal. Biochem.*, 112, 195, 1981.

11. **Bjerrum, O. J., Larsen, K. P., and Wilken, M.,** Some recent developments of the electroimmunochemical analyses of membrane proteins. Application of Zwittergent, Triton X-114 and Western blotting technique, in *Modern Methods in Protein Chemistry,* Tschesche, H., Ed., Walter de Gruyter, Berlin, 1983, 73.
12. **Schäfer-Nielsen, C. and Rose, C.,** Separation of nucleic acids and chromatin proteins by hydrophobic interaction chromatography, *Biochim. Biophys. Acta,* 696, 323, 1982.
13. **Clarke, H. G. M. and Freeman, T.,** A quantitative immunoelectrophoresis method (Laurell electrophoresis), in *Protides Biol. Fluids,* Peeters, H., Ed., Elsevier, Amsterdam, 1967, 503.
14. **Bhakdi, S., Jenne, D., and Hugo, F.,** Electroimmunoassay-immunoblotting (EIA-IB) for utilization of monoclonal antibodies in quantitative immunoelectrophoresis: the method and its applications, *J. Immunol. Methods,* 80, 25, 1985.
15. **Bütikofer, P., Frenkel, E. J., and Ott, P.,** Crossed immunoblotting: identification of proteins after crossed immunoelectrophoresis and electrotransfer to nitrocellulose membranes, *J. Immunol. Methods,* 84, 65, 1985.

4.3.3 ELECTROTRANSFER AT ACIDIC pH

Steffen U. Friis

INTRODUCTION

Immunoblotting of separated polypeptides is usually performed after electrophoresis at slightly alkaline pH, often after denaturation of the proteins by sodium dodecylsulfate (SDS). However, some proteins, e.g., the wheat prolamins, gliadins, a group of basic proteins rich in nonpolar amino acids, have to be electrophoresed at an acidic pH due to their solubility properties. Electrophoretic systems developed for this purpose, using lactate as the common counterion, separates gliadin into four groups of polypeptides, termed α-, β-, γ-, and ω-gliadin, depending on their electrophoretic mobility.[1] Other proteins such as histones, cytochromes, and lysozymes, are also most preferably electrophoresed at acidic pH due to their isoelectric point.

Electrotransfer to nitrocellulose at acidic pH has as yet not been described, and problems with binding to nitrocellulose might be expected. Below, a system is given where proteins after separation by electrophoresis in polyacrylamide gel at pH 3.1 are electrotransferred to nitrocellulose in a lactate containing transfer buffer at pH 2.7.

EQUIPMENT AND REAGENTS

Apparatus

A homemade plexiglass blotting cell with platinium electrodes in a 5 ℓ glass tank was used for transfer. The voltage applied came from a 12 V/4 A battery charger (Hella, West Germany) (Figure 1).

Materials

Sodium chloride, sodium acetate, glycine, glacial acetic acid, Tween® 20, hydrogen peroxide 30% (v/v), and Coomassie Brilliant Blue R 250 were obtained from Merck, Darmstadt, West Germany, while Trizma base, 3-amino-9-ethyl carbazole, and *N,N*-dimethylformamide were purchased from Sigma, St. Louis. BDH Chemicals, Poole, U.K. delivered D,L-lactic acid. Nitrocellulose (BA 85, pore size 0.45 μm) was from Schleicher & Schuell, Dassel, West Germany. AGF 138 filter paper was obtained from AGF, Ebeltoft, Denmark. Peroxidase-conjugated swine antirabbit immunoglobulin was a product of Dakopatts, Glostrup, Denmark. Gliadin was extracted from defatted Danish wheat (Vulca) according to the method described by Charbonnier.[2]

The primary antibodies were from rabbits immunized with gliadin. Gliadin at 1 mg was dissolved in 1 mℓ of 100 m*M* lactic acid and mixed with an equal volume of complete Freund's adjuvant. Over a period of 12 weeks, rabbits were injected intracutaneously with 200 μℓ of the antigen solution at 2 week intervals. One week after the sixth injection the rabbits were bled for 40 mℓ. The immunization was continued with a booster injection every fourth week and 40 mℓ of blood was collected 1 week after each injection. The IgG-fraction was isolated on Protein A-Sepharose® CL-4B[3] (Pharmacia Fine Chemicals, Uppsala, Sweden).

Solutions

1. Transfer buffer (pH 2.7)

100 m*M* lactic acid	22.5 mℓ
H_2O	ad 3000 mℓ

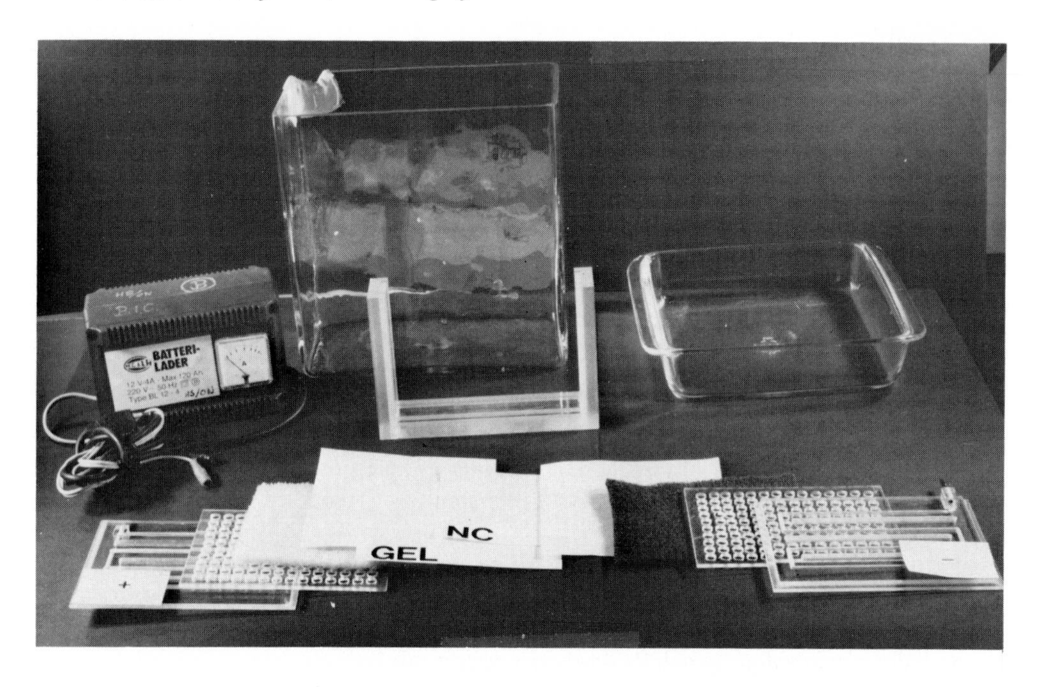

FIGURE 1. Equipment for electrotransfer at acidic pH. The blotting cell frame (in the middle) holds the blotting sandwich tight together. The sandwich (front) is symmetrically built up of layers of plexiglass grids with platinum electrodes, perforated plexiglass plates, acrylic sponges, and filter papers on either side of the gel slab and nitrocellulose sheet core. The layers are assembled submerged in a vessel (background to the right) containing transfer buffer and transfer is carried out in a 5 ℓ glass tank (background middle) with a 12 V battery charger (background left) as a power supply.

2. Washing buffer (pH 10.2)	
50 m*M* Tris	30.3 g
150 m*M* sodium chloride	43.8 g
H_2O	ad 5000 mℓ
3. Incubation buffer (pH 10.2)	
0.05% (w/v) Tween® 20	20.0 g
Washing buffer	ad 1000 mℓ
4. Blocking buffer (pH 10.2)	
2% (w/v) Tween® 20	20.0 g
Washing buffer	ad 1000 mℓ
5. Staining buffer (pH 5.5)	
50 m*M* sodium acetate	6.8 g
Glacial acetic acid	1.8 mℓ
H_2O	ad 1000 mℓ
6. Staining solution (pH 5.5)	
3-Amino-9-ethyl carbazole	25.0 mg
Dissolved in *N,N*-dimethylformamide	2.5 mℓ
Staining buffer	ad 50.0 mℓ
Hydrogen peroxide 30% (w/v)	25 μℓ

All buffers can be stored for weeks at 4°C. Staining solution shall be prepared immediately before use.

PROCEDURES

Gel Electrophoresis

Gliadin is separated by polyacrylamide gel (9.3% T; 3.2% C) electrophoresis at pH 3.1 as previously described.[1] The system is a telescope electrophoresis system in which the gel is cast at neutral pH and equilibrates to pH 3.1 during a 6.5 hr electrophoresis. Anode tank buffer: 100 m*M* lactic acid, pH 2.7. Cathode tank buffer: 50 m*M* Tris adjusted to pH 6.6 with lactic acid.

Assembly of Blotting Cell

The blotting cell is horizontally assembled submerged in a 22 × 22 × 6 cm vessel containing transfer buffer. All layers are well wetted before assembly and single-use gloves are used to avoid contamination of the nitrocellulose.

The layers of the sandwich are held tight together by the cell frame. At the plexiglass grid with platinum electrode, a 3-mm thick plexiglass plate, 15.7 × 11.0 cm and perforated with 140 holes with a diameter of 8 mm, protects the electrode from the next layer which is a 15 × 11 × 1 cm acrylic sponge. This is followed by one sheet of filter paper, the gel slab, and the nitrocellulose sheet, cut out to the size of the gel. The sandwich is closed by one sheet of filter paper, 1-cm thick acrylic sponge, a perforated plexiglass plate and finally a plexiglass grid with electrode.

Transfer

The transfer is performed in a 5 ℓ glass tank containing 3 ℓ transfer buffer at 4°C with a constant voltage of 7 V/cm, the distance between the electrodes being 1.7 cm. The proteins migrate against the cathode. Precooling of buffer and stirring during transfer is not necessary. After transfer the gel is stained for proteins to control elution efficiency.

Blocking and Washing

The blocking procedure is essentially as described by Bjerrum et al.[4] The nitrocellulose sheets are quenched by incubation in blocking buffer for 5 min followed by washing in incubation buffer.

Incubation with Primary Antibodies

The antibodies are suitable diluted with incubation buffer and incubated on a rocking table at 4°C for 16 hr. Excess antibodies are removed by washing the nitrocellulose 3 × 5 min in 3 × 25 mℓ washing buffer.

Incubation with Secondary Antibodies

Peroxidase conjugated swine antirabbit immunoglobulin antibodies are used as secondary antibodies. A 1:500 dilution in incubation buffer is used and incubation for 2 hr at 20°C on a rocking table is sufficient. Washing is performed as above but is terminated with a 5-min equilibration in staining buffer. The last step prevents precipitation in the subsequent staining solution.

Staining for Peroxidase Activity

Peroxidase activity is localized with staining solution prepared immediately before use. In order to minimize background staining, incubation is carried out in a tray impervious to light. The procedure is complete after 10 min and the nitrocellulose is thoroughly washed in tap water and dried between layers of filter paper in a similar tray. They can conveniently be stored in transparent plastic bags, preferably in the dark (see Figure 2).

COMMENTS ON THE PROCEDURE

As for other blotting experiments, each step in the procedure should be preceded by a dot-immunobinding assay (see Section 2); in this case, especially, to ascertain that the acidic pH does not interfere with binding of the protein to nitrocellulose or destroys antigen

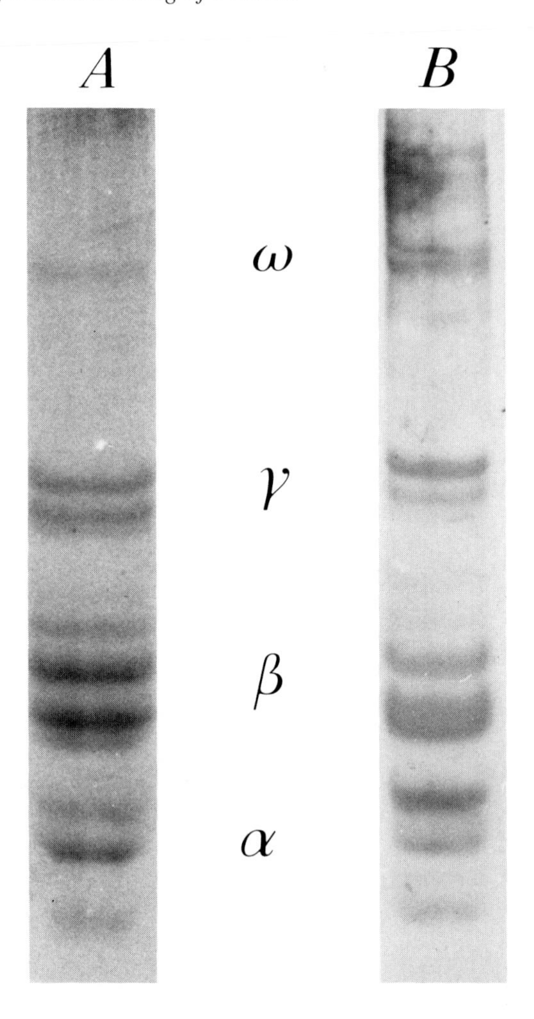

FIGURE 2. Transfer of gliadin at pH 2.7. (A) 100 μg gliadin separated into its four major groups of polypeptides: α-, β-, γ-, and ω-gliadin by polyacrylamide gel electrophoresis at pH 3.1 and stained with Coomassie Brilliant Blue R 250. Cathode at the bottom of the lane. (B) 100 μg gliadin separated as A and electrotransferred to nitrocellulose at pH 2.7. Immunoglobulin from a rabbit immunized with gliadin was used as a primary antibody (26 μg/mℓ) and peroxidase-conjugated swine antirabbit immunoglobulin as secondary antibody. Staining was performed with 3-amino-9-ethylcarbazole.

reactivity. To avoid denaturation of the proteins during transfer caused by heat generation between the electrodes, transfer from gel to membrane is performed in a weak lactic acid solution at low field strength for 16 hr, 4°C in a 5 ℓ tank.

Contrary to transfer at pH 8.6, methanol in transfer buffer is not necessary, since polyacrylamide gels do not swell significantly in weak, acidic solutions.

Due to the sensitive detection of transferred proteins significant background staining sometimes appears within slab lanes. It may be caused by trailing proteins in the polyacrylamide gel separation, though they are not usually detectable by ordinary protein staining in gels.

USE AND LIMITATIONS

With the purpose of decreasing transfer time, transfer in electroblotting equipment without buffer vessels (see Chapter 4.3.2) has been tried, but the temperature between the electrodes exceeded 80°C during transfer.

After transfer and quenching, the possibilities for storing and restaining are the same as for experiments performed at pH 8.6. The transferred proteins can be stored for at least 6 months, when quenched nitrocellulose is kept in incubation buffer at 4°C. The immunoreactions and staining can be repeated in order to test different antibodies. Dried immunoblots can be reincubated with primary and/or secondary antibodies and restained within at least 2 years from transfer.

The described procedure for transfer at pH 2.7 might be used without modifications for immunoblotting of other corn prolamins and basic proteins such at cytochromes, histones, and lysozymes.

REFERENCES

1. **Friis, S. U. and Schäfer-Nielsen, C.,** Separation of gliadin at pH 3.1 in a polyacrylamide gel suitable for blotting procedures, *J. Biochem. Biophys. Methods,* 10, 301, 1985.
2. **Charbonnier, L.,** Etude des proteines alcoolo-soluble de la farine de ble, *Biochemie,* 55, 1217, 1973.
3. **Hjelm, H., Hjelm, K., and Sjöquist, T.,** Protein A *Staphylococcus aureus:* its isolation by affinity chromatography and its use as an immunosorbent for isolation of immunoglobulins, *FEBS Lett.,* 28, 73, 1982.
4. **Bjerrum, O. J., Larsen, K. P., and Wilken, M.,** Some recent developments of the electrochemical analysis of membrane proteins. Application of Zwittergent Triton X-114 and Western blotting technique in *Modern Methods in Protein Chemistry,* Tschesche, H., Ed., Walter de Gruyter, Berlin, 1983, 79.

4.3.4 PROTEASE-FACILITATED PROTEIN TRANSFER

Wade Gibson and Donna Mia Benko

INTRODUCTION

This chapter describes a procedure that can be used to increase the efficiency of electrotransferring high-M_r proteins or protein complexes from polyacrylamide gels onto nitrocellulose membranes. The procedure, called "protease-facilitated transfer",[1] is a modification of the general protein transfer technique described by Towbin et al.,[2] and works as follows. A proteolytic enzyme is introduced into the gel during electrotransfer and cleaves the resident proteins. The smaller size of the resulting fragments enables them to move more readily from the gel. The procedure speeds electrotransfer in general and promotes a more quantitative recovery of high-M_r proteins without compromising the antibody-binding properties of the resulting nitrocellulose-bound molecules.

EQUIPMENT AND REAGENTS

This procedure requires no special equipment beyond that routinely used to separate the proteins of interest by polyacrylamide gel electrophoresis (PAGE), and electrotransfer them to the solid phase of choice. The electrotransfer unit used in the experiments described here consisted of a 12 V battery charger (Model 0122-06, Schauer Manufacturing Corp., Cincinnati, Ohio) that supplied an initial current of 1 to 1.5 A to the horizontal transfer assembly shown schematically in Figure 1 and described in the legend. The applicability of the procedure, however, is independent of the design of the transfer unit. The electrode buffer was the same as described by Towbin et al.;[2] the proteases used were Pronase (No. 537088, Calbiochem-Behring Corp., La Jolla, Calif.), and *Staphylococcus aureus* V8 protease (No. 39-900, Miles Scientific, Naperville, Ill.); and the SDS used during electrotransfer was from Bio-Rad (No. 161-0301, Richmond, Calif.). Nitrocellulose was from Schleicher & Schuell (BA85, 0.45 μm, Keene, N.H.).

Other reagents used were concanavalin A and wheat germ agglutinin, (Nos. L-1104 and L-2107, respectively) from E-Y Laboratories, San Mateo, Calif., and polyvinylpyrrolidone 40 (No. PVP40), aniline, sodium cyanoborohydride, and sodium metaperiodate, all from Sigma Chemical Co., St. Louis.

Densitometric measurements of autoradiographic and fluorographic images on XAR-5 film (Eastman Kodak Co., Rochester, N.Y.) were made using an E-C 910 densitometer (E-C Corp., St. Petersburg, Fla.) equipped with a 540 nm interference filter.

SOLUTIONS

Protease buffer:	
Tris (25 m*M*)	0.3 g
glycine (192 m*M*)	1.44 g
SDS (Bio-Rad, 0.1%, w/v)	0.1 g
Distilled water, add to final volume of	100 mℓ
crystalline pronase (10 to 20 μ/mℓ)	1 to 2 mg

The buffer is stable but protease solution is prepared fresh. Some proteases may require a preincubation step to eliminate contaminating enzymes.

The electrotransfer buffer was Tris (25 m*M*), glycine (192 m*M*), and methanol (20%, v/v), degassed prior to use. For the recipe, see appendix of Chapter 1.

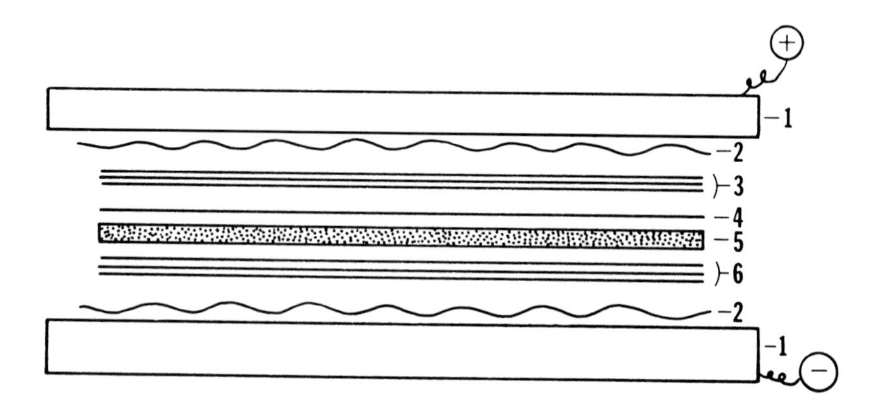

FIGURE 1. Diagram showing horizontal electrotransfer assembly. (1) Graphite slabs (Gra Far Corp., Detroit) (30 × 16 × 2 cm) were used as electrodes. (2) Neoprene matting (No. 9400, Cole-Parmer Instrument Co., Chicago) separated the gel sandwich (elements 3 to 6) from the electrodes. The gel sandwich was composed of three sheets of Whatman 3MM paper soaked in electrotransfer buffer (3), a sheet of BA85 nitrocellulose prewetted in electrotransfer buffer (4), the polyacrylamide gel section (5), and three sheets of Whatman 3MM paper soaked in protease buffer (6), and was assembled as described in the text. The electrode chamber, a plastic tray measuring 22 × 33 × 6 cm, was filled to a depth just beneath the top of the upper electrode (about 1.8 ℓ). Tight contact between the gel and the nitrocellulose was maintained by the weight of the upper electrode slab (about 1.7 kg).

PROCEDURES

Proteins are appropriately solubilized and subjected to slab gel SDS-PAGE.[3,4] Following electrophoresis, the gel is placed on top of three pieces of 3MM paper, (Whatman Ltd., Maidstone, England), previously cut to the size of the gel, and soaked in protease buffer (about 0.2 mℓ/cm^2 of gel surface area) containing the desired amount of proteolytic enzyme (typically 10 to 20 μg/mℓ). A piece of nitrocellulose, cut to the size of the gel and prewetted in electrotransfer buffer, is next applied to the top of the gel, followed by an uppermost pad of three pieces of 3MM paper, also prewetted in that buffer. The paper-gel sandwich should be assembled quickly to avoid dehydrating the gel, and care must be taken to squeeze any entrapped air bubbles from the sandwich before placing it between the electrodes of the transfer unit. The optimal transfer time must be determined empirically since it depends on such variables as the thickness and concentration of the gel, the composition of the electrotransfer buffer, and the strength of the electric field during transfer. As an approximation, the time required to transfer a 40,000 to 60,000 M_r protein in the absence of protease, will be adequate to transfer essentially all proteins in the presence of a protease (e.g., 45 to 90 min using the conditions described here). The resulting nitrocellulose replica is removed from the transfer unit and subsequently processed as usual (e.g., quenched in preparation for antibody-[5,6] or lectin-binding[7,8] assay). The polyacrylamide gel "remnant" is routinely stained with Coomassie brilliant blue to verify that all proteins were removed.

COMMENTS ON THE PROCEDURE

The following comments may be helpful. First, it is important that the SDS used contain as little tetra- and hexadecyl sulfate as possible since these contaminants inhibit the activity of some proteases.[9] Second, to align the gel with the paper elements of the electrotransfer sandwich, it is convenient to transfer the gel from the glass plate of the SDS-PAGE unit onto a thin, flexible sheet of waterproof material (e.g., Parafilm M, American Can Co.,

Greenwich, Conn., or wax paper). Then, after placing the protease-wetted 3MM paper strips on top of the gel, the assembly is inverted onto a second piece of the same waterproof material, and the initial support sheet (now on top) is carefully peeled away from the gel. The remaining layers are added, and the procedure continued as described earlier. Third, following protease-facilitated electrotransfer the polyacrylamide gel remnant may show some Coomassie brilliant blue staining due to the Pronase that has been introduced into it.

EVALUATION

Protease-facilitated electrotransfer has been shown to be an effective means of increasing the efficiency of transferring high-M_r proteins (i.e., >100,000) from polyacrylamide gels onto nitrocellulose.[1] The procedure is simple and the results are reproducible.

Improved transfer of high-M_r proteins is not at the expense of low-M_r species. Their binding was essentially the same in the presence or absence of protease. It has also been determined that the use of protease during transfer does not interfere with most post-transfer assays (e.g., antibody binding; Amido Black, India ink, or colloidal gold staining;[14] and trinitrophenyl-derivatization for immunodetection[15]). Instead, the assay sensitivity for high-M_r proteins is generally increased due to the greater amount of protein transferred (Figure 2B, D; Reference 1). Both polyclonal (e.g., Figure 2, Reference 6) and monoclonal (e.g., Figure 2B) antibodies have been tested, and all appeared to react at least equally well with proteins after protease-facilitated transfer. If a particular antigenic determinant proved sensitive to one protease, another having a different cleavage specificity could be used. *Staphylococcus aureus* V8 protease, for example, has been used as an effective alternative to Pronase (Figure 3; unpublished results). An alternate protease must be small enough to penetrate the polyacrylamide gel matrix, and it must work in the presence of SDS (used to render it negatively charged and ensure correct migration in the electric field), or have an appropriate intrinsic charge to ensure correct migration.

A parameter not investigated in the first report of this procedure was the time required for transfer. This has since been done and the results of the experiment are presented in Figure 2. Using the apparatus described here and Pronase at a concentration of 20 $\mu g/m\ell$, it was found that 70 to 80% of the comparatively high-M_r major capsid protein (MCP, 153,000) of cytomegalovirus was removed from the 10% gel during the first 15 min of transfer, and by 30 min more than 99% of the MCP was removed from the gel. The 20 to 30% increase of MCP on the nitrocellulose between 15 and 30 min corresponds well to the 20 to 30% lost from the gel during that same interval (data not presented for MCP alone). It was also shown in this experiment that the immunological reactivity of the nitrocellulose-bound matrix protein (MP) closely paralleled the pattern of MCP transfer. Thus, immunological activity was not lost even following exposure to transfer conditions for twice the necessary time. These results indicate that the protease-facilitated procedure reduces the time required for transfer (e.g., 99% of high-M_r protein transferred from the gel within 30 min) in addition to improving the efficiency of transferring high-M_r proteins.

One practical application of this technique is the transfer of very large disulfide-crosslinked complexes for immunoanalysis. The ability to transfer such complexes to nitrocellulose without destroying their disulfide crosslinks may aid studies of conformation-dependent epitopes. In instances where preservation of the disulfide crosslinks is not essential, and where the use of protease may be undesirable, an appropriately charged reducing agent such as mercaptopropionic acid could be used to reduce the disulfide bonds and thereby separate the protein complex into smaller components more easily removed from the gel. It is also noted that protein bands transferred by the protease-facilitated technique will contain an increased number of amino- and carboxy-terminal residues as a consequence of cleavage and, therefore, should exhibit an increased reactivity with reagents for those groups (e.g.,

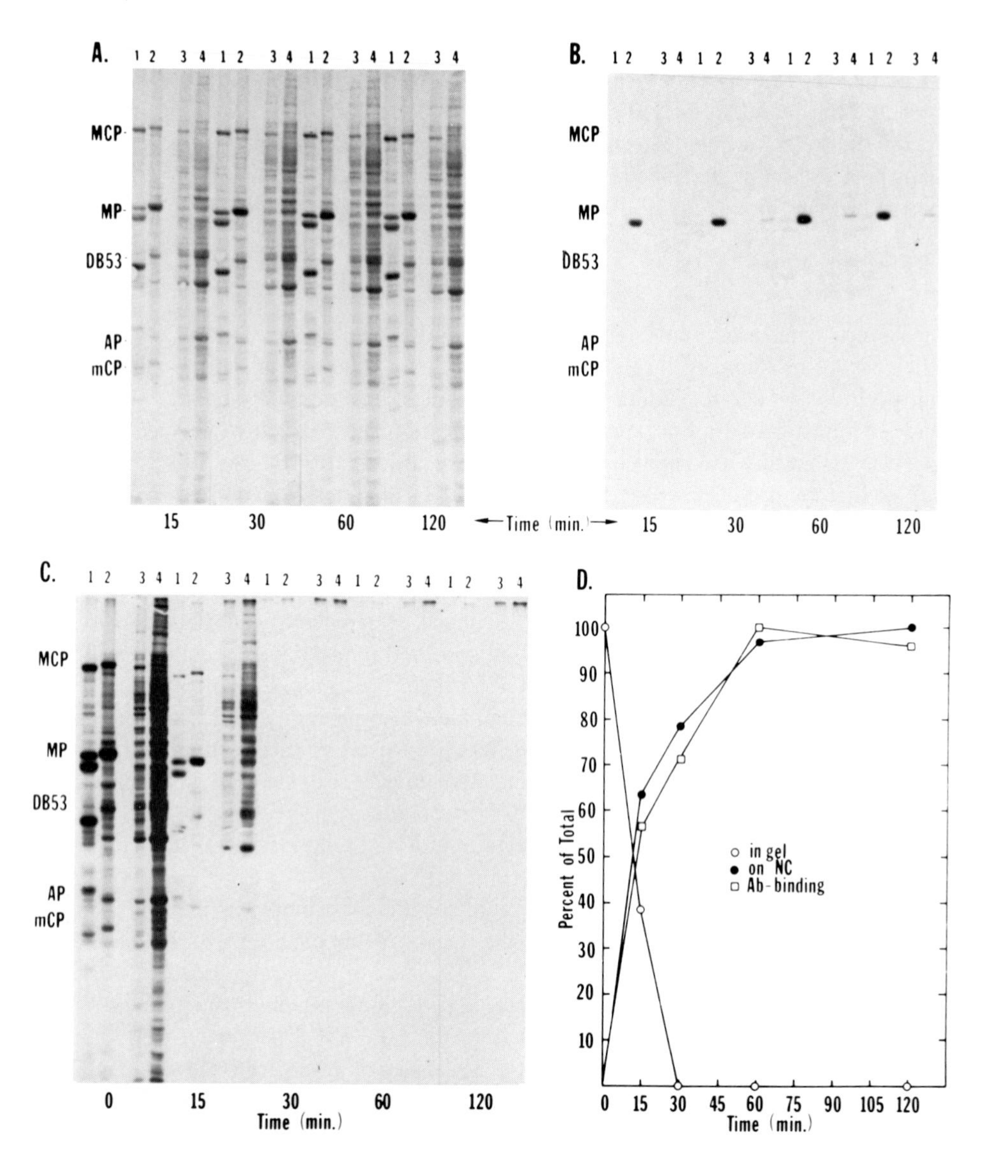

FIGURE 2. Time course of protease-facilitated transfer. Four preparations of ^{35}S-methionine labeled proteins were subjected to SDS-PAGE in a 10% gel: nuclear proteins from Colburn cytomegalovirus (1) or human cytomegalovirus (HCMV) (2) infected cells; cytoplasmic proteins from Colburn (3) or HCMV (4) infected cells. Five repeats were run. One was stained with Coomassie brilliant blue immediately after SDS-PAGE (0 min); the others were subjected to protease-facilitated transfer for the times indicated. All gel strips were stained with Coomassie brilliant blue and processed for fluorography[10,11] (panel C); all nitrocellulose replicas were dried and subjected to autoradiography (panel A). The nitrocellulose replicas were subsequently probed with a monoclonal antibody against the HCMV matrix protein,[12] followed by ^{125}I-protein A as an indicator,[5] and then subjected to fluorography[13] (panel B). Panel D shows the combined amounts of the proteins MCP, MP, and DB53 present at each time point in the gel remnants (''in gel'') and on the nitrocellulose strips (''on NC''), and the amount of antibody bound to the matrix protein (''Ab-binding'') after electrotransfer. Values presented are based on densitometric measurements of images shown (or shorter exposures) in panels A, B, and C, and are calculated as percent of the highest value for each. Abbreviations: major capsid protein (MCP, 153 kdaltons), matrix protein (MP, 69 kdaltons), DNA-binding protein (DB53, 53 kdaltons), assembly protein (AP, 36 kdaltons), and minor capsid protein (mCP, 34 kdaltons).

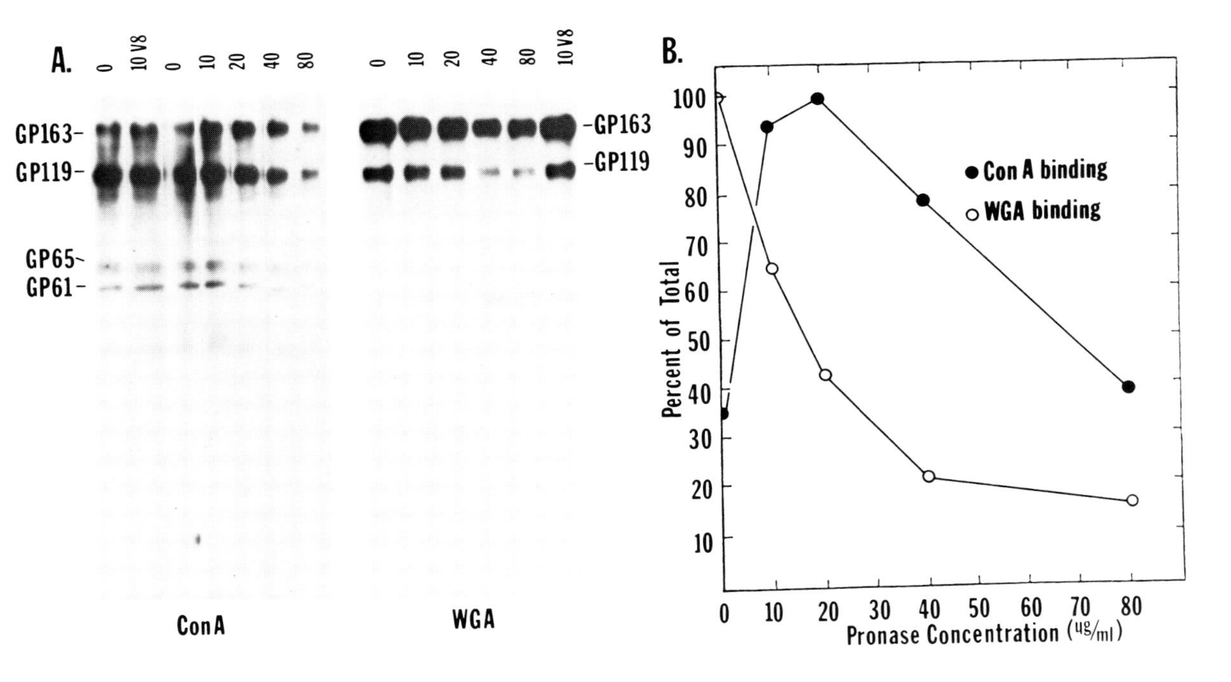

FIGURE 3. Lectin-binding after protease-facilitated transfer. Colburn cytomegalovirus virion proteins were separated by SDS-PAGE in 10% gels. The gels were sectioned into lanes, and the proteins in each lane were then electrotransferred to nitrocellulose using the indicated amounts of Pronase (0 to 80 μg/mℓ) or *S. aureus* V8 protease (0 to 10 μg/mℓ) (V8). Following electrotransfer the nitrocellulose strips were probed with ^{125}I-concanavalin A (Con A) or ^{125}I-wheat germ agglutinin (WGA), as described elsewhere.[7,8] Shown here are fluorograms of the lectin-probed nitrocellulose strips (panel A), and a graphic representation of the effect of each Pronase concentration on the reactivity of gp163 with these lectins (panel B), calculated as percent of the highest value for each. Glycoproteins (GP) are indicated by $M_r \times 10^{-3}$.

stains or radiolabeling compounds). One final consideration is that active protease may be transferred through the gel and onto the nitrocellulose. Although this possibility has not been tested directly, it does not appear to be a problem since immunological reagents (e.g., antibodies and ^{125}I-protein A, stored at $-70°$ between uses) and lectin probes (e.g., ^{125}I-wheat germ agglutinin, stored at 4° between uses) can be used repeatedly without apparent proteolytic damage.

USE AND LIMITATIONS

More recently, the protease-facilitated electrotransfer technique has been used in studies of cytomegalovirus glycoproteins.[8] Results of initial experiments indicated that when ^{3}H-glucosamine-labeled glycoproteins were electrotransferred to nitrocellulose in the presence of 20 μg/mℓ Pronase, two of the radiolabeled bands (gp163, gp119) were disproportionately weak, compared with their relative intensities in a "nontransferred" gel. This observation suggested that some glycosylated fragments were lost during protease-facilitated transfer. To test this possibility, virion glycoproteins were electrotransferred to nitrocellulose using different concentrations of Pronase, and then assayed with the lectins wheat germ agglutinin (WGA, binds sialic acid under conditions used) and concanavalin A (Con A, binds high mannose, N-linked oligosaccharides). Results of this experiment (Figure 3) showed that the amount of Con A binding was increased (e.g., 50 to 70% for gp163) using moderate concentrations of Pronase (e.g., 10 to 20 μg/mℓ) during electrotransfer, but decreased when higher concentrations (e.g., 40 to 80 μg/mℓ) were used. The WGA binding properties of the glycoproteins, however, were more severely affected. Even at the lowest concentration of Pronase tested (10 μg/mℓ) there was a reduction in WGA binding. This was most dramatic for gp163, whose WGA binding was decreased by 30 to 40% following transfer with 10 μg/mℓ Pronase, and by 80 to 85% following transfer with 80 μg/mℓ Pronase. Thus, while the use of low concentrations of Pronase to facilitate transfer of glycoproteins is compatible with the Con A binding assay, another enzyme may be more suitable if WGA binding is to be assayed following transfer.

It should be emphasized that this effect appears to be limited to glycoproteins. When proteins labeled with ^{14}C-amino acids,[1] ^{35}S-methionine (Figure 2), ^{32}P-orthophosphate (unpublished results), or ^{3}H-glucosamine[8] were compared following protease-facilitated transfer, only those labeled with ^{3}H-glucosamine showed a reduced fluorographic intensity. Further, when gp163 transfer was measured directly by the intensity of the nitrocellulose-bound, ^{35}S-methionine-labeled protein, rather than indirectly by its carbohydrate moiety (i.e., ^{3}H-glucosamine; lectin binding), it was found that the protein itself behaved like others. More was transferred in the presence of Pronase and there was no detected loss of bound protein radioactivity when higher concentrations of Pronase were used during transfer.

These results indicate that there is a selective loss of glycopeptides during protease-facilitated transfer (using Pronase), and that the effect is strongest on peptides bearing sialylated (i.e., WGA binding) oligosaccharides and magnified by decreasing their size (e.g., by using too much Pronase). One explanation of this effect is that the high ratio of carbohydrate to amino acids in small glycopeptides reduces their binding to nitrocellulose. If this interpretation is correct, it may be possible to eliminate the problem by using a less voracious protease such as *Staphylococcus aureus* V8. Such an enzyme would be expected to cleave less frequently, even at high concentrations, and yield larger glycopeptides with correspondingly lower ratios of carbohydrate to amino acids — hypothetically better able to bind to nitrocellulose.

ACKNOWLEDGMENTS

We acknowledge the excellent technical assistance of Aimee Wolitzky, and thank Naomi Swensson for help in typing the manuscript. This work was aided by Research Grants AI22711, AI13718, AI19373, and Institutional Research Grant RR5378, all from NIH. Donna Benko was supported by PHS T32 CA09243, awarded by the National Cancer Institute, DHHS.

REFERENCES

1. **Gibson, W.,** Protease-facilitated transfer of high-molecular-weight proteins during electrotransfer to nitrocellulose, *Anal. Biochem.*, 118, 1, 1981.
2. **Towbin, H., Staehelin, T., and Gordon, J.,** Electrophoretic transfer of proteins from polyacrylamide gels to nitrocellulose sheets: procedure and some applications, *Proc. Natl. Acad. Sci. U.S.A.*, 76, 4350, 1979.
3. **Laemmli, U. K.,** Cleavage of structural proteins during the assembly of the head of bacteriophage T4, *Nature (London)*, 277, 680, 1970.
4. **Gibson, W.,** Structural and nonstructural proteins of strain Colburn cytomegalovirus, *Virology*, 111, 516, 1981.
5. **Burnette, W. N.,** "Western blotting": electrophoretic transfer of proteins from SDS-polyacrylamide gels to unmodified nitrocellulose and radiographic detection with antibody and radioiodinated protein A, *Anal. Biochem.*, 112, 195, 1981.
6. **Weiner, D. and Gibson, W.,** Identification of a primate cytomegalovirus group-common protein antigen, *Virology*, 115, 182, 1981.
7. **Bartles, J. R. and Hubbard, A. L.,** ^{125}I-wheat germ agglutinin blotting: increased sensitivity with polyvinylpyrrolidone quenching and periodate oxidation/reductive phenylamination, *Anal. Biochem.*, 140, 284, 1984.
8. **Benko, D. and Gibson, W.,** Primate cytomegalovirus glycoproteins: Lectin-binding properties and sensitivities to glycosidases, *J. Virol.*, 59, 703, 1986.
9. **Lacks, S., Springhorn, S., and Rosenthal, A.,** Effect of the composition of sodium dodecyl sulfate preparations on the renaturation of enzymes after polyacrylamide gel electrophoresis, *Anal. Biochem.*, 100, 357, 1979.
10. **Bonner, W. and Laskey, R. A.,** A film detection method for tritium labeled proteins and nucleic acids in polyacrylamide gels, *Eur. J. Biochem.*, 46, 83, 1974.
11. **Laskey, R. and Mills, A.** Quantitative film detection of ^{3}H and ^{14}C in polyacrylamide gels by flurography, *Eur. J. Biochem.* 56,335, 1975.
12. **Weiner, D., Gibson, W., and Fields, K. L.,** Anti-complement immunofluorescence establishes nuclear localization of human cytomegalovirus matrix protein, *Virology*, 147, 19, 1985.
13. **Swanstrom, R. and Shank, P. R.,** X-ray intensifying screens greatly enhance the detection by autoradiography of the radioactive isotopes ^{32}P and ^{125}I, *Anal. Biochem.*, 86, 184, 1978.
14. **Moeremans, M., Daneels, G., and DeMey, J.,** Sensitive colloidal metal (gold or silver) staining of protein blots on nitrocellulose membranes, *Anal. Biochem.*, 145, 315, 1985.
15. **Wolff, M., Pfeifle, J., Hollman, M., and Anderer, A.,** Immunodetection of nitrocellulose-adhesive proteins at the nanogram level after trinitrophenyl modification, *Anal. Biochem.*, 147, 396, 1985.

SECTION 5: MATRICES

5.1 NONCOVALENT ATTACHMENT

5.1.1. NITROCELLULOSE MEMBRANES AS SOLID PHASE IN IMMUNOBLOTTING

Lise Nyholm and Jakob Ramlau

INTRODUCTION

The so-called "unspecific" binding of macromolecules to almost any kind of solid surface is a classical problem when working with dilute solutions. In techniques such as ELISA (enzyme-linked immunosorbent assay) and immunoblotting, this property has been turned to advantage, and the binding has proved to be so robust that these techniques have become widely used, although the nature of the binding has not been clearly elucidated. It is generally assumed that the binding of macromolecules to nitrocellulose membranes is due to hydrophobic interactions. These are very briefly described as associations between hydrophobic areas favored by a gain in entropy when the water adjoining the hydrophobic areas is extruded to the surrounding bulk water.

The following short review discusses how the assumption that hydrophobic interactions are involved in the binding to nitrocellulose membranes relates to the empirical findings.

NITROCELLULOSE MEMBRANES — STRUCTURE AND COMPOSITION

The nitrocellulose or, more correctly, the cellulose nitrate, consists of purified cellulose esterified with three nitrate groups per glucose unit. In the production of membranes this cellulose nitrate (hereafter denoted NC for brevity) is dissolved in a mixture of organic solvents. The solution is spread as a thin film on a smooth surface. Strict attention to the evaporation of the solvent mixture makes it possible to control the porosity of the membranes. Several pore sizes are available, the 0.45 μm membranes being most widely used.

The pore size stated in catalogs refers to an "equivalent" pore size (limiting diameter for challenge latex particles), but scanning electron microscope (SEM) pictures, which abound in brochures on sterile filtration, show the filter as a network with long tortuous channels through the membrane, presenting an enormous internal surface accessible for binding. It is difficult to construct a geometrical model for the internal surface, but rough estimates arrive at an internal surface area some 100 times larger than the outer surface. The large, inner surface undoubtedly contributes to the very high binding capacity of NC membranes compared to the plastic surfaces used in ELISA. Pure NC filters such as Schleicher & Schuell's are stated to bind 80 to 200 $\mu g/cm^2$,[1,2] compared to 100 ng/cm^2 in ELISA plastic.[3] Even weakly binding proteins may be retained within the NC matrix, because the large excess of binding sites will allow multiple dissociations and reassociations. The reported better binding of low-M_r proteins in NC membranes with pore size 0.2 μm[1,4] could reasonably be explained by such a better retention of weakly binding species in the more porous membrane. However, the use of membranes with smaller pore size is not a general solution to problems with weakly binding proteins. Thus, calmodulin did not bind to even 0.025 μm membranes under normal blotting conditions.[5]

Some NC membranes, e.g., Millipore, also contain cellulose acetate which probably reduces the binding capacity somewhat (found in Ref. 6 to be 15 $\mu g/cm^2$). Membranes designed for filtration are often treated with surfactants such as Triton® to improve wettability.

Such additives might interfere with binding (see later discussion of detergents). Residual solvent might remain in the membrane after processing and contribute to or interfere with binding.[7] The presence of such residues is suggested by the fact that prolonged baking of the NC membrane (as is routine in DNA blotting) increases its brittleness. Age might also influence membrane properties and has been reported to increase hydrophobicity of the membrane.[7]

ROLE OF BLOTTING TECHNIQUE

When evaluating the effects of various factors on the binding to or elution from NC membranes one should consider the experimental details of the various blotting techniques, or rather NC techniques since the very versatile so-called "dot-blot" (dot immunobinding) is actually not a blotting technique. Proteins spotted onto NC membranes, which are left for some time before blocking, have a better opportunity for irreversible binding than will the proteins in a capillary blot or an electroblot. In capillary blots or manifold dot-blots, which resemble simple filtration, the solvent flow will tend to remove any unbound fraction of the protein molecules. Virus retention on NC filters thus depends on solvent flow rate.[8] In electroblots the electric field tends to remove the unbound fraction and might eventually elute the protein from the membrane.[2,4]

In the following paragraphs on the influence of pH, salts, organic solvents and detergents, we have included results from some of our own experiments in which we employed a dot-immunobinding assay modified so as to resemble the genuine blotting techniques (experimental details given in the Appendix).

INFLUENCE OF pH ON BINDING

Hydrophobic interaction is favored when the pH is close to the pI of the protein.

There is no systematic study on the influence of pH on the binding of proteins to NC membranes, although good binding has been reported both at slightly basic pH and in 0.7% acetic acid.[6] The binding of small single-stranded DNA is stronger at pH 3 than at pH 6.[9]

Membrane filters (including NC) have been used for years for virus concentration. Therefore, the various factors affecting virus binding to filters have been studied rather thoroughly.[8,10,11] Viruses generally seem to bind best to NC membranes at low pH, and may even be eluted at pH 9 to 10. This pH-dependent variation in binding has been explained as being caused by electrostatic attraction due to postulated negative charges on the Millipore NC filter,[8] but hydrophobic interactions have also been implicated.[11]

We have investigated the binding to NC (Schleicher & Schuell) of three purified proteins: human neuron-specific enolase (pI ~5, M_r ~90,000), human IgG (pI ~7 to 8, M_r ~150,000), and avidin (pI ~10, M_r ~70,000) at pH 2 to 12 (ionic strength 0.1). All three proteins showed good binding at the pH extremes, denaturation probably exposing hydrophobic groups as suggested for a related finding in ELISA.[12] Within these extremes the binding varied with pH; the acidic enolase showed strongest binding at pH 4 and weakest binding at pH 8 to 10, while the opposite trend was seen for the more basic IgG and avidin. This observed variation in binding with varying pH is consistent with the notion of a hydrophobic contribution to binding.

The main point is that although a large number of proteins bind tenaciously to NC under a wide range of experimental conditions including extremes of pH, there are proteins which bind poorly under conditions generally used in electrophoresis: pH 8 to 9 and low ionic strength. Thus, attempts to make capillary blots of human neuron-specific enolase from agarose gels were unsuccessful, since nearly all the enolase passed the NC without binding. However, "touch-blots" (see Appendix) of such gels with NC moistened in 0.1 or 1 *M*

CH_3COOH or 0.1 *M* NaOH gave excellent results, especially if NaCl or $(NH_4)_2SO_4$ was also added to the NC equilibration buffer.

INFLUENCE OF SALTS ON BINDING

Hydrophobic interactions are strengthened with increasing concentrations of salts with salting-out potential. Ions with high-charge density such as PO_4^{3-} or SO_4^{2-} (water structure forming, antichaotropic ions of the Hofmeister series) have high salting-out potential and strengthen hydrophobic interaction, while ions with low-charge density such as SCN^- (water structure disturbing, chaotropic) destabilize hydrophobic interaction and can therefore be used for solubilization of, e.g., membrane proteins.[13,14]

The effects of salts on the binding of protein to NC has not been systematically studied, probably because the addition of salt to electrotransfer buffers is awkward due to increased joule heating. However, NC seems to behave like a hydrophobic matrix as regards the influence of salts. The binding of single-stranded DNA to NC requires high concentrations of NaCl in the buffer[15] and the binding of RNA requires even more NaCl.[16,17] Calmodulin has been found to bind to NC only if 0.1 *M* NaCl or 0.025 *M* phosphate (pH 6.5) was present in the electrotransfer buffer.[5] Poliovirus binding to NC was strengthened by anti-chaotropic ions (at high pH) and chaotropic ions reversed this effect, so hydrophobic interaction was implicated at high pH (but electrostatic at low pH).[11]

We have investigated the effect of the antichaotropic salt ammonium sulfate on the binding of the three aforementioned proteins in the pH range 2 to 8. As mentioned earlier, neuron-specific enolase showed poor binding at pH 8 at low ionic strength (0.1), and IgG and avidin showed relatively poor binding at pH 4 at low ionic strength. For each protein the addition of ammonium sulfate (to an ionic strength of 0.5) increased the binding, so that it became equally strong for all pH levels within the investigated range pH 2 to 8. Enhanced binding was also observed in "touch-blots", when the NC membrane was preequilibrated in 1 *M* solutions of NaCl or $(NH_4)_2SO_4$.

All the above observations indicate that hydrophobic interactions are involved in protein binding to NC membranes. On the other hand, Parekh et al.[18] found that chaotropes did not elute proteins from NC as expected from the "hydrophobic hypothesis" for binding.

INFLUENCE OF ORGANIC SOLVENTS

Proteins can often be eluted from hydrophobic matrices by decreasing the polarity of the medium with organic solvents.[13] In this respect NC does not behave like a typical hydrophobic matrix. Ethylene glycol at a concentration of 25%, which is often used as an efficient eluant in hydrophobic chromatography,[13] had no inhibitory effect on the binding of human IgG and avidin when tested in the described dot-immunobinding procedure. Another observation, which apparently does not fit very well with a hydrophobic binding mechanism, is the improved binding in the presence of 20% methanol.[2] Methanol was originally included in the electrotransfer buffer to counteract swelling of the gel during transfer, but was found to improve binding, although it slowed the speed of transfer.[2] It is tempting to ascribe these effects to an incipient precipitation of the proteins in the presence of methanol, since methanol is a well-known precipitant of proteins[13] and binding strength to NC often correlates with the M_r of the proteins.[18] Methanol might also enhance the binding by removing SDS from the proteins.[4] The finding that 5% pyridin or 40% acetonitrile can elute proteins from NC[18] is in accordance with the hypothesis that binding to NC is due to hydrophobic interactions.

NC itself dissolves in acetone, and blotted proteins may be recovered in this way.[23]

INFLUENCE OF DETERGENTS

Detergents are often used to elute recalcitrant proteins from hydrophobic columns.[19] It is therefore noteworthy that we, in accordance with others, (e.g., Reference 18) could not elute already bound proteins from NC with 1% solutions of either SDS (sodium dodecyl sulfate), Triton® X-100, or Tween® 20. However, preequilibration of the NC in 1% solution of each of the detergents completely blocked the binding (the only exception being avidin in Triton® X-100, which showed some residual binding). Similar observations have led to the use of detergents such as Tween® 20 as blocking agents in immunoblotting procedures.[20] There seems to be no blocking effect of 1% cetyltrimethylammonium bromide, deoxycholate, or Zwittergent 3-12.[21] Higher detergent concentration such as 5% SDS has been reported to elute bound proteins from NC.[1] At low detergent concentrations, the lack of effect seems to depend on detergent composition and on the protein.

In accordance with Erickson et al.[22] we have found that 0.1% SDS in the electrotransfer buffer did not interfere with binding. Lin and Kasamatsu[4] found that 0.05% Nonidet P-40 completely inhibited the binding of viral polypeptides to NC, and even eluted some of them. Parekh et al.[18] found that combinations of ionic and nonionic detergents (e.g., 0.1% SDS + 0.5% deoxycholate + 0.5% Triton® X-100) could elute proteins from NC, while none of the detergents alone (1%) were effective.

BINDING OF GLYCOLIPIDS

Lipopolysaccharides can be transferred to NC after separation in nondenaturing polyacrylamide gels.[24] This is in accordance with the hydrophobic hypothesis for binding. Glycosphingolipids have been transferred to NC in isopropanol/water (2:1).[25] It is noteworthy that these glycolipids could bind to the NC after blocking with bovine serum albumin (BSA), and that they were eluted if Triton® X-100 (percentage not stated) was present in washing buffers.

REVERSIBILITY OF BINDING TO NITROCELLULOSE

It has already been mentioned that some proteins will not be eluted by conditions inhibiting their binding. This indicates that some secondary alterations take place after binding, which enhance the binding to the membrane. The baking of NC membranes after transfer of DNA and RNA seems to bring this about.[26] Whether this "zippering into position"[4] of the proteins is explained solely by multipoint attachment via several hydrophobic patches on the molecule, or whether other mechanisms are involved is open to speculation. The main point is that a multitude of different proteins bind so tenaciously to NC membranes that they will resist washing and elution conditions as harsh as those used for covalently bound ligands. Olmsted[27] used covalent blotting for the affinity purification of antibodies reacting with transferred bands, and it is obvious that NC blotting can be used for the same purpose. Bound antibodies have been eluted with 8 *M* urea, 0.1 *M* mercaptoethanol,[22] 6 *M* NH_4SCN,[28] or acidic pH[29] without eluting the NC bound antigen, and even several hours incubation at 37° with the chaotropes 3 *M* guanidinium HCl or 6 *M* urea did not elute proteins from NC.[18]

An interesting, though rather hypothetical, aspect of the "zippering" or secondary folding into more stable configuration is the possible renaturation of proteins blotted from SDS gels. Many proteins have turned out to be remarkably antigenically active after blotting in spite of the prior harsh treatment.[28] One might speculate that hydrophobic groups on the NC could compete with SDS on the protein and thus help renature the protein into more or less native configuration in the same way as excess nonionic detergent will compete SDS away from proteins, resulting in at least partial restoration of antigenicity.[30]

TENTATIVE CONCLUSIONS AND FUTURE DEVELOPMENTS

Though a vast number of papers attest to the versatility of NC membranes in blotting procedures there is, to our knowledge, no thorough description of possible binding mechanisms, but rather a general consensus that hydrophobic interactions are involved. In this chapter we have used this hydrophobic model as a conceptual framework for presenting the different observations more than we have attempted a critical evaluation of the hypothesis. Those interested in theoretical aspects of hydrophobic interaction in general are referred to References 31 and 32.

The use of different NC membranes and techniques is a hindrance to the straightforward comparison between various findings, but the main obstacle is probably the simple fact that proteins are different. There is by now ample evidence that conditions under which some proteins will bind firmly to NC will elute others. It is thus not easy to extract a set of clear guidelines for NC blotting, but a few practical hints may be given. Concerning the choice of NC brand we have found Schleicher & Schuell's to be the best (highest binding capacity, highest signal:background ratio) with Millipore's very flexible HAHY membranes as a second choice. In case of marginal binding of a given protein one should consider reduction of solvent flow, voltage, or electrotransfer time, addition of methanol or salt, change of pH, or maybe other blocking conditions.

Although most proteins stick very firmly to the NC, there is still room for improvement. A matrix with predictable binding, preferably equal for all proteins, would be ideal. The brittleness of NC membranes is acceptable in the laboratory, but if pre-dotted NC membranes should come into use for field tests (see Chapter 9.1.4) or for "doctor's office/bedside" testing,[33] some improvement in physical structure would be essential. NC might also be of interest within ELISA technology. Its high binding capacity could make NC-coated ELISA tubes/plates worth considering.

New membranes allowing post-transfer covalent coupling would probably be very useful. Such matrices already exist, but the amino groups necessarily present in most gel buffers interfere with coupling. One might envisage a bifunctional membrane, containing both groups for the binding of proteins and primary amino groups for coupling with glutaraldehyde, carbodiimide, or other crosslinkers. A special application of blotting combined with covalent coupling might be solid phase sequencing of separated proteins or peptide fragments. Solid phase sequencing of proteins covalently coupled to aminoalkylsilan-modified glass beads has been described.[34] A similar approach might be imagined for peptides transferred from gels to modified glass fiber filters.

It is thus possible that the blotting technique, which was originally intended as an immunochemical second dimension to the PAGE, might develop into a tool for further molecular characterization of separated components.

APPENDIX: MATERIALS AND METHODS

Nitrocellulose: Schleicher & Schuell BA 85 (0.45 μm).

Antigens: Human neuron specific enolase and avidin, purified in-house. Human IgG, KabiVitrum AB, Stockholm, Sweden.

Antibodies: First layer — Rabbit antibovine neuron-specific enolase (A589), rabbit anti-avidin (Z 359), and rabbit antihuman IgG (A 090). Second layer — Alkaline phosphatase-conjugated swine antirabbit IgG (D 306). All from Dakopatts, Glostrup, Denmark.

Dot immunobinding procedure: A procedure was employed which intended to imitate conditions in blotting techniques. The NC was preequilibrated in the relevant buffer (see below), surplus buffer being removed with filter papers. 3 μℓ of protein were pipetted *in duplo* onto the moist NC membrane (150 ng to 9 pg/spot, 5-fold dilutions from 50 μg/mℓ to 3 ng/mℓ in the relevant equilibration buffer + 0.1 mg/mℓ BSA). The NC was washed immediately afterward in equilibration buffer for 30 min before routine blocking and further processing. All operations at room temperature (approximately 20°C).

Equilibration buffers: For testing influence of pH — Glycine-HCl, pH 2; sodium acetate, pH 4; sodium phosphate, pH 6; sodium barbital, pH 8; glycine-HCl, pH 10 and 12; all adjusted to ionic strength 0.1 with NaCl.[34] Influence of ammonium sulfate — As above pH 2 to 8, but with ammonium sulfate added to ionic strength 0.5. Influence of ethylene glycol (only IgG and avidin) — 0.025 *M* Tris, 0.193 *M* glycine, pH 8.3 with and without 25% (v/v) ethylene glycol. Influence of detergents (only IgG and avidin) — 0.025 *M* Tris, 0.193 *M* glycine, pH 8.3 with and without 1% (v/v) SDS, Triton® X-100, or Tween® 20. For investigating possible eluting effects of detergents on already bound protein 3 μℓ antigen dilution (only IgG and avidin) in 0.025 *M* Tris, 0.193 *M* glycine, pH 8.3 (+ 0.1 mg/mℓ BSA) were applied *in duplo* on NC which was then left overnight before washing (30 min) in 1% (v/v) SDS, Triton® X-100, or Tween® 20 in 0.025 *M* Tris, 0.193 *M* glycine, pH 8.3, followed by routine blocking and antibody incubation. In the same experiment was also included identically dotted NC which had been made immediately before washing. Parallel controls were washed in 0.025 *M* Tris, 0.193 *M* glycine, pH 8.3 without detergent.

Blocking, antibody incubation, and staining: Excess binding sites were blocked in 1% (w/v) BSA in washing buffer (0.05 *M* Tris, 0.15 *M* NaCl, 0.5% (v/v) Tween® 20, pH 9) before overnight incubation with first antibody (1:500 (anti-enolase) or 1:1000 in washing buffer with 0.1% (w/v) BSA. After 3 × 10 min washing the NC was incubated in second antibody for 3 hr (1:1000 in washing buffer with 0.1% BSA). After another 3 × 10 min washing staining was performed as described in the Appendix to Chapter 1.

Estimation of binding: The amount of bound protein was estimated visually from the developed color. Since this can only be regarded as a semiquantitative method, we have only referred to the results in semiquantitative terms in the preceding text. At maximal binding the detection limit was at 80 to 400 ng/mℓ (0.2 to 1.2 ng/spot). Detection limit varied with varying pH by a maximum of 3 to 4 dilution steps, depending on antigen. Addition of ammonium sulfate increased the binding corresponding to a maximum of two dilution steps in the detection limit. Effects of detergents ranged from no eluting effect (same detection limit) to full blockade when the NC was preequilibrated in detergent at all protein dilutions tested.

"Touch-blot": Agarose gel electrophoresis was performed in 1% (w/v) agarose (HSA, Litex, Glostrup, Denmark) in Tris-barbital, pH 8.6 (ionic strength 0.02) at 5 to 10 V/cm (10 to 100 ng of protein mixed with bromophenol blue as marker). After electrophoresis excess fluid was removed from the gel surface by brief contact with a filter paper. The NC was soaked in 0.1 *M* CH_3COOH, electrophoresis buffer or 0.1 *M* NaOH, alone or with 1 *M* NaCl (or $(NH_4)_2SO_4$) added, blotted semidry between filter papers, and placed on the gel surface, avoiding air bubbles. After 3 to 6 min, the NC was lifted off the gel and cut into strips before blocking and antibody incubation. (Sometimes five consecutive blots from the same gel was performed.) Subsequently the gel plate was fixed in picric acid and stained in Coomassie brilliant blue as described in Reference 36.

REFERENCES

1. **Burnette, W. N.,** "Western blotting"; electrophoretic transfer of proteins from sodium dodecyl sulphate-polyacrylamide gels to unmodified nitrocellulose and radiographic detection with antibody and radioiodinated protein A, *Anal. Biochem.*, 112, 195, 1981.
2. **Gershoni, J. M. and Palade, G. E.,** Electrophoretic transfer of proteins from sodium dodecyl sulphate-polyacrylamide gels to a positively charged membrane filter, *Anal. Biochem.*, 124, 396, 1982.
3. **Cantarero, L. A., Butler, J. E., and Osborne, J. W.,** The adsorptive characteristics of proteins for polystyrene and their significance in solid-phase immunoassays, *Anal. Biochem.*, 105, 375, 1980.
4. **Lin, W. and Kasamatsu, H.,** On the electrotransfer of polypeptides from gels to nitrocellulose membranes, *Anal. Biochem.*, 128, 302, 1983.
5. **Rochette-Egly, C. and Daviaud, D.,** Calmodulin binding to nitrocellulose and Zetapore membranes during electrophoretic transfer from polyacrylamide gels, *Electrophoresis*, 6, 235, 1985.
6. **Towbin, H., Staehelin, T., and Gordon, J.,** Electrophoretic transfer of proteins from polyacrylamide gels to nitrocellulose sheets: procedure and some applications, *Proc. Natl. Acad. Sci. U.S.A.*, 76, 4350, 1979.
7. **Schneider, Z.,** Aliphatic alcohols improve the adsorptive performance of cellulose nitrate membranes — application in chromatography and enzyme assays, *Anal. Biochem.*, 108, 96, 1980.
8. **Wallis, C., Melnick, J. L., and Gerba, C. P.,** Concentration of viruses from water by membrane chromatography, *Ann. Rev. Microbiol.*, 33, 413, 1979.
9. **Smith, M. R., Devine, C. S., Cohn, S. M., and Lieberman, M. W.,** Quantitative electrophoretic transfer of DNA from polyacrylamide or agarose gels to nitrocellulose, *Anal. Biochem.*, 137, 120, 1984.
10. **Shields, P. A. and Farrah, S. R.,** Influence of salts on the electrostatic interaction between poliovirus and membrane filters, *Appl. Environ. Microbiol.*, 45(2), 526, 1983.
11. **Farrah, S. R., Shah, D. O., and Ingram, L. O.,** Effects of chaotropic and antichaotropic agents on elution of poliovirus adsorbed on membrane filters, *Proc. Natl. Acad. Sci. U.S.A.*, 78, 1229, 1981.
12. **Conradie, J. D., Govender, M., and Visser, L.,** ELISA solid phase: partial denaturation of coating antibody yields a more effective solid phase, *J. Immunol. Methods*, 59, 289, 1983.
13. **Scopes, R.,** *Protein Purification*, Springer-Verlag, Berlin, 1982.
14. **Hatefi, Y. and Hanstein, W. G.,** Destabilization of membranes with chaotropic ions, *Methods Enzymol.*, 31, 770, 1974.
15. **Southern, E. M.,** Detection of specific sequences among DNA fragments separated by gel electrophoresis, *J. Mol. Biol.*, 98, 503, 1975.
16. **Popovic, D. A.,** Hydrophobic chromatography of 28 S and 18 S ribosomal RNAs on a nitrocellulose column, *J. Chromatogr.*, 236, 234, 1982.
17. **Thomas, P. S.,** Hybridization of denatured RNA and small DNA fragments transferred to nitrocellulose, *Proc. Natl. Acad. Sci. U.S.A.*, 77, 5201, 1980.
18. **Parekh, B. S., Mehta, H. B., West, M. D., and Montelaro, R. C.,** Preparative elution of proteins from nitrocellulose membranes after separation by sodium dodecyl sulphate-polyacrylamide gel electrophoresis, *Anal. Biochem.*, 148, 87, 1985.
19. **Bjerrum, O. J. and Gianazza, E.,** in *Electrochemical Analysis of Membrane Proteins*, Bjerrum, O. J., Ed., Elsevier, Amsterdam, 1983, 142.
20. **Batteiger, B., Newhall, V. W. J., and Jones, R. B.,** The use of Tween 20 as a blocking agent in the immunological detection of proteins transferred to nitrocellulose membranes, *J. Immunol. Methods*, 55, 297, 1982.
21. **Bjerrum, O. J., Larsen, K. P., and Wilken, M.,** Some recent developments of the electroimmunochemical analysis of membrane proteins. Application of Zwittergent, Triton X-114 and Western blotting technique, in *Modern Methods in Protein Chemistry — Review Articles*, Tschesche, H., Ed., Walter de Gruyter, Berlin, 1983, 79.
22. **Erickson, P. F., Minier, L. N., and Lasher, R. S.,** Quantitative electrophoretic transfer from SDS polyacrylamide gels to nitrocellulose sheets: a method for their reuse in immunoautoradiographic detection of antigens, *J. Immunol. Methods*, 51, 241, 1982.
23. **Anderson, P. J.,** The recovery of nitrocellulose-bound protein, *Anal. Biochem.*, 148, 105, 1985.
24. **Sidberry, H., Kaufman, B., Wright, D. C., and Sadoff, J.,** Immunoenzymatic analysis by monoclonal antibodies of bacterial lipopolysaccharides after transfer to nitrocellulose, *J. Immunol. Methods*, 76, 299, 1985.
25. **Towbin, H., Schoenenberger, C., Ball, R., Braun, D. G., and Rosenfelder, G.,** Glycosphingolipid-blotting: an immunological detection procedure after separation by thin layer chromatography, *J. Immunol. Methods*, 72, 471, 1984.
26. **Seed, B.,** Attachment of nucleic acids to nitrocellulose and diazonium-substituted supports, in *Genetic Engineering, Principles and Methods*, Vol. 4, Setlow, J. K. and Hollaender, A., Eds., Plenum Press, New York, 1982, 91.

27. **Olmsted, J. B.,** Affinity purification of antibodies from diazotized paper blot of heterogeneous protein samples, *J. Biol. Chem.,* 256(23), 11955, 1981.
28. **Anderson, N. L., Nance, S. L., Pearson, T. W., and Anderson, N. G.,** Specific antiserum staining of two-dimensional electrophoretic patterns of human plasma proteins immobilized on nitrocellulose, *Electrophoresis,* 3, 135, 1982.
29. **Legocki, R. P. and Verma, D. P. S.,** Multiple immunoreplica technique: screening for specific proteins with a series of different antibodies using one polyacrylamide gel, *Anal. Biochem.,* 111, 385, 1981.
30. **Schäfer-Nielsen, C. and Bjerrum, O. J.,** Immunoelectrophoretic analysis of sodium dodecylsulphate treated proteins, *Scand. J. Immunol.,* Suppl. 2, 73, 1975.
31. **Tanford, C.,** *The Hydrophobic Effect: Formation of Micelles and Biological Membranes,* John Wiley & Sons, New York, 1973.
32. **Lewin, S.,** *Displacement of Water and its Control of Biochemical Reactions,* Academic Press, London, 1974.
33. **Gordon, J., Staehelin, T., and Towbin, H.,** International patent application PCT/EP/80/00018 from Ciba-Geigy.
34. **Machleidt, W. and Wachter, E.,** New supports in solid phase sequencing, *Methods Enzymol.,* 47, 263, 1977.
35. **Dawson, R. M. C., Ed.,** *Data for Biochemical Research,* 2nd ed., Clarendon Press, Oxford, 1969.
36. **Axelsen, N. H., Ed.,** Handbook of Immunoprecipitation-in-Gel Techniques, Blackwell Scientific, Oxford, 1983.

5.1.2 PROTEIN BINDING TO CHARGE-DERIVATIZED MEMBRANES IN BLOTTING PROCEDURES — A CRITICAL STUDY

Jakob Ramlau

INTRODUCTION

Nitrocellulose membranes have proven their worth empirically in blotting, but the nature of the binding of macromolecules to the nitrocellulose remains an open question (see Chapter 5.1.1). It has thus been tempting to develop other matrices with more well-defined binding properties. One such type of binding is the electrostatic binding predominant in ion exchange chromatography.

Filter paper can easily be derivatized to DEAE paper, but this positively charged material has never been popular as a blotting medium, probably due to its weakness when wet.[1]

It has been observed that asbestos filters retain particles and soluble components much smaller than the macrostructure of the filter should allow. This has been found to be due to the presence of charged groups on the asbestos. Other filters with fixed positive charges have therefore been produced to replace the banned asbestos.

Some of these charge-derivatized filters have been used in DNA blotting[2] as the nucleic acids are polyanions and bind well to positively charged matrices. Gershoni and Palade[3] used the positively charged nylon membrane, Zetabind, for protein blotting utilizing its very high capacity for protein (up to 400 μg/cm^2) with its attendant disadvantage of blocking problems.

The electrostatic nature of the binding might limit the use of high salt concentrations in incubation and washing buffers and native proteins might not bind well at transfers taking place near the pI of the protein. An important advantage of the nylon membranes is their mechanical strength.

Several firms offer these charge-derivatized nylon membranes: Zetabind® (AMF-CUNO, Le Attaques, 62730 Marck, France), Zetaprobe® (Bio-Rad) and Biodyne® (Pall, address is found in Reference 5). Other nylon membranes are sold for blotting (especially with nucleic acids) without detailed information on the nature of the membrane: Hybond N® (Amersham), GeneScreen® (New England Nuclear), and Nytran® (Schleicher & Schuell).

MATERIALS AND PROCEDURES

An earlier study (Chapter 5.1.1) had shown that various parameters did not influence the binding of proteins to nitrocellulose in a simple manner compatible with a straightforward hydrophobic binding. It was thus of interest to test whether a charge-modified nylon membrane exhibited a more predictable ion exchange type of binding.

Membranes

The membranes used were Pall Biodyne A® (a nylon membrane-derivatized with 50% amino and 50% carboxyl groups), pore size 1.2 μm.

Samples and Application

A simple immunobinding system (dot blotting) was used (see Chapter 2). Samples, 3 μℓ, of normal rabbit immunoglobulin (X 903 from Dakopatts, DK-2600, Glostrup, Denmark) in fivefold serial dilutions from 40 μg/mℓ down to 13 ng/mℓ (corresponding to an applied amount of protein from 120 ng to 40 pg/dot) in buffers containing 0.1 mg/mℓ bovine serum albumin (BSA, Sigma A 4503) were dotted onto the dry membranes and allowed to air dry at room temperature.

The serial dilutions were prepared in the buffers described later and the individual blots were incubated for 30 min in the same buffers prior to blocking and subsequent processing.

The following buffers were used: 0.1 *M* sodium citrate buffer at pH 3 and 5, 0.1 *M* sodium phosphate buffer at pH 7, 0.1 *M* Tris/HCl at pH 9, and 0.1 *M* ethanolamine/HCl at pH 11. Each application/incubation was performed with the buffer alone and with the buffers supplemented with 0.25, 0.5, and 2.0 *M* NaCl and 0.2 and 1% of the two detergents Triton® X-100 (v/v) and sodium dodecyl sulfate (SDS) (w/v).

Blocking Procedure

Preliminary experiments had shown that blocking could be performed with 10% (w/v) nonfat dry milk[4] (commercial food grade from a chain store) in PBS-Tween® 20 (10 m*M* sodium phosphate buffer, pH 7.4, 0.145 *M* NaCl, 0.1% (v/v) Tween® 20) for 1 hr at room temperature.

Immunochemical Detection

Alkaline phosphatase conjugated swine antirabbit immunoglobulins was used in a direct technique. The blots were incubated for 2 hr in the antibody (D 306 from Dakopatts) diluted 1:500 in PBS-Tween® 20 containing 0.1% (w/v) BSA.

The blots were developed with 5-bromo-4-chloro-3-indolyl phosphate and Nitro Blue Tetrazolium as described in the Appendix to Chapter 1.

RESULTS

The blotting performed at pH 7 (dilutions prepared and blots washed in pH 7 buffer without any additives) was used as a standard for comparisons. Under these conditions a 3-mm dot containing 40 pg of rabbit immunoglobulin could be detected.

The blocking with reconstituted skim milk was efficient insofar as no background was observed immediately after the staining. Nevertheless, the membranes apparently retained some staining reagents in spite of the washing since on later exposure to light they developed some background staining. This phenomenon has never been troublesome with nitrocellulose membranes in our laboratory.

The sensitivity obtained under the standard circumstances was not affected by the addition of NaCl (0.25 to 2 *M*) nor was there any significant difference in sensitivity in the pH range tested (pH 3 to 11). However, both 0.2 and 1.0% Triton® X-100 reduced the binding substantially, increasing the detection limit to 1 and 5 ng/dot. Similar reduction was observed with both percentages of SDS. This detergent-induced reduction in binding was somewhat less marked at pH 9 and 11.

DISCUSSION

The experiments demonstrate the high sensitivity obtainable in even a direct immunoenzymatic system. The problems in blocking caused by the high capacity of membranes[3] was conquered by using instant milk.[4] The binding of the rabbit immunoglobulin to the derivatized nylon is probably not only electrostatic as it is difficult to imagine an ion exchange type of interaction taking place at 2 *M* NaCl. The detailed technical information provided by Pall[5] recommends transfer of DNA onto Biodyne A by ''capillary rise'' blotting (''Southern blotting''[6]) (see Chapter 4.1) in the presence of 3 *M* NaCl, 0.3 *M* sodium citrate, pH 7, followed by baking at 80°C. This procedure is an almost exact replica of the classical nitrocellulose blotting.

It is interesting to note that Reed and Mann[7] comment on the tendency to use ''nitrocellulose recipes'' unchanged for nylon membranes and that they further demonstrate that salt concentration does not influence binding of DNA to Zetaprobe® and that transfers can be performed both at neutral pH and in 0.4 *M* NaOH.

The pH independence found in our experiments is in accordance with Gershoni and Palade,[3] who found no pH influence in the range 2 to 8, although the proteins might have been present as SDS complexes under the conditions used. The influence of pH on ion exchange type interaction would be expected to be as follows: at very low pH all protein in question will be positively charged as will the amino groups on the membrane, while the carboxyl groups will be minimally charged (if the pK_a of the immobilized groups is 5, only 1% of them will be negatively charged and able to interact with the positively charged immunoglobulin at pH 3).

The same type of argument holds for very basic pH values. The observed independence of pH on the binding is thus not easily compatible with a purely electrostatic interaction.

That other forces than ionic attraction take part in the binding is suggested by the experiments with detergents. Thus, the nonionic detergent Triton® X-100 should not interfere with electrostatic binding and the concentrations of SDS used add only 7 and 35 m*M* to the concentration of ions. It is thus more reasonable to assume that hydrophobic forces are involved in the binding to the membranes and that these are counteracted by the detergents. The experience from the experiments with nitrocellulose indicate that detergents do not influence the immunochemical detection per se (see Chapter 5.1.1).

CONCLUSIONS

Our findings suggest that the binding of at least rabbit immunoglobulin to the Biodyne® A membrane is not solely dependent on electrostatic forces, but also involves hydrophobic forces.

These findings do not necessarily detract from the practical value of the membranes. They possess a large capacity for proteins (up to 400 μg/cm² for BSA) which make them attractive in applications requiring quantitative binding or when working with weakly binding proteins where a surplus of capacity might ensure binding. The mechanical strength of the membranes is attractive as they can be cut, folded, and rolled up in any arrangement, making them well suited for any odd laboratory application and for rough use in the field test formats.

Several papers describe the use of nylon membranes in protein blotting (see Chapter 6.1.3), but their chief advantages will probably be exploited in work with nucleic acids since they bind DNA regardless of size and single/double strandedness as well as RNA.

REFERENCES

1. **Danner, D. B.,** Recovery of DNA fragments from gels by transfer to DEAE-paper in an electrophoresis chamber, *Anal. Biochem.*, 125, 139, 1982.
2. **Chomczynski, P. and Quasba, P. K.,** Alkaline transfer of DNA to plastic membrane, *Biochem. Biophys. Res. Commun.*, 122, 340, 1984.
3. **Gershoni, J. M. and Palade, G. E.,** Electrophoretic transfer of proteins from sodium dodecylsulphate-polyacrylamide gels to a positively charged membrane filter, *Anal. Biochem.*, 124, 396, 1982.
4. **Spinola, S. M. and Cannon, J. G.,** Different blocking agents cause variation in the immunologic detection of proteins transferred to nitrocellulose membranes, *J. Immunol. Methods*, 81, 161, 1985.
5. Protocol Guide for DNA Transfer to Pall Biodyne A Nylon Membranes, Pall Tech. note, Code SD 933, from Pall, Europa House, Havant Street, Portsmouth, PO1 3PD, England.
6. **Southern, E. M.,** Detection of specific sequences among DNA fragments separated by gel electrophoresis, *J. Mol. Biol.*, 98, 503, 1975.
7. **Reed, K. C. and Mann, D. A.,** Rapid transfer of DNA from agarose gels to nylon membranes, *Nucl. Acids Res.*, 13, 7207, 1985.

5.2 IMMUNOREACTIVITY OF PROTEINS BOUND COVALENTLY TO DIAZOBENZYLOXYMETHYL PAPER

Bodil Norrild and Ruth Feldborg

INTRODUCTION

The blotting method developed by Southern[1] was created for transfer of DNA to nitrocellulose by elution of DNA from agarose gels directly onto the membranes. Later it became desirable to study the hybridization of DNA probes to bound RNA, but it proved difficult to bind RNA to the nitrocellulose. Alwine et al.[2] therefore prepared diazobenzyloxymethyl-activated Whatman paper which could bind single-stranded RNA through diazo-bonds.

As proteins could also be bound through diazo-bonds, it was obvious that we should try to adapt the method for analysis of the immunoreactivity of immobilized denatured proteins.[3]

This chapter mainly describes the immunologic reactivity of proteins covalently linked to chemically activated paper through diazobenzyloxymethyl-reactive groups.

METHODS

Preparation of Diazobenzyloxymethyl (DBM) Paper

The paper was chemically activated in a two-step procedure[2]: **Preparation of aminobenzyloxymethyl (ABM) paper**

1. 14 × 25 cm Whatman 540 paper (A. G. Frisenette, Ebeltoft, Denmark) was soaked in 10 mℓ of an aqueous solution of 0.8 g 1-(*m*-nitrobenzyloxy)methyl pyridinium chloride and 0.25 g sodium acetate. Use glass-distilled water throughout the procedure.
2. The paper was dried at 60°C in an incubator.
3. Heated to 130 to 135°C for 35 min in an oven.
4. Washed twice for 20 min in water.
5. Dried at 60°C as in step 2.
6. Washed twice for 20 min in toluene and air dried under a hood.
7. The paper was then treated with 150 mℓ 20% (w/v) sodium dithionite in a plastic tray for 30 min at 60°C. The pan with the paper was shaken every 10 min.
8. Washed for 20 min in water.
9. Washed for 20 min in 30% (w/v) acetic acid.
10. Washed 3 to 4 times in water until the odor of hydrogen sulfide disappeared.
11. The ABM paper was dried and stored in a sealed bag at 4°C. The ABM paper could be stored for 1 to 2 months. We did not use it beyond this point.

Diazobenzyloxymethyl (DBM) paper

1. The ABM paper was diazotized just prior to use in a solution made up of 3.2 mℓ freshly made sodium nitrite ($NaNO_2$) solution (10 mg/mℓ), 80 mℓ 1.8 *M* HCl, and 40 mℓ water. Incubation was performed in a plastic tray for 30 min at 4°C.
2. Washed 5 times for 5 min in 100 mℓ ice-cold water.
3. Washed once for 10 min in ice-cold transfer phosphate buffer (50 m*M* sodium phosphate buffer, pH 7.4).

Note: The diazotized paper should be used immediately.

SDS-Polyacrylamide-Agarose Gels

Regular 9.25% (w/v) SDS (sodium dodecyl sulfate)-polyacrylamide gels[4] crosslinked with 0.24% (w/v) *N,N*-diallyltartardiamide (Bio-Rad, Richmond, Calif.) were too dense for the passive transfer of the separated proteins by elution from the gel to the DBM paper. The 9.25% (w/v) gel solution was therefore supplemented with 1% agarose (type HSB Litex, Copenhagen). The gel mixture was boiled before addition of both 0.1% (w/v) SDS, 0.035% (w/v) ammonium persulfate, and *N,N*-diallyltartardiamide. After the addition of these components the gel was poured when the solution was still warm. The gels were 1.5-mm thick. A 3% (w/v) stacking gel was poured on top of the separation gel and used with or without slots, depending on the experiment.[5,6] The electrophoresis was done in a Bio-rad slab-gel electrophoresis cell for 18 to 20 hr at 10 mA.

Blotting of Proteins onto the DBM Paper

The acrylamide-agarose gel was treated twice with 100 mℓ of a 2% (w/v) periodic acid for 30 min at room temperature, followed by one 15- to 30-min wash in a 0.5 *M* sodium phosphate buffer, pH 7.5, and two 15-min washes in the blotting buffer (0.05 *M* sodium phosphate, pH 7.5). The gel was immediately covered with the freshly prepared DBM paper, and the elution of the proteins and transfer to the paper were done as described by Southern[1] for the blotting of DNA. Basically, the proteins were eluted from the gel by soaking buffer through the gel and the DBM paper. The gel was placed on filter paper which functioned as wicks, and on top of the DBM paper a pile of paper towels allowed the suction of liquid through the gel. The proteins were covalently bound to the paper as long as the reactive diazo-groups maintained their activity. The transfer was done either at room temperature or at 4°C for 20 hr. The paper maintained its reactivity for a longer time at 4°C. Remaining diazo-groups were blocked by incubation of the DBM paper in a 0.1 *M* Tris-HCl buffer (pH 9.0) supplemented with 10% (v/v) ethanolamine and 0.25% (w/v) gelatin (buffer 1).[7]

Antigen Preparation

Herpes simplex virus type 1 (HSV-1) or type 2 (HSV-2) proteins were used in our studies. Vero cells were grown as monolayer cultures and infected with 10 plaque-forming units (PFU) of virus under standard conditions.[7] The infected cell proteins were extracted 18 hr after infection by solubilization of the cells in a disruption buffer with 2% (w/v) SDS and 5% (v/v) 2-mercaptoethanol.[10] Radioactively labeled HSV proteins were prepared by changing the culture medium 4 hr after infection to a medium containing either 2 μCi/mℓ of the ^{14}C-labeled amino acids leucine, isoleucine, and valine,[6] or 20 μCi/ℓ of ^{35}S-methionine.[5,8] Radioisotopes were purchased from NEN, Dreieich, West Germany. The infected cell proteins were labeled from 4 to 18 hr postinfection and then solubilized in disruption buffer. The nomenclature of the infected cell proteins (ICPs) is as introduced by Morse et al.[4]

Antibodies

Rabbit antibodies to individual HSV proteins were used as detailed by Norrild et al.[5,6] The monospecific rabbit antibodies against the glycoproteins gB, gC, and gD[9] were used for identification of the glycoproteins on the transfers. It should be noted that the rabbit antibodies to gC did not react in immunoblotting.[6]

Human antibodies from HSV-seropositive donors were used to screen for the spectrum of antibodies reactive with individual proteins.[10] Only selected sera are presented.

Immunoreaction

Strips of DBM paper with the covalently bound, transferred HSV proteins, blocked with gelatin were in accordance with the original procedure[5] incubated for 5 to 6 hr at 37°C with antibody. Incubation buffer: 0.05 *M* Tris-HCl buffer (pH 7.4) containing 0.15 *M* NaCl,

0.005 *M* EDTA, 0.25% (w/v) gelatin, and 0.05% Triton® X-100 (buffer 2). Rabbit antibodies were diluted 1/10, or 1/50, and human sera were diluted 1/100.[5,10] After extensive washing in the dilution buffer for 18 hr at 37°C, the strips were incubated for 2 hr at 37°C in the same buffer containing 0.25 μCi of ^{125}I-labeled protein A (Amersham, Buckinghamshire, England or NEN). Unbound protein A was washed off in the incubation buffer supplemented with 1 *M* NaCl and 0.4% (w/v) sarkosyl (buffer 3).[5,7] The strips were blotted dry and the autoradiographic image of the reactive proteins was obtained by exposure of the strips to Kodak® XPR-1 film at 4°C.

IMMUNOBLOTTING WITH RABBIT SERA

Electrophoretic separation of HSV-1 proteins was possible in agarose-polyacrylamide gels, although the bands appeared more diffuse than in regular polyacrylamide gels (Figure 1, slot B).[11] All radioactive proteins were transferred under conditions used as illustrated in Figure 1, slots B and C.

The immunoreaction with hyperimmune polyvalent rabbit serum made against all the HSV proteins and used in the dilution 1/10 reacted with most of the immobilized proteins. As an exception, the proteins designated ICP 20 did not bind antibodies, either because the hyperimmune serum did not have antibodies against the protein or because the denatured ICP 20 was unable to bind antibodies (Figure 1, slot E). Antisera made against the glycoproteins gB or gD reacted selectively with the two glycoproteins as illustrated in Figure 1, slot F, where a mixture of the sera was used. Each of the antibody preparations reacted with only gB or gD, respectively (data not shown). Antibodies made against gC did not bind to the glycoprotein, although the protein did bind to the paper (Figure 1, slot G). The gC is highly glycosylated and contains large amounts of o-linked carbohydrate.[12,13] The denaturation of the protein by SDS and boiling is likely to destroy the antigen-binding sites as the antibodies immunoprecipitate gC from an infected cell extract made with Triton® X-100 containing buffer.[14]

Because of the limited resolution of the agarose-polyacrylamide gels and perhaps also because of small quantities of the precursor molecules to the gB and gD, these were not visible in the transfers. When post-translational processing of the glycoproteins was inhibited by addition of the drug tunicamycin to the cell cultures, the precursors accumulated and could easily be identified in the described immunoblotting assay.[6]

IMMUNOBLOTTING WITH HUMAN ANTIBODIES

The immunoblotting was used by several investigators for screening of human sera.[15,16] In our original screening the DBM paper was used,[10] and Figure 2 shows the data obtained with selected sera. The differences in the subset of reactive antibodies from various donor sera should be noted. The reactivity of the human sera to HSV-1 and -2 proteins could be of interest for typing purposes, although other methods, such as enzyme-linked immunosorbent assay (ELISA), using purified HSV-1 and HSV-2-specific proteins have been developed recently for that very purpose.[17]

The screening of human sera by means of immunoblotting with nitrocellulose as described, for example, in Chapter 9.1.1 and 9.3.1, was more sensitive and identified more proteins than obtained with the presently described method. Another advantage of the nitrocellulose procedure was the use of peroxidase-coupled antihuman IgG instead of the iodinated protein A-binding, as the iodinated product was both expensive and had a limited "lifetime".[18] The immunoblotting has proven useful in several serum-screening studies for HSV antibodies,[16,18,19] and also the analysis of human cerebrospinal fluids for HSV-antibodies has been done.[20]

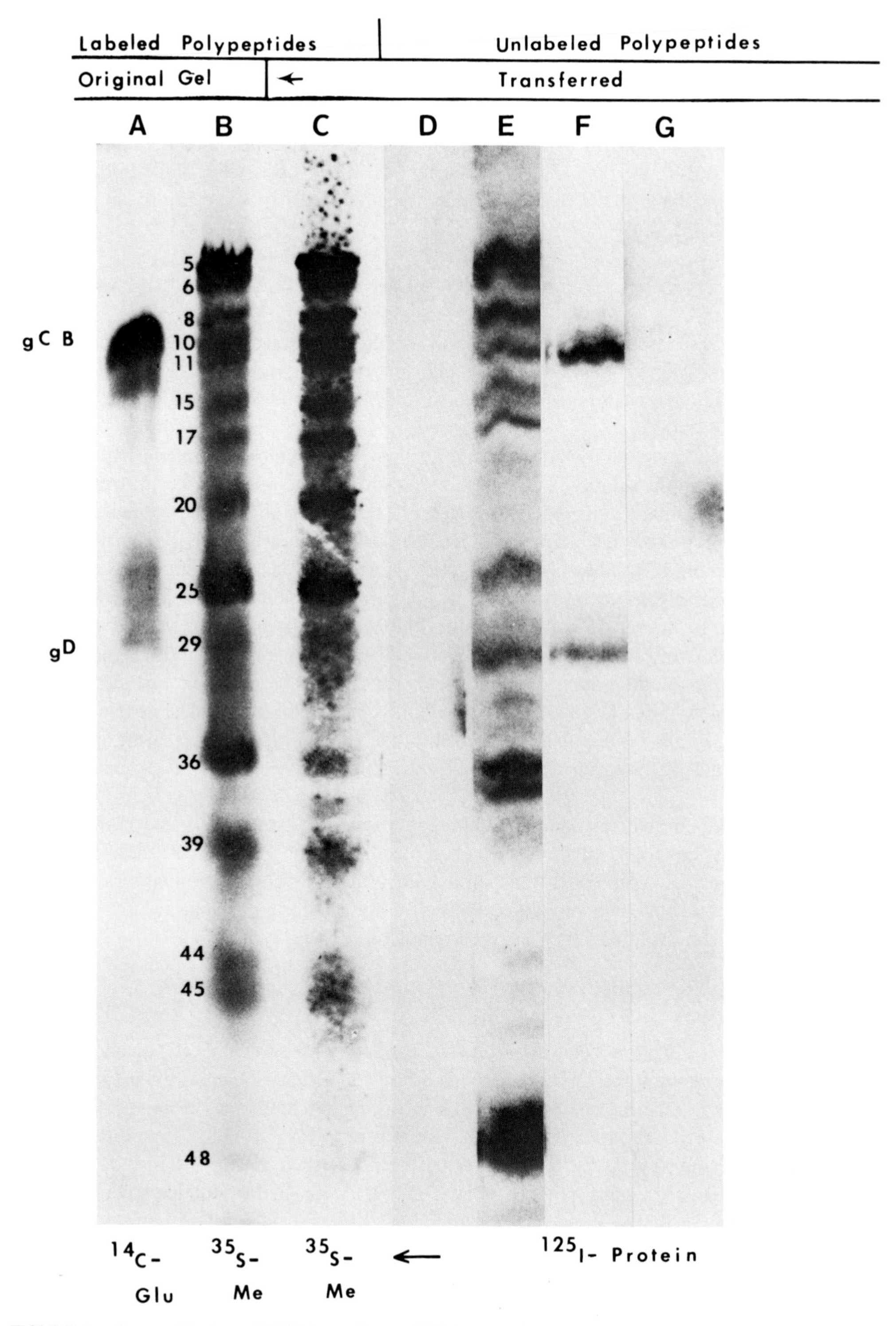

FIGURE 1. Immunoblotting of HSV-1 proteins on DBM paper after electrophoretic separation of the proteins in 9.25% SDS-polyacrylamide gels containing 1% agarose. Autoradiographs: ^{14}C-glucosamine-HSV proteins (A) ^{35}S-methionine-labeled HSV-1 proteins, (B) transferred ^{35}S-methionine-labeled HSV-1 proteins, (C) Lanes D to G: unlabeled HSV-1 proteins were separated and transferred to the DBM paper and immunoreacted with, (D) preimmune rabbit serum, (E) rabbit immune serum reactive with all HSV-1 proteins, (F) rabbit immune serum reactive with gB and gD, and (G) rabbit immune serum reactive with gC. Bound IgG was detected with ^{125}I-protein A. The numbers in lane B and the ICP numbers as defined by Morse et al.[4]

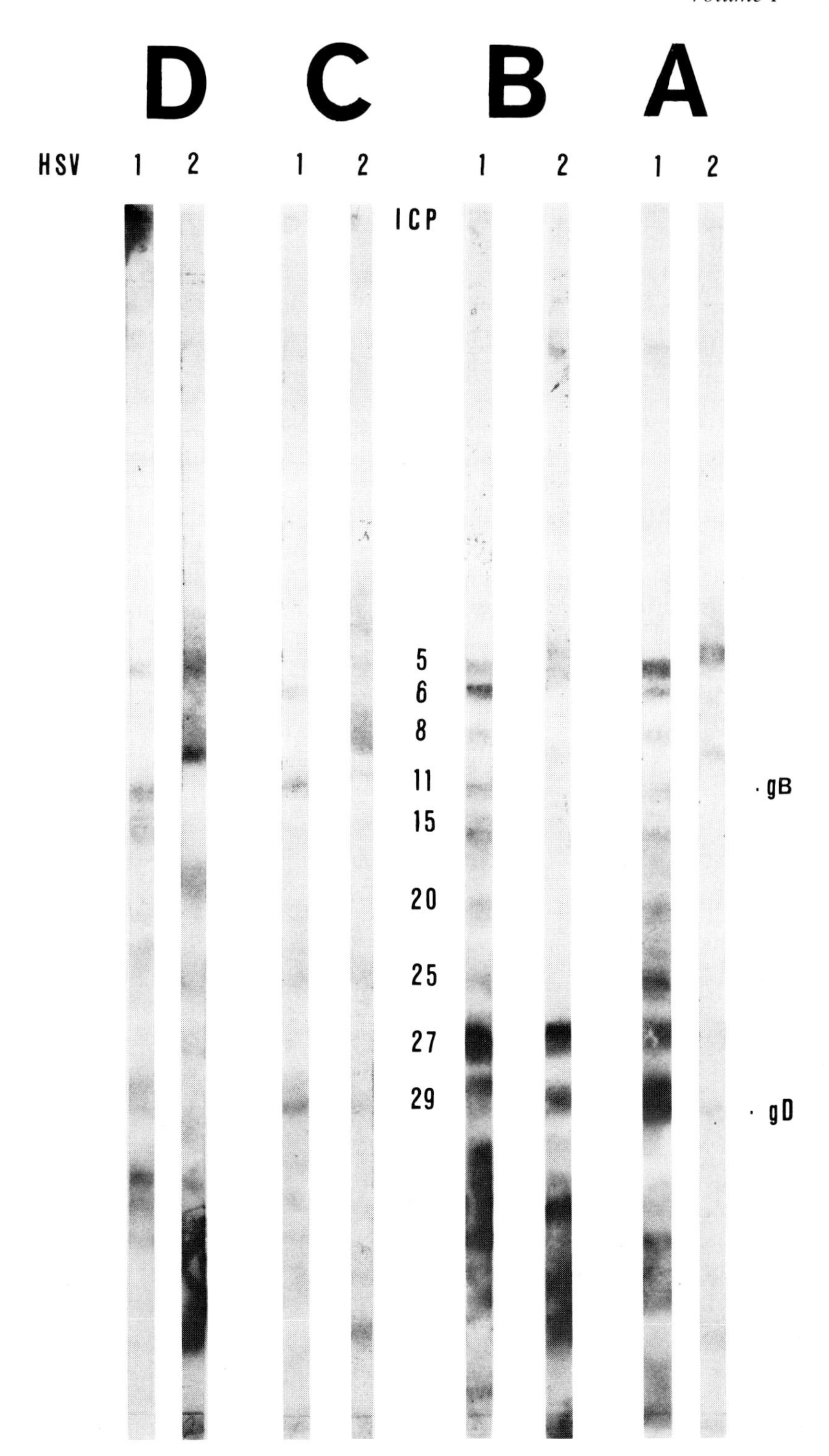

FIGURE 2. Immunoblotting of HSV-1 and -2 proteins on DBM paper. Unlabeled HSV proteins immunoreacted with human sera (1/100) from: (panel A) a seropositive healthy donor, (panel B) a donor with cervical carcinoma *in situ*, (panels C and D) donors with invasive cervical carcinoma.

It should be noted that immunoblotting with nitrocellulose is important in screening of human sera for antibodies against other viruses such as, e.g., HIV[21] (see Chapter 9.1.2).

EVALUATION

The major advantage of the method where proteins are covalently bound to paper is the possibility of analyzing the immunoreactivity of proteins extracted under denaturing conditions, which was not possible by the immunoprecipitation methods used previously. The spectrum of proteins available for analysis was thus increased as many proteins were only brought into solution by extraction with SDS and 2-mercaptoethanol. This treatment separated polypeptides associated with disulfide bonds and the following electrophoresis and immunoreaction with specific antibodies made against the native protein made it possible to analyze the precursor/product relationship between, e.g., the various polypeptides of the HSV glycoproteins.[3,5] The identification of the HSV "35-family" of proteins is another example of proteins which were only identified after the development of the immunoblotting assay, as the proteins are soluble only in the presence of SDS.[22] Six proteins showed immunological cross-reactivity with monoclonal antibodies applied in an immunoblotting assay.[22]

Another advantage of binding the proteins covalently to DBM paper is that the proteins are not washed off during the procedure. There have been reports of a minor loss of protein from the nitrocellulose membranes, expecially when the transferred proteins were "renatured" before incubation with antibodies by a wash in Triton® X-100. This phenomenon was not a problem in our experiments on DBM paper.

The limitations of either one of the immunoblotting techniques are the risk of destroying immunoreactivity of the proteins by the denaturation process, both during the extraction procedure and during the electrophoresis. To date this has not been a major problem when rabbit sera were used in the test, but we have reported that the denatured glycoprotein gC of HSV-1 does not bind the antibodies made by immunization of rabbits with immunoprecipitates cut from agarose gels.[5] Other investigators have made rabbit antibodies which react in the blotting test, but these antibodies were made against denatured proteins.[23] The development of monoclonal antibodies to a certain protein also showed that some of the monoclonals did not react in immunoblotting whereas others did (unpublished results). Apparently, specific antigen determinant sites were destroyed by the denaturation, as many of the monoclonal antibodies reacted in other tests such as ELISA, immunofluorescence, and immunoprecipitation.

APPLICATION AND FUTURE ASPECTS

The possibility of binding proteins covalently to DBM paper makes the technique useful for the development of diagnostic methods where recycling of the paper-bound proteins is wanted or where small amounts of specimens are available. We have tried such an application for the analysis of tear samples for their content of antibodies to HSV proteins. The virus proteins were covalently linked to small discs of DBM paper and the tear samples could, if necessary, be sucked into the paper directly from the capillary tubes used for the collection of tears. The technique is basically a solid-phase immunosorbent assay which allows a quantitation of specific antibodies.[24,25]

More handy ELISA assays have been developed for similar purposes,[26] and only where special reasons would make the use of paper-bound proteins an advantage should the more complicated assay be used. If small amounts of a purified protein are used as antigen it could be desirable to reuse the paper-bound protein for analysis of several specimens, as the bound antibodies can be removed from the paper without loss of the covalently bound protein (see Chapter 8.8).

The applications described have emphasized the potential of the method, and when similar problems are raised in future research projects, the technique might be useful.

REFERENCES

1. **Southern, E. M.,** Detection of specific sequences among DNA fragments separated by gel electrophoresis, *J. Mol. Biol.*, 98, 503, 1975.
2. **Alwine, J. C., Kemp, D. J., and Stark, G. R.,** Method for detection of specific RNAs in agarose gels by transfer to diazobenzyloxymethyl-paper and hybridization with DNA probes, *Proc. Natl. Acad. Sci. U.S.A.*, 74, 5350, 1977.
3. **Braun, D. K., Pereira, L., Norrild, B., and Roizman, B.,** Application of denatured, electrophoretically separated, and immobilized lysates of herpes simplex virus-infected cells for detection of monoclonal antibodies and for studies of the properties of viral proteins, *J. Virol.*, 46, 103, 1983.
4. **Morse, L. S., Pereira, L., Roizman, B., and Schaffer, P. A.,** Anatomy of herpes simplex virus (HSV) DNA. X. Mapping of viral genes by analysis of polypeptides and functions specified by HSV-1 x HSV-2 recombinants, *J. Virol.*, 26, 389, 1978.
5. **Norrild, B., Pedersen, B., and Roizman, B.,** Immunological reactivity of herpes simplex virus 1 and 2 polypeptides electrophoretically separated and transferred to DBM paper, *Infect. Immun.*, 31, 660, 1981.
6. **Norrild, B. and Pedersen, B.,** The effect of tunicamycin on the synthesis of herpes simplex virus type 1 glycoproteins and on their expression on the cell surface, *J. Virol.*, 43, 395, 1982.
7. **Renart, J., Reiser, J., and Stark, G. R.,** Transfer of proteins from gels to diazobenzyloxymethyl-paper and detection with antisera: a method for studying antibody specificity and antigen structure, *Proc. Natl. Acad. Sci. U.S.A.*, 76, 3116, 1979.
8. **Roizman, B., Norrild, B., Chan, C., and Pereira, L.,** Identification and preliminary mapping of a unique herpes simplex virus 2 glycoprotein with monoclonal antibodies, *Virology*, 133, 242, 1984.
9. **Spear, P. G.,** Membrane proteins specified by herpes simplex viruses. I. Identification of four glycoprotein precursors and their products in type 1-infected cells, *J. Virol.*, 17, 991, 1976.
10. **Vass-Sørensen, M., Abeler, V., Berle, E., Pedersen, B., Davy, M., Thorsby, E., and Norrild, B.,** Prevalence of antibodies to herpes simplex virus and frequency of HLA-antigens in patients with preinvasive and invasive cervical cancer, *Gynecol. Oncol.*, 18, 349, 1984.
11. **Norrild, B.,** Membrane-bound proteins of herpes simplex virus, in *Electroimmunochemical Analysis of Membrane Proteins*, Bjerrum, O. J., Ed., Elsevier, Amsterdam, 1983, 395.
12. **Olofsson, S., Norrild, B., Andersen, A. B., Pereira, L., Jeansson, S., and Lycke, E.,** Populations of herpes simplex virus glycoprotein gC with and without affinity for the N-acetyl-galactosamine specific lectin of helix pomatia, *Arch. Virol.*, 76, 25, 1983.
13. **Johnson, D. C. and Spear, P. G.,** O-linked oligosaccharides are acquired by herpes simplex virus glycoproteins in the Golgi apparatus, *Cell*, 32, 987, 1983.
14. **Norrild, B.,** Immunochemistry of herpes simplex virus glycoproteins, *Curr. Top. Microbiol. Immunol.*, 90, 67, 1980.
15. **Eberle, R. and Siao-Wen Mou,** Relative titers of antibodies to individual polypeptide antigens of herpes simplex virus type 1 in human sera, *J. Infect. Dis.*, 148, 436, 1983.
16. **Ashley, R. and Corey, L.,** Association of herpes simplex virus polypeptide specific antibodies and the natural history of genital herpes infections, in *Herpesvirus, UCLA Symposia on Molecular and Cellular Biology*, Vol. 21, Rapp, F., Ed., Alan R. Liss, New York, 1984, 37.
17. **Nahmias, A. J.,** personal communication.
18. **Stubbe-Teglbjærg, C., Feldborg, R., and Norrild, B.,** Immunological reactivity of human sera with individual herpes simplex proteins: a comparative study including sera from patients with preinvasive or invasive cervical cancer and from control persons, *J. Med. Virol.*, in press.
19. **Bernstein, D. I., Lovett, M. A., and Bryson, Y. J.,** The effects of acyclovir on antibody response to herpes simplex virus in primary genital herpetic infections, *J. Infect. Dis.*, 150, 7, 1984.
20. **Forsgren, M., Sköldenberg, B., Norrild, B., Grandien, M., and Jeansson, S.,** Transient antibody response after early treatment of herpes simplex encephalitis with acyclovir, *Antiviral Res.*, in press.
21. **Hirsch, M. S., Wormser, G. P., Schooley, R. T., Ho, D. D., Felsenstein, D., Hopkins, C. C., Joline, C., Duncanson, F., Sarngadharan, M. G., Saxinger, C., and Gallo, R. C.,** Risk of nosocomial infection with human T-cell lymphotropic virus III (HTLV-III), *N. Engl. J. Med.*, 312, 1, 1985.

22. **Braun, D. K., Roizman, B., and Pereira, L.,** Characterization of post-translational products of herpes simplex virus gene 35 proteins binding to the surface of full capsids but not empty capsids, *J. Virol.*, 49, 142, 1984.
23. **Compton, T. and Courtney, R. J.,** Evidence for post-translational glycosylation of a nonglycosylated precursor protein of herpes simplex virus type 1, *J. Virol.*, 52, 630, 1984.
24. **Norrild, B., Pedersen, B., and Møller-Andersen, S.,** Herpes simplex virus-specific secretory IgA in lacrimal fluid during herpes keratitis, *Scand. J. Lab. Clin. Invest.*, 42(Suppl. 161), 29, 1982.
25. **Pedersen, B., Møller-Andersen, S., Klauber, A., Ottoway, E., Prause, J. U., and Norrild, B.,** Secretory IgA specific for herpes simplex virus in lacrimal fluid from patients with herpes keratitis, *Br. J. Ophthalmol.*, 66, 648, 1982.
26. **Meiland, H. T., Hebjørn, S., Møller-Andersen, S., Larsson, A. M., and Norrild, B.,** Herpes simplex virus specific IgA in cervico-vaginal secretions, and amnion fluid in pregnancy, in preparation.

SECTION 6: VISUALIZATION OF TRANSFER

6.1 STAINING OF PROTEINS ON THE BLOTTING MEMBRANE

6.1.1. SILVER STAINING OF BLOTS

Kai-Chung L. Yuen

INTRODUCTION

The biological and biochemical studies of proteins have been aided tremendously by the recent advent of new techniques that allow a sensitive and specific detection such as silver staining and immunoblotting. The former detects nanograms of proteins, while the latter method identifies minute quantities of antigens using specific antibodies. A wealth of information on silver staining of proteins on polyacrylamide gels,[1-9] isoelectric focusing gels,[10,11] and acetic acid-urea gels[12] has been published in recent years. Interestingly, the same staining technique was found applicable for detection of DNA and RNA in polyacrylamide gels.[13-16]

Immunoblotting is able to demonstrate the complexity of the entire protein species present, and yet is able to specifically identify a few proteins of interest from the protein mixtures. Most general staining techniques for proteins on matrices interfere with the subsequent detection of antigens using antibodies.[17-20] In this chapter, a silver staining method that allows detection of nanograms of proteins transferred from polyacrylamide gels to nitrocellulose, and subsequent use of antibodies to identify antigens on the same membrane is described.

PROCEDURES

Silver Staining

Careful handling of the nitrocellulose membrane with clean forceps is essential for satisfactory results. The nitrocellulose membrane (0.45 μm) was obtained from Schleicher & Schuell. All solutions should be made fresh in deionized distilled water.

1. Immediately following electroblotting of proteins to nitrocellulose, the membrane was washed in 10 m*M* Tris-HCl, pH 7.4 for 15 min in a glass tray on a shaker at room temperature. The membrane was then washed with two changes of deionized water for 5 min each.
2. The nitrocellulose was submerged in 100 mℓ of 0.5% (w/v) potassium ferricyanide for 5 min, then rinsed for 10 sec with deionized water.
3. The membrane was then incubated for 20 min in a 100-mℓ solution containing 0.2 g silver nitrate, 0.2 g ammonium nitrate, and 0.5 mℓ 37% (v/v) formaldehyde.
4. The nitrocellulose was rinsed twice with deionized water for 5 sec each.
5. Following the washes, the membrane was transferred to a clean glass tray containing 100 mℓ of 3% (w/v) sodium carbonate, and 0.25 mℓ of 37% (v/v) formaldehyde for at least 3 hr.
6. After rinsing in deionized water for 15 min, the nitrocellulose was soaked for 15 min in a 100-mℓ solution containing 1.4 mℓ ammonium hydroxide, 70 mg sodium hydroxide, and 0.2 g silver nitrate.
7. The nitrocellulose was then rinsed briefly in deionized water. A 100-mℓ solution containing 0.005% (w/v) citric acid and 0.019% (v/v) formaldehyde was added. Pro-

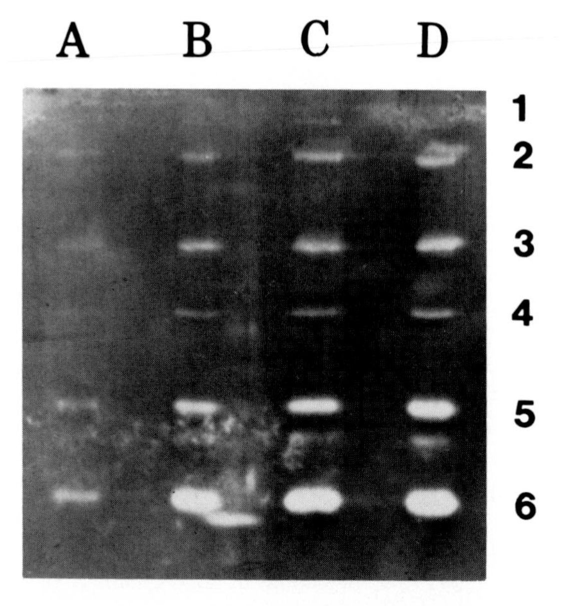

FIGURE 1. Silver staining of proteins on nitrocellulose membrane. Protein M_r standards were separated by electrophoresis on a 15% SDS-polyacrylamide gel, electrotransferred to nitrocellulose membrane, and then silver stained as described in the text. Lane A: (1) 3.20 ng of phosphorylase b, (2) 4.15 ng of BSA, (3) 7.35 ng of ovalbumin, (4) 4.15 ng of carbonic anhydrase, (5) 4 ng of soybean trypsin inhibitor, and (6) 6.05 ng of α-lactalbumin. Lane B: 5 × the amounts of proteins of lane A. Lane C: 10 × the amounts of proteins of lane A. Lane D: 50 × the amounts of proteins of lane A.

teins should start to appear as white bands on a yellowish-brown background (Figure 1). For long-term storage, the nitrocellulose membrane can be transferred to a tray containing deionized water.

Immunodetection of Proteins after Silver Staining

1. After silver staining, the nitrocellulose was equilibrated in 100 mℓ of 10 m*M* Tris-HCl, pH 7.4, for at least 2 hr. It was then incubated overnight at room temperature with constant shaking in 100 mℓ of buffer A: 50 m*M* NaCl, 2 m*M* EDTA, 10 m*M* Tris, pH 7.4, 4% (w/v) bovine serum albumin (BSA) (fraction V), and 0.1% (w/v) sodium azide.
2. The buffer was then discarded and replaced with 50 mℓ of buffer A containing an appropriate amount of antibodies. Incubation was carried out for 16 hr.
3. The membrane was washed at least five times (10 min each) with 50 mℓ buffer B (50 m*M* NaCl, 2 m*M* EDTA, and 10 m*M* Tris-HCl, pH 7.4).
4. After the washings, 50 mℓ of fresh buffer A was added containing 2×10^5 cpm of ^{125}I-labeled *Staphylococcus aureus* protein A (Miles Scientific).
5. After a 3- to 4-hr incubation, the nitrocellulose was washed six times with buffer B (50 mℓ and 10 min each), then air dried, and exposed overnight against a Kodak XAR-5 X-ray film at −70°C. Autoradiograph of such immunodetection technique using polyoma virus infected mouse kidney cells and specific antibody made against the polyoma major structural protein (VP1) is shown in Figure 2.

COMMENTS ON THE PROCEDURES

This silver staining technique is capable of detecting nanograms of proteins (Figure 1); however, different proteins appear not to be stained equally well. As little as 4 ng of carbonic

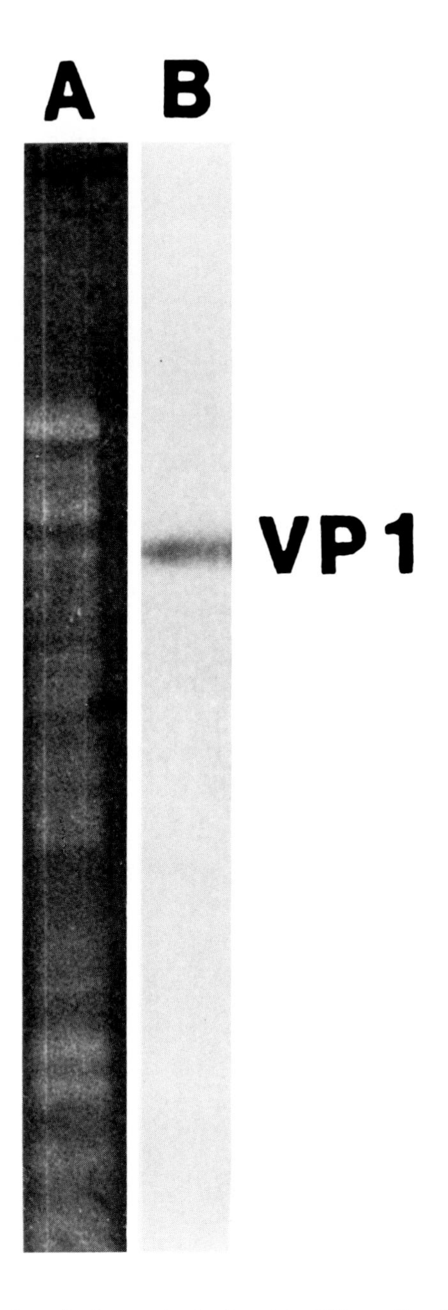

FIGURE 2. Immunodetection of polyoma structural protein VP1 on nitrocellulose membrane after silver staining. Primary mouse kidney cells infected with polyoma virus for 48 hr were scraped from the plates and the cells were pelleted and solubilized in sample buffer (0.062 *M* Tris, pH 6.8, 2% SDS, 5% 2-mercaptoethanol, 10% glycerol, and 0.002% bromophenol blue). Proteins at 15 μg were separated by 15% SDS-PAGE (polyacrylamide gel electrophoresis), transferred to the nitrocellulose membrane, silver stained, and then radioimmunodetected using rabbit antipolyoma VP1 antibody prepared as previously described.[22] Lane A: silver stain pattern of proteins from polyoma-infected mouse kidney cells. Lane B: antibody blot of lane A using antipolyoma VP1 antibody.

anhydrase, soybean trypsin inhibitor and BSA, 6 ng of α-lactalbumin, and 7 ng of ovalbumin can be readily detected (Figure 1). On the other hand, 32 ng seemed to be the limit of detection for phosphorylase b. However, the relatively low sensitivity of detection for phosphorylase b may be primarily due to the low transfer efficiency of proteins with such high M_r (= 94,000).

The proteins appear as white bands on a dark background, suggesting that it was the membrane, not the proteins, that was stained with silver. This observation prompted us to look into the ability of the proteins to bind to antibodies after being silver stained. When polyoma and granulosis viral proteins were silver stained and then immunoblotted, the silver stained proteins reacted equally well with the antibodies as proteins that were transferred to nitrocellulose and immunoblotted without prior staining.[21]

The combination of silver staining of proteins on nitrocellulose, followed by immunodetection, provides a powerful tool for specifically detecting antigens from complex mixtures of proteins. When cell extracts infected with polyoma virus were separated on SDS (sodium dodecyl sulfate)-polyacrylamide gel, transferred to nitrocellulose, silver stained, and then probed with antibodies to the major viral structural protein VP1, only one band lit up among the hundreds of cellular proteins present (Figure 2).

It should be noted that the staining process can be stopped at various steps in the scheme. Immediately after protein transfer, the nitrocellulose membrane can be stored in 10 m*M* Tris-HCl overnight (point 1). In addition, incubation in sodium carbonate solution (point 5) can be lengthened to an overnight step. We found that the most critical step in the whole staining procedure was the rinsing of the membrane with deionized water after the first silver nitrate reaction (point 4). It was observed that brief rinsing (5 sec, twice) produced a dark brown background, while longer washings resulted in a lighter background. A darkly stained membrane provides better contrast between the proteins and the background.

Unlike gels, nitrocellulose membranes are relatively easy to handle and they do not occupy much space. These qualities allow simultaneous staining of multiple membranes. We had successfully stained three to four membranes in the same tray. After staining, the membranes can be stored in deionized water for months without observable change. Nitrocellulose membranes do not change in size appreciably after the staining manipulations, thus making comparison of proteins an easy task.

The combination of the ultrasensitive silver staining and immunodetection provides a powerful means for the detection of antigens present in minute quantities. It is especially useful in clinical studies for detecting antigens present in infected cells amid numerous cellular proteins. This silver staining should also be applicable for two-dimensional (2D) gel electrophoresis systems, thus adding another dimension to the usefulness of this technique.

REFERENCES

1. **Switzer, R. C., Merril, C. R., and Shifrin, S.,** A highly sensitive silver stain for detecting proteins and peptides in polyacrylamide gels, *Anal. Biochem.*, 98, 231, 1979.
2. **Oakley, B. R., Kirsch, D. R., and Morris, N. R.,** A simplified ultrasensitive silver stain for detecting proteins in polyacrylamide gels, *Anal. Biochem.*, 105, 361, 1980.
3. **Wray, W., Boulikas, T., Wray, V. P., and Hancock, R.,** Silver staining of proteins in polyacrylamide gels, *Anal. Biochem.*, 118, 197, 1981.
4. **Merril, C. R., Dunau, M. L., and Goldman, D.,** A rapid sensitive silver stain for polypeptides in polyacrylamide gels, *Anal. Biochem.*, 110, 201, 1981.
5. **Sammons, D. W., Adams, L. D., and Nishizawa, E. E.,** Ultrasensitive silver-based color staining of polypeptides in polyacrylamide gels, *Electrophoresis*, 2, 135, 1981.
6. **Ochs, D. C., McConkey, E. H., and Sammons, D. W.,** Silver stains for proteins in polyacrylamide gels: a comparison of six methods, *Electrophoresis*, 2, 304, 1981.
7. **Eschenbruch, M. and Burk, R. R.,** Experimentally improved reliability of ultrasensitive silver staining of proteins in polyacrylamide gels, *Anal. Biochem.*, 125, 96, 1982.
8. **Berson, G.,** Silver staining of proteins in polyacrylamide gels: increased sensitivity by a blue toning, *Anal. Biochem.*, 134, 230, 1983.
9. **Ohsawa, K. and Ebata, N.,** Silver stain for detecting 10 femtogram quantities of protein after polyacrylamide gel electrophoresis, *Anal. Biochem.*, 135, 409, 1983.

10. **Confavreux, C., Gianazza, E., Chazoti, G., Lasne, Y., and Aranud, P.,** Silver stain after isoelectric focusing of unconcentrated cerebrospinal fluid: visualization of total protein and direct immunofixation of immunoglobulin G, *Electrophoresis,* 3, 206, 1982.
11. **Lasne, F., Benzerara, D., and Lasne, Y.,** A fast silver staining method for protein detection after isoelectric focusing in agarose gels, *Anal. Biochem.,* 132, 338, 1983.
12. **Mold, D. E., Weingart, J., Assaraf, J., Lubahn, D. B., Kelner, D. N., Shaw, B. R., and McCarty, S. R.,** Silver staining of histones in triton-acid urea gels, *Anal. Biochem.,* 135, 44, 1983.
13. **Boulikas, T. and Hancock, R.,** A highly sensitive technique for staining DNA and RNA in polyacrylamide gels using silver, *J. Biochem. Biophys. Methods,* 5, 219, 1981.
14. **Somerville, L. L. and Wang, K.,** The ultrasensitive silver "protein" stain also detects nanograms of nucleic acids, *Biochem. Biophys. Res. Commun.,* 102, 53, 1981.
15. **Herring, A. J., Inglis, N. F., Ojeh, C. K., Snodgrass, D. R., and Menzies, J. D.,** Rapid diagnosis of rotavirus infection by direct detection of viral nucleic acid in silver-stained polyacrylamide gels, *J. Clin. Microbiol.,* 16, 473, 1982.
16. **Iglol, G. L.,** A silver stain for the detection of nanogram amounts of tRNA following two-dimensional electrophoresis, *Anal. Biochem.,* 134, 184, 1983.
17. **Towbin, H., Staehelin, T., and Gordon, J.,** Electrophoretic transfer of proteins from polyacrylamide gels to nitrocellulose sheets: procedure and some applications, *Proc. Natl. Acad. Sci. U.S.A.,* 76(9), 4350, 1979.
18. **Burnette, W. N.,** "Western blotting": electrophoretic transfer of proteins from sodium dodecyl sulfate-polyacrylamide gels to unmodified nitrocellulose and radiographic detection with antibody and radioiodinated protein A, *Anal. Biochem.,* 112, 195, 1981.
19. **Reinhart, M. P. and Malamud, D.,** Protein transfer from isoelectric focusing gels: the native blot, *Anal. Biochem.,* 123, 229, 1982.
20. **Hancock, K. and Tsang, V. C. W.,** India ink staining of proteins on nitrocellulose paper, *Anal. Biochem.,* 133, 157, 1983.
21. **Yuen, L. K. C., Johnson, T. K., Denell, R. E., and Consigli, R. A.,** A silver staining technique for detecting minute quantities of proteins on nitrocellulose paper: retention of antigenicity of stained proteins, *Anal. Biochem.,* 126, 1982.
22. **McMillen, J. and Consigli, R. A.,** Immunological reactivity of antisera to sodium dodecyl sulfate-derived polypeptides of polyoma virions, *J. Virol.,* 21, 1113, 1977.

6.1.2 INDIA INK STAINING OF PROTEINS ON NITROCELLULOSE

Kathy Hancock and Victor C. W. Tsang

INTRODUCTION

Since the development of protein blotting,[1] investigators have needed a sensitive way to detect the blotted proteins directly on the nitrocellulose, thus enabling one to locate molecular weight markers as well as to directly visualize resolved protein mixtures. Immunological detection methods[2] only reveal the proteins with epitopes recognized by the antibodies used. A portion of the gel can be cut and silver stained, or a companion gel can be run and silver stained, but the change in gel size that occurs during silver staining makes direct comparison between the gel and blot difficult. To help solve this problem, we developed a simple India ink stain.[3] With this stain, blotted proteins can be easily detected directly on the nitrocellulose with a sensitivity of approximately 1 ng/mm, nearly equal to the sensitivity of the silver stain.[4]

REAGENTS

Phosphate-buffered saline (PBS)

- 0.15 *M* NaCl in 0.01 *M* Na_2HPO_4/NaH_2PO_4, pH 7.20
- PBS containing 0.3% (v/v) Tween® 20* (PBS-TW)
- 3 mℓ of Tween® 20 (polyoxyethylene sorbitan monolaurate from Sigma Chemical Co., St Louis) in 1000 mℓ PBS. Dispense Tween® directly into PBS using a 14-gauge cannula and syringe while the PBS is stirring. Prepare fresh daily.

India ink

- Pelikan fount India drawing ink for fountain pens (Pelikan AG, D-3000, Hannover 1, West Germany) diluted 1:1000 in PBS-TW. Prepare fresh daily.

PROCEDURE

The optimum conditions for India ink staining proteins that have been resolved by SDS-PAGE (sodium dodecyl sulfate polyacrylamide gel electrophoresis)[2] and electrophoretically transferred[5] onto a nitrocellulose sheet are as follows:

1. Wash the blotted nitrocellulose sheet (18 × 20 cm) (0.20 μm, BA-83, Schleicher & Schuell, Keene, N.H.) at room temperature, 4 times for 5 min each, with approximately 250 mℓ of PBS-TW. Agitate during washes by using a rocker platform or an orbital rotator (both from Bellco Biotechnology, Vineland, N.J.). If necessary, remove adherent gel by wiping with dry gauze.
2. Stain with a 1:1000 dilution of India ink in PBS-TW for 2 hr, or overnight at room temperature with agitation.
3. Following staining, rinse briefly with deionized water. Rinsing for 1 min or less is usually sufficient. Allow the nitrocellulose to air dry.

* Use of trade names is for identification only and does not imply endorsement by the Public Health Service or by the U.S. Department of Health and Human Services.

COMMENTS ON PROCEDURE

One of the most important variables is the brand of India ink used for staining. Of the four brands we have tested, Pelikan fount India drawing ink for fountain pens is a more sensitive protein stain than Pelikan drawing ink, Higgins waterproof black India drawing ink. (Faber-Castell Corp., Newark), or Winsor & Newton Indian ink (England). The optimum concentration of India ink is 1 $\mu\ell$ ink/mℓ PBS-TW. Lower concentrations of ink do not stain the protein bands sufficiently, while higher concentrations cause increased background staining. Tween® 20 is necessary in the stain solution to help prevent background staining, but the concentration is not critical. Concentrations from 0.05 to 5.0% are all equally suitable.[3] Since we use PBS containing 0.3% (v/v) Tween® for all of the washing steps of the enzyme-linked immunoelectrotransfer blot technique,[2,5] it is convenient to use the same concentration of Tween® for the India ink stain. These wash steps are necessary to remove any unbound protein, adherent polyacrylamide gel particles, and residual SDS which may interfere with the stain. It is important to dispense the Tween® directly into the PBS that is being stirred to prevent clumping and sticking of the Tween® to the walls and bottom of the vessel. Dispensing the Tween® with a syringe fitted with a large gauge cannula provides an accurate way to measure the volume of Tween® and introduce it directly into the PBS.

Stained bands begin to appear within the first 5 min, but overnight staining is preferable for maximum sensitivity. However, for most applications, such as staining molecular weight standards, all bands are readily visible within 2 hr. India ink also stains proteins blotted onto nylon membranes such as GeneScreen (New England Nuclear, Boston), but the sensitivity has not yet been evaluated.

The India ink stain is useful for staining the proteins from a complex mixture resolved and electrotransferred onto nitrocellulose. The India ink stain is also useful for staining molecular weight markers resolved in a well adjacent to the antigens being examined. Prestained molecular weight markers are commercially available (Bethesda Research Laboratories, Gaithersburg, Md.) and are useful for locating blotted lanes so that the nitrocellulose can be cut and exposed to different reagents. The proteins resolve into rather broad bands, however, and their molecular weights differ from those published by the manufacturer. All lots must be recalibrated,[6] which hinders their usefulness as molecular weight standards.

EVALUATION

The India ink stain is nearly as sensitive as the silver stain[4] of the same proteins in a polyacrylamide gel. In Figure 1, all protein bands that stain with India ink are easily detected in the lane loaded with 6 μg (or 1.0 μg/mm well width) of the protein mixture CY, a complex mixture of cytosolic antigens from the adult *Schistosoma mansoni*.[7] In fact, the same bands are also detectable in the lane loaded with 1.5 μg of the mixture CY (or 0.25 μg/mm well width), although the staining intensity of most bands is considerably less. A silver stained gel (Figure 2) was also loaded with 1.5 μg of the cytosolic antigens CY. The silver staining is more intense than the India ink stain of the same amount of protein. However, when comparing both stains visually, nearly all the bands detectable by the silver stain are detectable with the India ink stain. Some of the band sharpness may be lost with the India ink stain. These observations are confirmed by the laser densitometer scan of the silver stain and the 1.5 μg India ink stain (Figures 3A and 3C).

The minimum detectable amounts each of six proteins resolved by SDS-PAGE and electrotransferred onto nitrocellulose was determined (Table 1). These amounts are from 88- to 180-fold less than the values we previously reported.[3] The difference can be explained by increased transfer efficiency due to the use of a higher ionic strength blot buffer (0.212 *M* Tris-HCl, 20% (v/v) methanol, pH 9.18 rather than 0.025 *M* Tris-HCl, 0.193 *M* glycine,

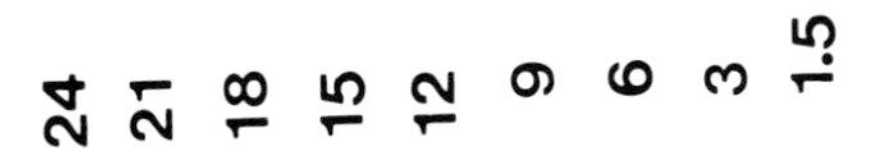

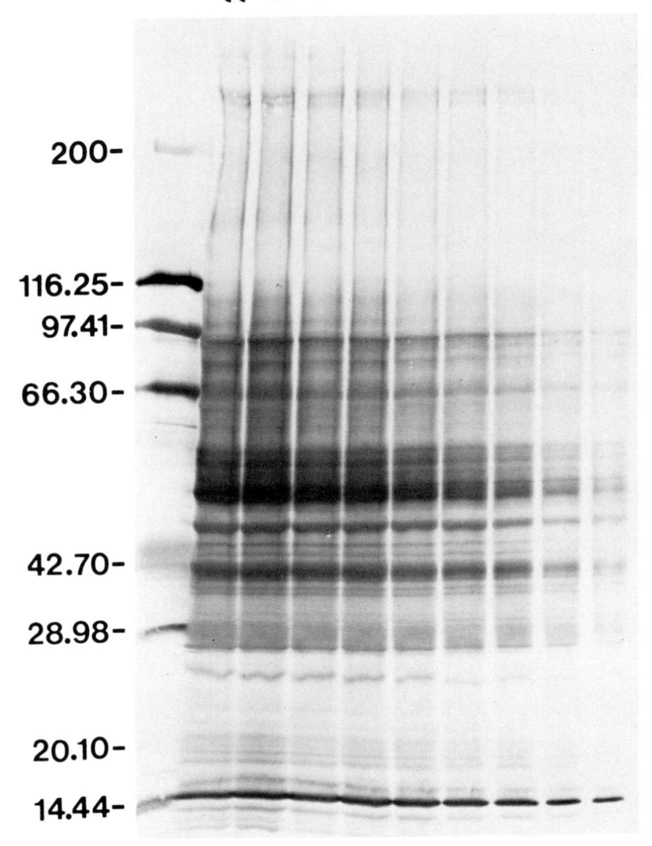

FIGURE 1. India ink stain of proteins electrotransferred onto nitrocellulose — evaluation of sensitivity. Adult *Schistosoma mansoni* worms were fractionated,[7] and a portion of the cytosolic fraction (CY) was treated with 20 μg SDS/μg protein using a solution of 10% SDS in 9 *M* urea buffered with 0.01 *M* Tris-HCl, pH 8.00. Bromophenol blue tracking dye was added (2 μℓ/100 μℓ sample) and the sample heated at 65°C for 15 min. The numbers across the top indicate the amount of treated CY loaded, in micrograms, in a 0.75-mm thick, 5 to 20% polyacrylamide gradient gel and electrophoresed.[2] The width of each well is 6 mm; therefore, 1.5 μg of protein loaded is equivalent to 0.25 μg/mm, 3 μg is equivalent to 0.5 μg/mm, etc. After electrophoresis (Pharmacia GE-2/4 LS, Piscataway, N.J.), the proteins were electrotransferred (Bio-Rad Trans-Blot cell) onto nitrocellulose using 0.212 *M* Tris-HCl, pH 9.18, with 20% (v/v) methanol as the blot buffer. The initial blotting conditions were 1.0 A, 124 Volts DC, buffer at 4°C. After 45 min, when the current had reached 2.0 A, the current was held at 2.0 A for the remaining 15 min. Final conditions were 2.0 A, 116 Volts DC, 36°C. The electrodes were 3.5-cm apart. India ink staining was done as described in the procedure section. The numbers down the side indicate molecular weight $\times$ 10^3. The molecular weight marker proteins are an equal combination of Bio-Rad's high molecular weight markers and Pharmacia's low molecular weight markers. The proteins and their molecular weights are as follows: myosin, 200,000;[9] β-galactosidase, 116,248;[10] phosphorylase b, 97,412;[11] bovine serum albumin, 66,296;[12] ovalbumin, 42,699;[13] carbonic anhydrase, 28,980;[14] soybean trypsin inhibitor, 20,100;[15] and α-lactalbumin, 14,437.[16]

20% (v/v) methanol, pH 8.35); blotting with current at 1.0 A minimum, voltage usually above 100 V DC; and transferring from a 10% gel rather than a gradient gel. Under these conditions, virtually all of the proteins are transferred. For comparison, the detection limits of our silver stain[4] for these same proteins are listed in Table 1.

The India ink stain is much more sensitive than the other stains we have tested for staining proteins on nitrocellulose (Figure 4). More protein bands are visible and the staining intensity is darker with India ink as compared to Amido Black and Fast Green. Previous tests with Coomassie Blue showed unacceptably high background staining.[3] Silver staining,[4] by the

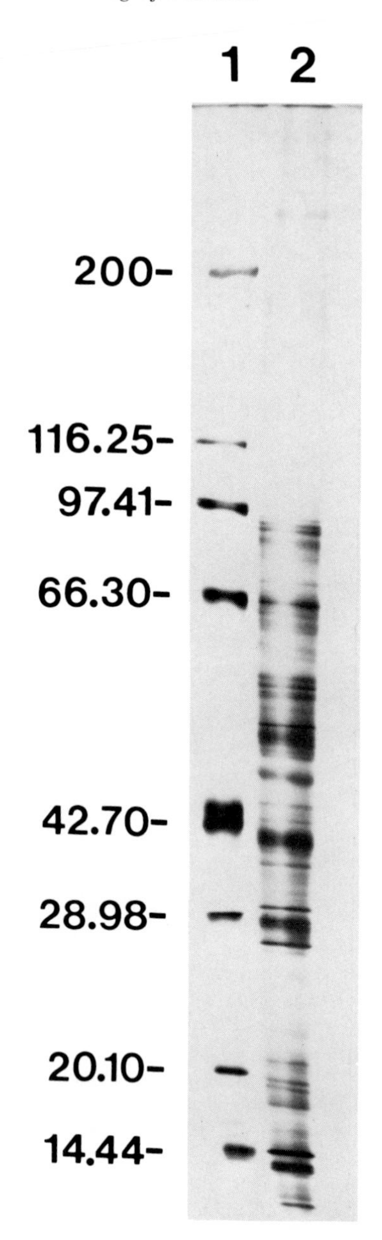

FIGURE 2. Silver stain of proteins in polyacrylamide gel. 1.5 μg of the protein mixture CY (lane 2) were electrophoresed in a 6-mm wide well on one side of the gel described in Figure 1. This portion of the gel was cut off and silver stained after electrophoresis.[4] Molecular weights $\times$ 10^3 of the standards are indicated in lane 1. Standards the same as described in Figure 1, but 10 times less, were loaded for the silver stain.

method we use for polyacrylamide gels, did not reveal any protein bands. Yuen et al.[8] have described a negative silver stain (background is stained, protein bands are unstained) and are able to detect nanogram amounts of proteins on nitrocellulose. The authors report that the stain is unsuitable when proteins are transblotted from gels such as ours. Our stacking and resolving gels do not contain SDS. SDS is present only in the upper reservoir buffer. Proteins are SDS treated before electrophoresis.[2]

Two important variables affect the India ink stain. One is the transfer efficiency of the protein from the gel to the nitrocellulose and the other is the affinity between the protein and the India ink. The transfer efficiency of the complex mixture of cytosolic proteins of

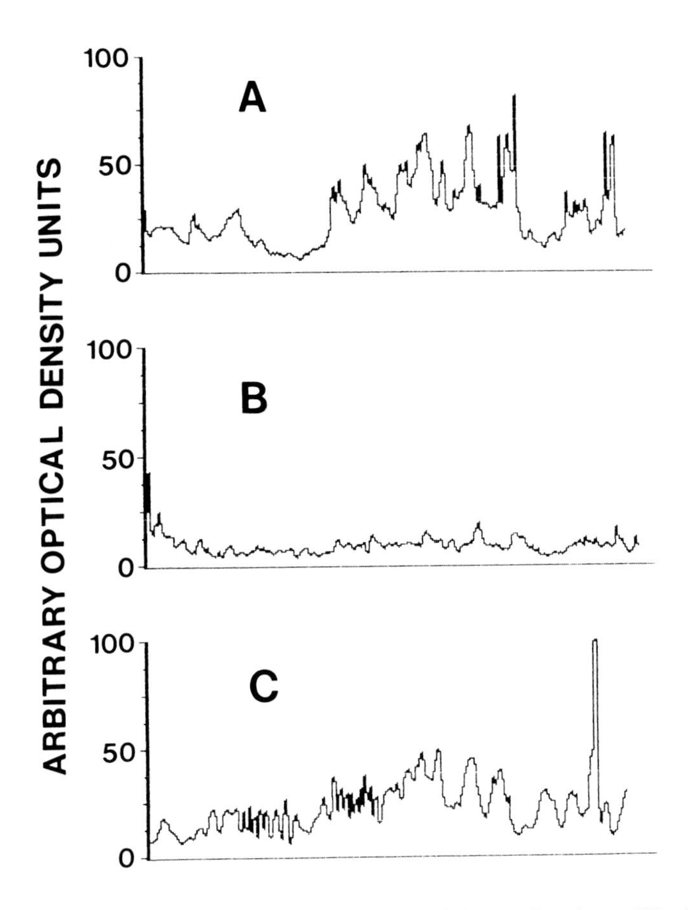

FIGURE 3. Densitogram of silver stains and India ink stain of the protein mixture CY. A scanning laser densitometer (Biomed Instruments, Inc., Fullerton, Calif.) was used to scan (A) the silver stained CY gel (Figure 2, lane 2), (B) the 1.5 μg lane of the gel which was silver stained after transfer of the CY proteins onto the nitrocellulose (not shown), and (C) the 1.5 μg lane of the India ink stained blot (Figure 1). A is the "before" blotting scan of the proteins in the polyacrylamide gel, B the "after" blotting scan of the proteins remaining in the gel, and C is the scan of the proteins that transferred to the nitrocellulose and stained with India ink. Gels and blot were scanned from the 5% (on the left) to the 20% portion of the gel.

Schistosoma mansoni, is illustrated in Figures 3A and 3B. The first scan represents the amount of protein in the gel before blotting, the second scan the amount of protein remaining in the gel after blotting. The transfer efficiency of eight molecular weight marker proteins is illustrated in Figures 5A and 5B. Note that B originally contained ten times more protein than A. By looking at individual proteins, it is easy to see that some proteins transfer better than others. For example, myosin (peak 1) and carbonic anhydrase (peak 6) do not transfer as efficiently as soybean trypsin inhibitor (peak 7) and α-lactalbumin (peak 8). The relative density of each protein band was determined before and after blotting and used to calculate the percentage of protein transferred (Table 2). The lowest transfer was 72% for myosin and the highest was 99% for β-galactosidase, ovalbumin, and α-lactalbumin.

The second important variable affecting the India ink stain is the affinity of the protein for the India ink. One protein in the cytosolic protein mixture CY, molecular weight approximately 15,000, has a high affinity for India ink (Figures 1 and 3C). Looking at the molecular weight marker proteins (Figures 1 and 5C), one can see that soybean trypsin inhibitor (M_r 20,100, peak 7), with a transfer efficiency of 98%, stains poorly with India ink. Myosin (M_r 200,000, peak 1) also stains poorly, although this may be partly attributable to low transfer efficiency. The causes for the variations of India ink binding found among

Table 1
SENSITIVITY OF THE INDIA INK STAIN VS. THE SILVER STAIN

	Minimum amt of protein detectable[a]	
Protein	**India ink stain (ng)**	**Silver stain (ng)**
Phosphorylase b	1.3	0.3
BSA	1.6	0.4
Ovalbumin	2.9	0.7
Carbonic anhydrase	1.6	0.4
Soybean trypsin inhibitor	50.0	0.4
α-Lactalbumin	9.5	0.6

Note: Twofold dilutions (diluent contained 2.5 m*M* Tris-HCl, 0.1% SDS, 50 m*M* dithiothreitol, 0.25 mg/mℓ bromophenol blue, pH 8.00) of low molecular weight standards (Pharmacia lot #5A101A, concentration of each protein given by manufacturer) were electrophoresed on two 10% polyacrylamide gels (10% acrylamide, 0.25% bis-acrylamide), 80-mm long, 160-mm wide, 0.75-mm thick. Each well was 4-mm wide. The proteins electrophoresed in one gel (Bio-Rad Model 360 mini-vertical cell, Richmond, Calif.) were electrotransfer-blotted (Bio-Rad Trans-Blot cell) onto nitrocellulose using 0.212 *M* Tris-HCl, 20% (v/v) methanol, pH 9.18, as the blot buffer. The electrodes were 3.5 cm apart. Initial blotting conditions were 1.0 A, 140 Volt DC, 1°C. After 40 min, the temperature of the buffer had increased to 40°C and the current to 2.8 A. Current was reduced and held at 1.0 A for the remainder of the 1-hr blotting time. The nitrocellulose was stained with India ink overnight. The other gel was silver stained.[4]

[a] Electrophoresed in a well 4-mm wide and 0.75-mm thick.

different proteins is difficult to comment upon due to lack of specific data. We did observe that total molecular "hydrophobicity"[17] of the various marker proteins in this chapter did not appear to have any correlation with their degree of stainability. Other charge distributions, surface localized hydrophobicities, and/or van der Waals forces may very well determine the stainability of proteins by India ink, which after all, is nothing more than carbon particles.

The India ink stain for proteins on nitrocellulose is easy to use and is far more sensitive than other stains used for detecting proteins on nitrocellulose. The India ink stain is capable of detecting as little as 1 to 2 ng of protein electrophoresed in a well 4-mm wide. For comparison, our silver stain[4] is capable of detecting as little as 0.3 ng of protein electrophoresed in a well 4-mm wide.

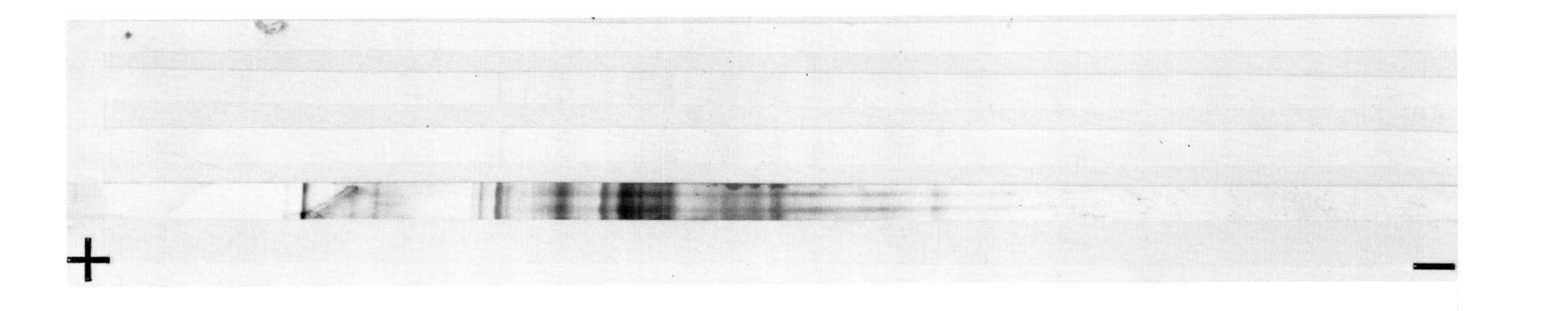

FIGURE 4. Comparisons of stains used for detecting proteins on nitrocellulose. A single-well gel was loaded with 1.5 μg CY/mm well width. The gel was electrophoresed and blotted as described in Figure 1. Strips were cut from the nitrocellulose sheet after washing 4 times for 10 min each with PSB-TW. The first strip was stained with Silver,[4] the second with Amido Black, the third with Fast Green (Amido Black and Fast Green from Bio-Rad, Richmond, Calif.), and the fourth with India ink. Both the Amido Black and Fast Green stains were 0.1% (w/v) solutions in 45% (v/v) methanol and 7% (v/v) acetic acid. Strips were stained in these two stains for 5 min and then destained for 15 min in 70% (v/v) methanol and 7% (v/v) acetic acid. Staining with Amido Black or Fast Green for 30 min instead of 5 and destaining for 15 min did not increase the sensitivity of the intensity of the stained hands.

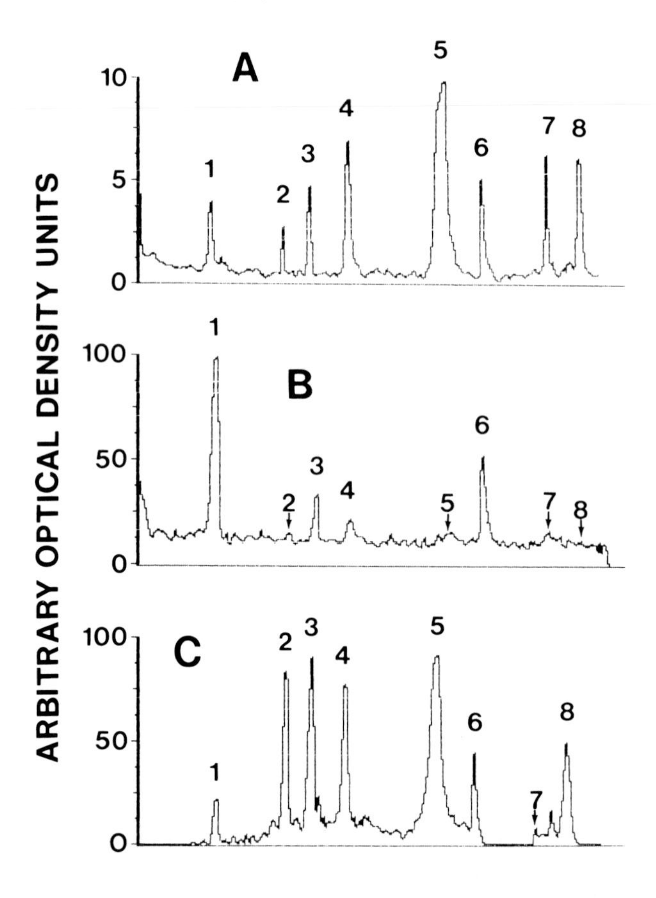

FIGURE 5. Densitogram of silver stains and India ink stains of molecular weight marker proteins. A scanning laser densitometer was used to scan (A) the silver stained molecular weight marker proteins shown in Figure 2, lane 1, (B) the silver stained molecular weight marker proteins in the gel following blotting (not shown), and (C) the India ink stained molecular weight marker proteins on the nitrocellulose (Figure 1). Ten times as much protein was loaded in the lane of the gel that was blotted as compared to the lane of the gel that was silver stained only. Peaks are indicated as follows: myosin (1), β-galactosidase (2), phosphorylase b, (3) bovine serum albumin (4), ovalbumin (5), carbonic anhydrase (6), soybean trypsin inhibitor (7), and α-lactalbumin (8). Gels and blot were scanned from the 5% (on the left) to the 20% portion of the gel.

Table 2
TRANSFER EFFICIENCY OF EIGHT PROTEINS ELECTROTRANSFERRED FROM A 5—20% GRADIENT POLYACRYLAMIDE GEL

	Relative density in silver stained gel			
Protein	**Peak no.**	**Before blot**	**After blot[a]**	**% of protein transferred[b]**
Myosin	1	57	15.7	72
β-Galactosidase	2	88	0.7	99
Phosphorylase b	3	100	4.0	96
BSA	4	123	3.7	97
Ovalbumin	5	165	1.3	99
Carbonic anhydrase	6	53	7.1	87
Soybean trypsin inhibitor	7	96	1.5	98
α-Lactalbumin	8	103	0.9	99

[a] The gel used for electrotransfer contained 10 times the amount of each protein as compared to the protein in the gel which was silver stained only. Data under this column were adjusted to reflect the differential in sample load.

[b] Estimated from density data measured by a scanning laser densitometer (Biomed Instruments Inc., Fullerton, Calif.).

REFERENCES

1. **Towbin, H., Staehelin, T., and Gordon, J.,** Electrophoretic transfer of proteins from polyacrylamide gels to nitrocellulose sheets: procedure and some applications, *Proc. Natl. Acad. Sci. U.S.A.*, 76, 4350, 1979.
2. **Tsang, V. C. W., Peralta, J. M., and Simons, A. R.,** The enzyme-linked immunoelectrotransfer blot techniques (EITB) for studying the specificities of antigens and antibodies separated by gel electrophoresis, *Methods Enzymol.*, 92, 377, 1983.
3. **Hancock, K. and Tsang, V. C. W.,** India ink staining of proteins on nitrocellulose paper, *Anal. Biochem.*, 133, 157, 1983.
4. **Tsang, V. C. W., Hancock, K., Maddison, S. E., Beatty, A. L., and Moss, D. M.,** Demonstration of species-specific and cross-reactive components of the adult microsomal antigens from *Schistosoma mansoni* and *S. japonicum* (MAMA and JAMA), *J. Immunol.*, 132, 2607, 1984.
5. **Tsang, V. C. W., Bers, G. E., and Hancock, K.,** Enzyme-linked immunoelectrotransfer blot (EITB), in *Enzyme-Mediated-Immunoassays,* Ngo, T. T. and Lenhoff, H. M., Eds., Plenum Press, New York, 1985.
6. **Tsang, V. C. W., Hancock, K., and Simons, A. R.,** Calibration of prestained protein molecular weight standards for use in the "Western" or enzyme-linked immunoelectrotransfer blot techniques, *Anal. Biochem.*, 143, 304, 1984.
7. **Tsang, V. C. W., Tsang, K. R., Hancock, K., Kelly, M. A., Wilson, B. C., and Maddison, S. E.,** *Schistosoma mansoni* adult microsomal antigens, a serologic reagent. I. Systematic fractionation, quantitation, and characterization of antigenic components, *J. Immunol.*, 130, 1359, 1983.
8. **Yuen, K. C. L., Johnson, T. K., Denell, R. E., and Consigli, R. A.,** A silver-staining technique for detecting minute quantities of proteins on nitrocellulose paper: retention of antigenicity of stained proteins, *Anal. Biochem.*, 126, 398, 1982.
9. **Woods, E. F., Himmelfarb, S., and Harrington, W. F.,** Studies on the structure of myosin in solution, *J. Biol. Chem.*, 238, 2374, 1963.
10. **Fowler, A. V. and Zabin, I.,** The amino acid sequence of beta-galactosidase of *Escherichia coli, Proc. Natl. Acad. Sci. U.S.A.*, 74, 1507, 1977.
11. **Titani, K., Koide, A., Hermann, J., Ericsson, L. H., Kumar, S., Wade, R. D., Walsh, K. A., Neurath, H., and Fischer, E. H.,** Complete amino acid sequence of rabbit muscle glycogen phosphorylase, *Proc. Natl. Acad. Sci. U.S.A.*, 74, 4762, 1977.

12. **Dayhoff, M. O., Ed.,** *Atlas of Protein Sequence and Structure,* Vol. 5(Suppl. 2), National Biomedical Research Foundation, Silver Spring, Md., 1976, 267.
13. **Nisbet, A. D., Saundry, R. H., Moir, A. J. G., Fothergill, L. A., and Fothergill, J. E.,** The complete amino-acid sequence of hen ovalbumin, *Eur. J. Biochem.,* 115, 335, 1981.
14. **Dayhoff, M. O., Ed.,** *Atlas of Protein Sequence and Structure,* Vol. 5(Suppl. 3), National Biomedical Research Foundation, Silver Spring, Md., 1978, 128.
15. **Koide, T. and Ikenaka, T.,** Studies on soybean trypsin inhibitors. I. Fragmentation of soybean trypsin inhibitor (Kunitz) by limited proteolysis and by chemical cleavage, *Eur. J. Biochem.,* 32, 401, 1973.
16. **Brew, K., Vanaman, T. C., and Hill, R. L.,** Comparison of the amino acid sequence of bovine alpha-lactalbumin and hens egg white lysozyme, *J. Biol. Chem.,* 242, 3747, 1967.
17. **Hopp, T. P. and Woods, K. R.,** Prediction of protein antigenic determinants from amino acid sequences, *Proc. Natl. Acad. Sci. U.S.A.,* 78, 3824, 1982.

6.1.3. COLLOIDAL METAL STAINING OF BLOTS

Marc Moeremans, Guy Daneels, Marc De Raeymaeker, and Jan De Mey

INTRODUCTION

In immunoblotting, the efficient transfer of electrophoretically separated polypeptides from the gel matrix to the membrane (blotting) is crucial. In order to judge the transfer efficacy, general protein staining methods directly on the membrane are required. Silver staining of a duplicate gel does not give accurate information because the transfer of certain proteins may be problematic. In addition, in most cases, protein staining of gels leads to dimension changes, which make a correlation of a specific detected protein band (on the membrane) with the overall protein pattern (in the gel) difficult.

Proteins can be transferred to different immobilizing matrices.[1,2] The most widely used membrane is nitrocellulose, but in some circumstances the use of nylon based membranes such as GeneScreen® or charge modified (cationized) ones like Zetaprobe® may be advantageous[3,4] (see Chapters 5.1.2 and 5.2). It is important to realize that the initial choice of membrane determines the choice of the subsequent staining method. This means that either negatively charged colloidal metal particles (gold) or positively charged particles (iron) must be used.

The colloidal gold staining (AuroDye®*) [5,6] consists in treating the protein blots (nitrocellulose) at low pH (proteins become cationized) with 20 nm gold particles (anionic) supplemented with Tween® 20. The accumulation of the particles results in red to dark red bands.

The colloidal iron staining (FerriDye®*)[7] consists in treating the protein blots (nitrocellulose, nylon membranes, positively charged nylon membranes) of SDS (sodium dodecyl sulfate)-polyacrylamide gels (SDS proteins are anionic) with iron particles (cationic) supplemented with Tween® 20. The binding of the particles to the negatively charged proteins (due to residual SDS) produces faint yellow brownish bands. This signal can be intensified by the Perls' reaction with potassium ferrocyanide which results in a deep blue color.

Although FerriDye is less sensitive than AuroDye, both methods are substantially more sensitive than conventional staining methods and match the sensitivity of the different subsequent immunostainings. In addition, FerriDye is the only general protein stain that works on nylon membranes.

GOLD STAINING WITH AURODYE

Equipment and Reagents

Apparatus

Magnetic stirrer with heating, Erlenmeyer flask, reflux cooler, stirring bars, glass Petri dishes, incubator at 37°C, pH meter, rocking table, gloves and tweezers, electrophoresis and blotting equipment.

Materials

Microfilters (0.22 μm, Millipore, SLGS 025BS), nitrocellulose membranes (e.g., Bio-Rad, Schleicher & Schuell), filter paper (e.g., Schleicher & Schuell, 604A-24343), sponge pads (6 mm, Bio-Rad).

Chemicals

Tetrachlorogold acid trihydrate, citric acid monohydrate, trisodium citrate dihydrate, di-

* AuroDye and FerriDye are trademarks from Janssen Life Sciences Products, a division of Janssen Pharmaceutica.

sodium hydrogen phosphate dihydrate, and sodium dihydrogen phosphate monohydrate were obtained from Merck, Tween® 20 from Bio-Rad.

Solutions

Tetrachlorogold acid solution	
1% (w/v) tetrachlorogold acid trihydrate	1.16g
Distilled water	ad 100 mℓ
Store at 4°C, protected from light	
Trisodium citrate solution	
1% (w/v) trisodium citrate dihydrate	1.14 g
Distilled water	ad 100 mℓ
Prepare fresh daily on, microfilter 0.2 μm	
Phosphate buffered saline (PBS, pH 7.2)	
1.47 m*M* potassium dihydrogen phosphate	0.2 g
8.1 m*M* disodium hydrogen phosphate	1.15 g
136.9 m*M* sodium chloride	8.00 g
2.68 m*M* potassium chloride	0.2 g
Distilled water	ad 1000 mℓ
Washing and blocking buffer	
0.3% (v/v) Tween® 20	3.0 mℓ
PBS	ad 1000 mℓ
Tween® 20 stock solution	
2% (v/v) Tween® 20	2 mℓ
Distilled water	ad 100 mℓ
Citrate buffer (pH 3.0)	
9 m*M* trisodium citrate dihydrate	2.65 g
41 m*M* citric acid monohydrate	8.62 g
Distilled water	ad 500 mℓ

Procedures

Preparation of AuroDye

In a scrupulously cleaned Erlenmeyer flask equipped with a reflux cooler, 247.5 mℓ bidistilled water is brought to boil and 2.5 mℓ of a 1% $HAuCl_4$ is added. Under vigorous stirring 7.5 mℓ trisodiumcitrate 1% is added. This mixture is heated for an additional 30 min, cooled, and stored at 4°C until use.[8] To 75 mℓ of this colloidal gold solution 5 mℓ 2% (v/v) Tween® 20 is added. After mixing, 20 mℓ of a 50 m*M* citrate buffer pH 3.0 is added. It is important to mix Tween® 20 and the solution very well before adding the citrate buffer. If this colloidal gold solution is stored properly, i.e., at 4°C and under sterile conditions, it can be stored for more than 6 months.

Staining Procedure[5]

After transfer the blot is washed extensively with excess PBS, pH 7.2, supplemented with 0.3% (v/v) Tween® 20 for 30 min at 37°C and 3 times for 15 min at room temperature.[5,9] About 100 mℓ (3 times) is sufficient for a blot of 15 × 10 cm. This extensive washing is necessary to remove adherent polyacrylamide gel particles and residual SDS. The blot is rinsed in excess distilled water (200 mℓ) for 3 min at room temperature.

The blot is incubated in colloidal gold stain at room temperature for at least 4 hr, preferably overnight. It is recommended to use about 0.2 mℓ/cm² blotting membrane. With high protein loads, 0.5 mℓ/cm² may be needed.

After staining, the blot is washed several times with excessive water and air dried.

Comments on the Procedure

This staining method is very sensitive and all possible precautions to avoid impurities must be employed. Handle membranes with clean tweezers.

The protein quantities loaded on the gel should be in the same range as normally used for silver staining, otherwise individual bands will not be visible.

Because it is practically impossible to avoid proteinaceous impurities in the transfer apparatus it is advisable to sandwich the gel between two nitrocellulose membranes. One will give the exact replica of the gel, the other will adsorb impurities.

High quality filter paper should be used. If other brands of filter paper are used than the one indicated, they should be checked. Low quality filter paper may result in the formation of a disturbing background.

If the colloidal gold stain turns blue it is necessary to replace it by a new solution, after a brief wash in excess water. This shift in color from red to blue is probably caused by the release of proteins from the blot. This release is more of concern when high amounts of proteins have been transferred. The capacity of nitrocellulose is limited to about 0.8 $\mu g/mm^2$.[1]

During incubation in the colloidal gold stain the background is initially slightly stained, but this is only temporary. After incubation for 4 hr the protein bands should be clearly visible.

It is advisable not to reuse the AuroDye solution. The concentration of the particles is probably too low to give optimal results. Gold solutions made as described earlier are just concentrated enough to give good staining results with 4 hr to overnight incubation.

Evaluation

To assess the sensitivity a twofold serial dilution series of Bio-Rad high M_r standards (1000 to 3.5 ng per protein band) was subjected to SDS-electrophoresis.[10] The last well was loaded with sample buffer alone. One gel was silver stained according to Morissey[11] (Figure 1a), a duplicate was electrotransferred to nitrocellulose membrane and stained with AuroDye (Figure 1b). By silver staining a duplicate gel and the gel after transfer, the blotting efficacy of myosin was shown to be low. Therefore, the detection limit for myosin is in the order of subnanogram amounts per square millimeter. The selective binding of the gold particles to protein bands is shown in the same figure since sample buffer was loaded only in the last well. The staining is clearly absent, except for the inevitable keratin bands.[12]

The method was also tested on a complex protein mixture. Human pleural fluid was analyzed on a 2D gel. The gel was silver stained (Figure 2a) and a duplicate was transferred to nitrocellulose and processed with AuroDye (Figure 2b), which in this experiment had a higher sensitivity than silver staining. Some proteins are not detected at all with silver staining but become clearly visible with AuroDye.

Colloidal gold staining at different pH shows that the interaction of negatively charged collolidal gold particles with polypeptides are of both hydrophobic and electrostatic nature.[5,6] The latter may result in different staining intensities of proteins differing in isoelectric point.[6] This is to our knowledge a common feature of all protein staining methods (e.g., silver staining, Coomassie blue, etc.); however, AuroDye stains more proteins on the 2D-blot than could be shown in the gel with silver staining.[13]

Due to the binding characteristics of AuroDye this method is limited to nitrocellulose membranes. The method works well on blots of both denatured and nondenatured proteins.

IRON STAINING WITH FERRIDYE

Equipment and Reagents

Apparatus

The equipment used for preparation of FerriDye is the same as for AuroDye, although one does not need a reflux cooler for the preparation of the solution.

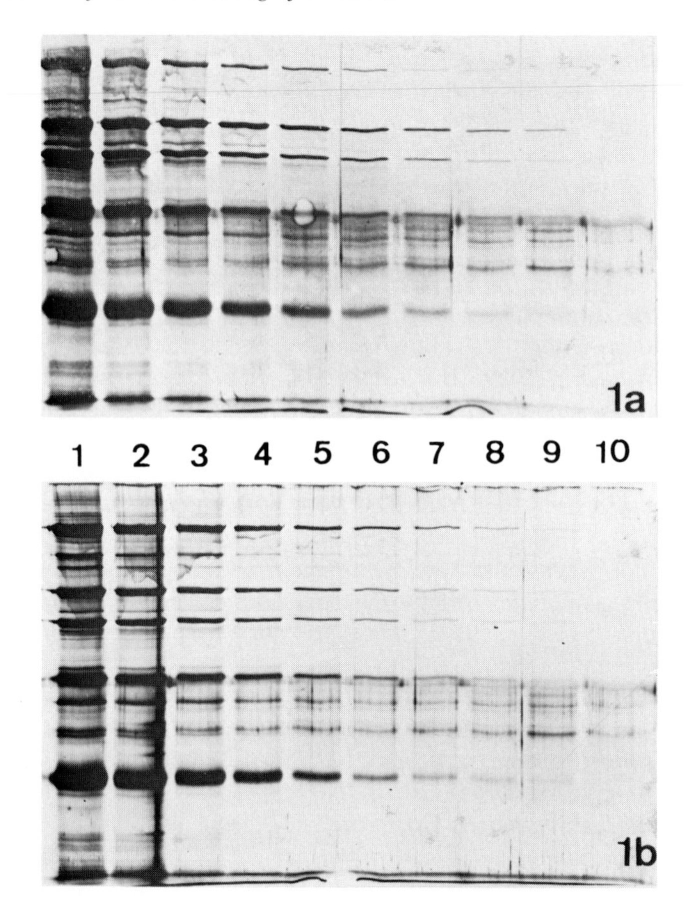

FIGURE 1. Comparison of the colloidal gold stain on nitrocellulose (1a) with silver staining[11] of SDS-polyacrylamide gel (1b). Nine dilutions (2-fold) of Biorad high molecular weight standards separated on a 7.5% gel. Amount per protein: Lane 1, 1000 ng; 2, 500 ng; 3, 250 ng; 4, 125 ng; 5, 62.5 ng; 6, 31.3 ng; 7, 15.6 ng; 8, 7.8 ng; 9, 3.9 ng. The last well (10) was loaded with sample buffer alone.

Materials

Microfilters (Millipore, 0.8 μm SLAA 025BS), nylon membranes, e.g., GeneScreen (NEN), Zetaprobe (Bio-Rad).

Chemicals

Ferric chloride hexahydrate and potassium hexacyanoferrate were obtained from Janssen Chimica. Sodium cacodylate was obtained from BDH and Tween® 20 was from Bio-Rad. Hydrochloric acid (Titrisol) was obtained from Merck.

Solutions

Ferric chloride hexahydrate	
0.5 *M* ferric chloride hexahydrate	33.79 g
Distilled water	ad 25 mℓ
Prepare fresh daily on microfilter (0.8 μm)	
Sodium cacodylate buffer	
0.1 *M* sodium cacodylate buffer	16.0 g
Distilled water	ad 900 mℓ

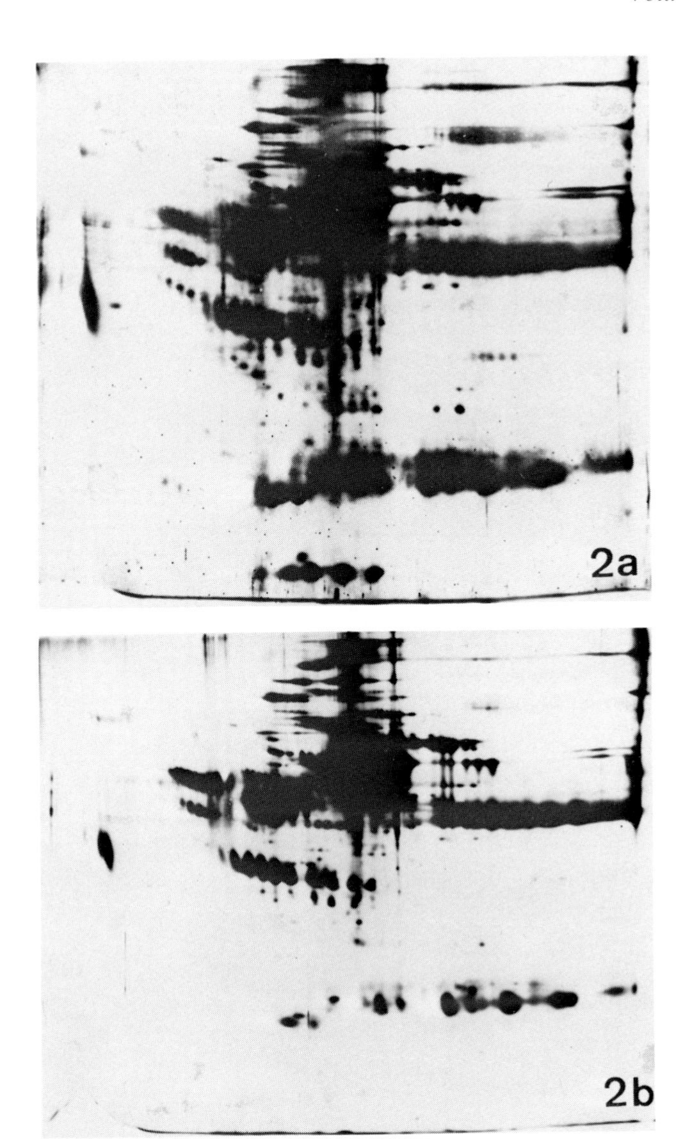

FIGURE 2. Comparison of AuroDye on nitrocellulose (2a) with silver staining[14] of polyacrylamide gel (2b). Human pleural fluid (30 μg) was run on a 2D gel. Proteins were separated in the first dimension with IEF using 2% LKB Ampholines pH 3.5 to 10 and 5 to 8 in a ratio 3:1 and in the second dimension with SDS-electrophoresis using a 10% T 2.6% C polyacrylamide gel. A duplicate gel was electrotransferred to nitrocellulose. (Courtesy of Professor Rabaey and Dr. Segers, State University of Gent, Belgium.)

Adjust the pH with 0.1 N HCl
Distilled water — ad 1000 mℓ
Sodim cacodylate is a toxic; use protective gloves and work under a fume hood

Potassium hexacyanoferrate
0.05 *M* potassium hexacyanoferrate(II)trihydrate — 2.11 g
Distilled water — ad 100 mℓ

Tween® 20 stock solution
10% (v/v) Tween® 20 — 10 mℓ
Distilled water — ad 100 mℓ

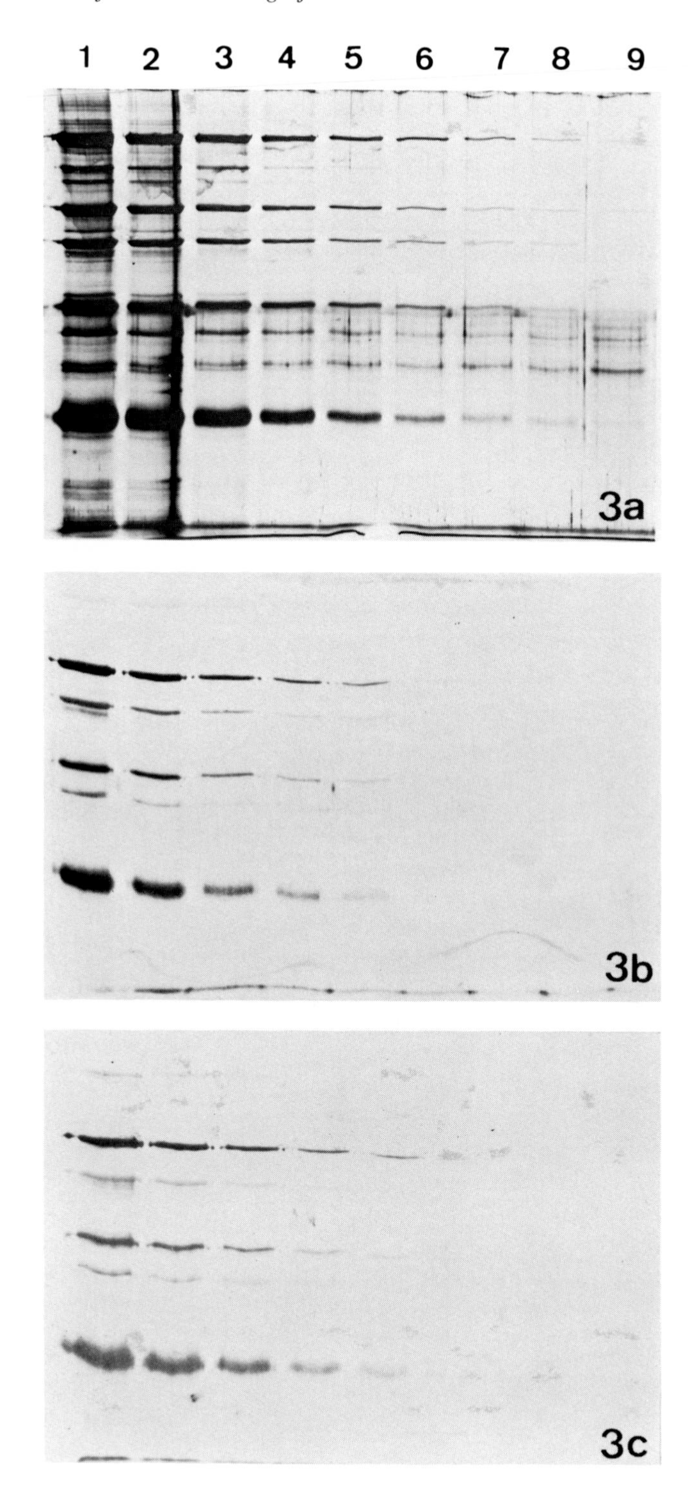

FIGURE 3. Nine dilutions (2-fold) of Bio-rad high molecular weight protein standards separated on 7.5% SDS-polyacrylamide gel. Amount per protein: Lane 1, 1000 ng; 2, 500 ng; 3, 250 ng; 4, 125 ng; 5, 62.5 ng; 6, 31.3 ng; 7, 15.6 ng; 8, 7.8 ng; 9, 3.9 ng. Silver staining[11] of polyacrylamide gel (3a). FerriDye staining of duplicate gels, transferred to either nitrocellulose (3b) or Zetaprobe (3c).

Procedure

Preparation of FerriDye

The cationic cacodylate colloid is prepared as follows[15,16]: 10 mℓ of 0.5 *M* $FeCl_3 \cdot 6\ H_2O$ is added drop by drop to 60 mℓ of boiling and stirring distilled water. The colloid is cooled to room temperature and 1 vol is mixed with 9 vol of 0.1 *M* sodium cacodylate buffer. The stock solution is diluted to an $OD_{460\ nm}$ = 0.5 with buffer. Tween® 20 (1 mℓ 10% (v/v)/49 mℓ diluted colloid) is added to this solution and brought to a final concentration of 0.2% (v/v).

Staining Procedure[7]

The protein blot is washed extensively with distilled water (200 mℓ for a 10 × 15 cm blot), 3 × 10 min in a large Petri dish. The blot is then incubated in the cationic colloid for 1 hr in sealed plastic bags. At least 0.3 mℓ FerriDye per square centimeter blotting membrane must be used. The blot is washed thoroughly with water (200 mℓ), 3 × 2 min, and then incubated in the developer solution.[17] The developer is prepared fresh by mixing 1 vol of 0.05 *M* $K_4Fe(CN)_6$, 2 vol of distilled water, and 2 vol of 1 N HCl.

It is recommended to prepare this solution and to carry out the reaction in a fume hood. After reaction for 1 min, the blot is rinsed in a large excess of water and air dried.

Comments on Procedure

Variable moderate background staining occurs.

FerriDye is less sensitive than AuroDye but it is still advisable to put an additional membrane (same sort) at the cathodic side during electroblotting. Work as cleanly as possible (clean glassware, handle membranes with tweezers), otherwise background staining may become problematic.

Evaluation

To assess the sensitivity a twofold serial dilution series of Bio-Rad high M_r standards (1000 to 3.5 ng per protein band) was subjected to SDS-PAGE. One gel was silver stained according to Morissey[11] (Figure 3a), a duplicate was electrotransferred to a nitrocellulose membrane and stained with FerriDye (Figure 3b). FerriDye is not as sensitive as silver staining or AuroDye. However, the lower sensitivity can be of some advantage, e.g., when higher amounts of protein have to be loaded. For positively charged membranes, this is not a problem in view of their higher binding capacity (4.8 $\mu g/mm^2$).[3]

The same high M_r standards were also transferred to the Zetaprobe membrane and processed for FerriDye as described earlier (Figure 3c).

FerriDye only stains SDS-denatured proteins; however, compared with AuroDye, FerriDye can be used on membranes such as nitrocellulose, unmodified and charge modified nylon-based membranes. Actually, FerriDye was developed for staining proteins transferred to positively charged nylon-based membranes. To our knowledge there was, until recently, no simple alternative for direct staining of proteins on such membranes.

REFERENCES

1. **Gershoni, J. M. and Palade, G. E.,** Protein blotting: principles and applications, *Anal. Biochem.*, 131, 1, 1983.
2. **Towbin, H. and Gordon, J.,** Immunoblotting and dot immunobinding — current status and outlook, *J. Immunol. Methods*, 72, 313, 1984.
3. **Gershoni, J. M. and Palade, G. E.,** Electrophoretic transfer of proteins from sodium dodecyl sulphate polyacrylamide gels to a positively charged membrane filter, *Anal. Biochem.*, 124, 396, 1982.

4. **Langanger, G., Moeremans, M., Daneels, G., Sobieszek, A., De Brabander, M., and De Mey, J.,** The molecular organization of myosin in stress fibers of cultured cells, *J. Cell Biol.*, 102, 200, 1986.
5. **Moeremans, M., Daneels, G., and De Mey, J.,** Sensitive colloidal metal (gold or silver) staining of protein blots on nitrocellulose membranes, *Anal. Biochem.*, 145, 315, 1985.
6. **Rohringer, R. and Holden, D. W.,** Protein blotting: detection of proteins with colloidal gold, and of glycoproteins and lectins with biotin-conjugated and enzyme probes, *Anal. Biochem.*, 144, 118, 1985.
7. **Moeremans, M., De Raeymaeker, M., Daneels, G., and De Mey, J.,** FerriDye: colloidal iron binding followed by Perls' reaction for the staining of proteins transferred from SDS-gels to nitrocellulose and positively charged nylon membranes, *Anal. Biochem.*, 153, 18, 1986.
8. **Frens, G.,** Controlled nucleation for the regulation of particle size in monodisperse gold suspensions, *Nature (London)*, 241, 20, 1973.
9. **Hancock, K. and Tsang, V. C. W.,** India ink staining of proteins on nitrocellulose paper, *Anal. Biochem.*, 133, 157, 1983.
10. **Laemmli, U. K.,** Cleavage of structural proteins during the assembly of the head of bacteriophage T4, *Nature (London)*, 227, 680, 1970.
11. **Morissey, J. H.,** Silver stain for proteins in polyacrylamide gels: a modified procedure with enhanced uniform sensitivity, *Anal. Biochem.*, 117, 307, 1981.
12. **Ochs, D.,** Protein contaminants of sodium dodecyl sulfate — polyacrylamide gels, *Anal. Biochem.*, 135, 470, 1983.
13. **Segers, J. and Rabaey, M.,** *Protides of Biological Fluids*, Vol. 33, Peeters, H., Ed., Pergamon Press, Oxford, 1985, 589.
14. **Wray, W., Boulikas, T., Wray, V. P., and Hancock, R.,** Silver staining of proteins in polyacrylamide gels, *Anal. Biochem.*, 118, 197, 1981.
15. **Seno, S., Tanaka, A., Urata, M., Hirata, K., Nakatsuka, H., Yamamoto, S.,** Phagocytic response of rat liver capillary endothelial cells and Kupffer cells to positive and negative charged iron colloid particles, *Cell Struct. Funct.*, 1, 119, 1975.
16. **Seno, S., Tsujii, T., Ono, T., and Ukita, S.,** Cationic cacodylate iron colloid for the detection of anionic sites on cell surface and the histochemical stain of acid mucopolysaccharides, *Histochemistry*, 78, 27, 1983.
17. **Perls, M.,** Nachweis von Eisenoxid in gewissen Pigmente, *Virchows Arch.*, 39, 42, 1867.

6.2 BLOCKING OF IMMUNOBLOTS

Byron E. Batteiger

INTRODUCTION

Blocking, or quenching, of immunoblots is the process whereby unoccupied protein binding sites on the solid phase matrix are saturated so that detection reagents (probes), such as antibodies or protein A, do not bind nonspecifically to the matrix. In addition, evidence suggests that blocking agents are important in reducing protein-protein interactions that could result in nonspecific adherence of probes to matrix-bound macromolecules, yielding false-positive reactions. Some data indicate that blocking solutions may also assist in partially renaturing proteins separated in gel systems utilizing protein denaturants such as sodium dodecyl sulfate (SDS).[1]

Examples of considerations in choosing blocking conditions follow. Some blocking agents, particularly detergents, are more likely to disrupt noncovalent forces that bind macromolecules to nitrocellulose and other membranes, resulting in loss of antigen, and hence specific binding, yielding false negative results.[2] On the other hand, the poorer overall background frequently obtained with some protein blocking procedures may reduce the sensitivity of visualizing weak, but specific binding, also yielding false negative results. Thus, it is clear that each blocking reagent has advantages and disadvantages, and no single procedure is clearly superior for all applications. However, for a given application, a best method can usually be found by directly comparing several standard procedures rather than arbitrarily using a single method.

The choice of blocking method depends on the type of matrix used, the detection probes, the nature of the macromolecule to be detected (e.g., protein or lipopolysaccharide), whether the results are to be quantitated, and even on whether or not the procedure is performed using small strips of membrane vs. larger sheets of membrane. This chapter is largely limited to the discussion of options for blocking or quenching nitrocellulose membrane. A standard approach to blocking charge-modified nylon membranes is also presented. However, the special blocking conditions required for other matrices are further mentioned in the chapters devoted to them (nylon membranes, Chapter 5.1.2; covalent binding matrices, Chapter 5.2). Blocking methods for dot immunobinding are included in Chapter 2.

The two major classes of blocking reagents are solutions of proteins and detergents, either nonionic or ionic. Examples of useful protein blocking agents are bovine serum albumin (BSA),[3,4] whole animal sera (e.g., horse serum),[5,6] hemoglobin,[7] and casein.[8] Recent experience would suggest that preparations of nonfat dry milk, initially developed for DNA blotting,[9,10] can also be used as inexpensive, effective blocking agents. Gelatin alone is not an adequate blocking agent, and does not improve results when used with detergents.[11] Tween® 20[11] has been used successfully as a blocking agent alone and appears to be the most commonly used of detergents. However, Nonidet P-40,[12] Triton® X-100,[13] as well as Tween® 20,[11] have also been used in procedures employing both proteins and detergents. For more extensive references to the many variations of blocking and washing, see recent review papers on immunoblotting.[2,6]

The procedures detailed in this chapter include both representative detergent and protein procedures.

REAGENTS AND PROCEDURES

Blocking Nitrocellulose Membranes

Blocking of nitrocellulose membranes can be accomplished either with detergents or proteins. We use either phosphate- or Tris-buffered saline (PBS, TBS) at pH 7.3 for our blocking buffers. Obviously, other buffer compositions and pH would serve equally well. In addition, we routinely use radioiodinated probes as detection reagents rather than enzyme detection systems. The relative merits of each system are discussed below.

Detergent Blocking

Blocking and Washing Buffer

PBS with Tween® 20	(pH 7.3)
4 m*M* KH_2PO_4	1.08 g
16 m*M* Na_2HPO_4	4.54 g
115 m*M* NaCl	13.4 g
0.05% (v/v) Tween® 20	1.0 mℓ
Distilled water	ad 2000 mℓ

Store at room temperature. We generally make fresh solutions every 1 to 2 weeks, and have the impression that older solutions do not block as efficiently as fresh solutions. Tween® 20 is obtained from Sigma Chemical Co., St. Louis.

Procedure

Immerse nitrocellulose in a glass or plastic container of appropriate size containing washing buffer for 60 min at room temperature. Nitrocellulose membranes so treated may be frozen and subsequently used for immunoblotting without re-blocking for 1 to 2 months. All primary and secondary antibody dilutions and incubations are then done in washing buffer. Washes between immunological detection steps also employ PBST with two changes of buffer over 30 min. Nitrocellulose processed in the manner described above can be stained with Amido Black 10 B after immunological reactions with excellent results. If radioiodinated reagents are used for the detection of specific antibody binding, autoradiograms of excellent quality can be obtained after such protein staining.[11]

Another variant of the above procedure that uses Tween® 20 as the sole blocking agent is described in the Appendix of Chapter 1.

Protein Blocking

These methods employ blocking and incubation buffers containing various proteins but no detergents. The washes consist of multiple changes of a saline solution. In experiments where protein blockers are required, we use either horse serum (procedure b) or nonfat dry milk (procedure c). BSA is included because it has been used extensively in the past as a standard method.

BSA

Blocking buffer: TBS with 3% (w/v) BSA (pH 7.3)	
10 m*M* Tris	1.2 g
150 m*M* NaCl	8.8 g
5 m*M* sodium azide	0.32 g
BSA (3% w/v)	30.0 g
Distilled water	ad 1000 mℓ

Adjust pH to 7.3 with 5 N HCl, then store at 4°C. BSA is radioimmunoassay grade, obtained from Sigma Chemical Co., St. Louis. Washing solution: 150 m*M* NaCl.

Procedure

Nitrocellulose strips are immersed in blocking buffer for at least 1 hr at 37 to 40°C.[3] Some investigators have suggested that longer blocking periods (e.g., 2 to 12 hr) improve background.[2,6] We have blocked at both room temperature and 37°C with good results, and routinely block for 1 hr. All antibody dilutions are carried out in blocking buffer so that it serves as the incubation buffer as well. The washing step between each incubation consists of five changes of 150 m*M* NaCl over 30 min, according to Towbin et al.[3]

In our experience, BSA as a blocker is not as effective as horse serum or nonfat dry milk, but has the advantages that it can be stored dry at 4°C and can be used with simple detection reagents such as protein A. Most serum-based blocking solutions cannot be used with protein A, since protein A will be bound by immunoglobulin G contained in such blocking solutions.

Whole Animal Sera

TBS with 10% (v/v) horse serum	
10 m*M* Tris	0.6 g
150 m*M* NaCl	4.4 g
5 m*M* sodium azide	0.16 g
10% (v/v) whole horse serum	50 mℓ
Distilled water	ad 500 mℓ

Adjust pH to 7.3 with 5 N HCl, store at 4°C. Horse serum is obtained from Gibco Laboratories, Chagrin Falls, Ohio.

Procedure

This procedure is patterned after that of Gordon et al.[5] Nitrocellulose is immersed in blocking buffer for at least 1 hr at 40°C. We obtain comparable results if the blocking step incubation is at room temperature or 37°C. Again, the duration of blocking may be varied (e.g., 2 to 12 hr). All antibody and probe dilutions are carried out in blocking buffer. The wash steps between each incubation with immunological reagents consist of five changes of 150 m*M* NaCl over 30 min.

In our hands, this procedure gives good overall results, with two exceptions. First, the overall background is higher than we generally observe with Tween® 20 or nonfat dry milk. As a result, we find that the background matrix binding of radiolabeled anti-antibodies (e.g., affinity-purified radioiodinated goat antirabbit IgG, antimouse IgG and IgM, and antihuman IgG) is excessively high when multiple antigen lanes are probed on single large (9 by 13 cm) nitrocellulose sheets. In these particular situations, we generally use a detergent-based (Tween® 20) blocking and incubation buffer which gives excellent background. Second, we find that although horse serum is excellent for eliminating nonspecific protein-protein binding sometimes seen with detergents,[2] it may reduce specific binding as well.

Nonfat Dry Milk

The use of nonfat dry milk as a blocking agent was first described by investigators in molecular genetics,[9,10] although casein had previously been used in lipopolysaccharide immunoblotting on nitrocellulose membranes.[8] We have had limited experience with nonfat dry milk as a blocking agent, but results of our preliminary experiments indicate that it may combine the advantages of BSA, being easy to store and compatible with any detection reagent, and whole serum, providing adequate background with good elimination of nonspecific protein-protein reactions. In addition, nonfat dry milk is cheap, widely available, and easier to handle than other proteins.

In recent immunoblotting experiments in our laboratory using several overlay steps, nonfat dry milk was directly compared to the other blocking procedures discussed in this chapter. Bacterial proteins were sequentially probed with immune guinea pig serum, hyperimmune rabbit antiguinea pig IgG serum, and radioiodinated affinity-purified goat antirabbit antibodies. In terms of overall background, elimination of nonspecific binding to individual polypeptides, and overall signal-to-noise ratio, 0.5% (w/v) nonfat dry milk appeared to be the best blocker.

TBS with 0.5% (w/v) nonfat dry milk	
10 m*M* Tris	0.6 g
150 m*M* NaCl	4.4 g
5 m*M* sodium azide	0.16 g
0.5% (w/v) nonfat dry milk	2.5 g
Distilled water	ad 500 mℓ
Adjust pH to 7.3 with 5 N HCl, store at 4°C	

Procedure

The procedure is identical to those used for BSA- and serum-containing blocking solutions. Blocking proceeds for 60 min at room temperature. Washes between incubation steps again consist of five changes of 150 m*M* NaCl over 30 min. Dry milk-containing blocking solutions for protein blotting have previously used 5.0% (w/v) concentrations.[9] Preliminary experiments using concentrations of dry milk ranging from 0.5 to 4.0% (w/v) indicated that the lower concentration yielded excellent results. Higher concentrations (e.g., 2 to 4%) reduced background further, but appeared to result in a generalized, small reduction in specific binding as well.

Blocking Charge-Modified Nylon Membranes

Blocking/quenching of charge-modified nylon membranes is more difficult than nitrocellulose membranes. The major advantages of nylon membranes are that they are more durable, have a higher binding capacity for proteins and nucleic acids, and may provide increased sensitivity.[2] These attributes also result in avid nonspecific binding of probes to unoccupied membrane sites unless vigorous blocking conditions are used. The blocking/quenching procedure given here is patterned after that of Gershoni and Palade.[7] Recent experience with nonfat dry milk in blocking DNA blots on charge-modified nylon membranes[9,10] (Chapter 5.1.2) suggests that such a procedure could potentially be useful for protein immunoblotting as well.

Blocking Buffer: PBS with 10% (w/v) BSA	
4 m*M* KH_2PO_4	0.11 g
16 m*M* Na_2HPO_4	0.45 g
115 m*M* NaCl	1.34 g
10% (w/v) BSA	20 g
Distilled water	ad 200 mℓ
Adjust pH to 7.3	
Dilution buffer: PBS with 2% (w/v) BSA	
Exactly as above except with 4 g BSA/200 mℓ solution	
Washing solution: PBS, pH 7.3	
Prepare PBS as above without BSA	

Procedure

Charge-modified nylon membranes bearing resolved polypeptides are incubated with blocking buffer at 45 to 50 °C for at least 12 hr. Time and temperature are apparently more critical here than with nitrocellulose. Probe dilutions are made in dilution buffer. Each wash step consists of five changes of washing solution, each change lasting 20 min.

USE AND LIMITATIONS

There are obviously inherent advantages and disadvantages in using either detergents or proteins as blocking agents with nitrocellulose membranes. There are two major disadvantages with detergents. One is the tendency to disrupt hydrophobic binding to nitrocellulose with loss of protein or other macromolecules and therefore specific binding. We have found that ionic detergents like sodium *N*-lauroyl sarcosinate (Sarkosyl), and the nonionic detergent Triton® X-100 cause considerable overall loss in specific binding.[11] Subtle overall losses of bound protein from nitrocellulose occur with Tween® 20 and Nonidet P-40 as well. In addition, all commonly used detergents disrupt the binding of bacterial lipopolysaccharides to nitrocellulose, so that detergent blockers, or combinations of detergents and proteins, are not suitable where retention of lipopolysacchride on the nitrocellulose is important.

The other disadvantage that we have noted with Tween® 20 is that this detergent is not as efficient as protein blockers in eliminating nonspecific binding to protein bands. For example, we have observed binding of human immunoglobulin to several outer membrane proteins of the bacterium *Chlamydia trachomatis* in serum from patients unlikely to have been infected,[14] and similar binding with preimmune serum from guinea pigs. These reactions were eliminated entirely when either 10% (v/v) horse serum or 0.5 % (w/v) nonfat dry milk were used as blocking agents.

However, detergents are exceedingly convenient, inexpensive, and produce excellent elimination of background matrix binding compared to most protein blocking agents. In our hands, protein blockers routinely give higher background binding, and in some cases this higher background may obscure minor bands and thus may reduce sensitivity. However, use of nonfat dry milk appears to give the best combination of low background along with elimination of nonspecific binding to individual protein bands. In experiments performed on small nitrocellulose strips using horse serum, the degree of background binding did not obscure reactive bands. However, in experiments where multiple antigens were probed with a single antibody preparation in sealed bags or Petri dishes, blocking with BSA or horse serum gave unacceptably high background binding. Under these circumstances, Tween® 20 gives far lower background and thus a superior result. Nonfat dry milk as the blocking reagent also gives good results in these types of experiments.

Another advantage of detergents in general, and Tween® 20 in particular, is that staining for protein, for example with Amido Black 10 B (Appendix, Chapter 1) after immune reactions is possible, while it is not with protein blockers. Such staining is useful for precise alignment of autoradiograms and stained strips, and also for excision of specific polypeptide bands for quantitation of bound radioactivity.[11]

In our laboratories, we use a Tween® 20-based procedure as a standard. This procedure is simple, cheap, very sensitive with excellent background, and allows protein staining after immunological probing. The major disadvantage may be false positive reactions, thought due to nonspecific binding of hydrophobic probes to hydrophobic domains of denatured polypeptides bound to nitrocellulose.[2] Since we are interested in bacterial membranes and in detecting antibodies to bacterial lipopolysaccharide, we have occasionally substituted a protein-based procedure, most often with horse serum. In addition, when screening sera obtained from animals and humans infected with *Chlamydia trachomatis,* we have routinely used horse serum blocking to eliminate what we believe to be false positive binding to certain bacterial membrane proteins. However, our group and others have recently obtained excellent results in protein immunoblotting with nonfat dry milk as a blocking agent. If such procedures prove suitable on further evaluation, they may prove to be the protein blocking agent of choice.

REFERENCES

1. **Petit, C., Sauron, M. E., Gilbert, M., and Theze, J.,** Direct detection of idiotypic determinants on blotted monoclonal antibodies, *Ann. Immunol. (Inst. Pasteur)*, 133D, 77, 1982.
2. **Gershoni, J. M. and Palade, G. E.,** Protein blotting: principles and applications, *Anal. Biochem.*, 131, 1, 1983.
3. **Towbin, H., Staehelin, T., and Gordon, J.,** Electrophoretic transfer of proteins from polyacrylamide gels to nitrocellulose sheets: procedure and some applications, *Proc. Natl. Acad. Sci. U.S.A.*, 76, 4350, 1979.
4. **Burnette, W. N.,** "Western blotting": electrophoretic transfer of proteins from sodium dodecyl sulfate-polyacrylamide gels to unmodified nitrocellulose and radiographic detection with antibody and radioiodinated protein A, *Anal. Biochem.*, 112, 195, 1981.
5. **Gordon, J., Towbin, H., and Rosenthal, M.,** Antibodies directed against ribosomal protein determinants in the sera of patients with connective tissue diseases, *J. Rheumatol.*, 9, 247, 1982.
6. **Towbin, H. and Gordon, J.,** Immunoblotting and dot immunobinding — current status and outlook, *J. Immunol. Methods*, 72, 313, 1984.
7. **Gershoni, J. M. and Palade, G. E.,** Electrophoretic transfer of proteins from sodium dodecyl sulfate-polyacrylamide gels to a positively charged membrane filter, *Anal. Biochem.*, 124, 396, 1982.
8. **Sidberry, H., Kaufman, B., Wright, D. C., and Sadoff, J.,** Immunoenzymatic analysis by monoclonal antibodies of bacterial lipopolysaccharides after transfer to nitrocellulose, *J. Immunol. Methods*, 76, 299, 1985.
9. **Johnson, D. A., Gatsch, J. W., Sportsman, J. R., and Elder, J. H.,** Improved technique utilizing nonfat dry milk for analysis of proteins and nucleic acids transferred to nitrocellulose, *Gene Anal. Technol.*, 1, 3, 1984.
10. **Reed, K. C. and Mann, D. A.,** Rapid transfer of DNA from agarose gels to nylon membranes, *Nucl. Acids Res.*, 13, 7207, 1985.
11. **Batteiger, B., Newhall, W. J. V., and Jones, R. B.,** The use of Tween® 20 as a blocking agent in the immunological detection of proteins transferred to nitrocellulose membranes, *J. Immunol. Methods*, 55, 297, 1982.
12. **Cohen, M. L. and Falkow, S.,** Protein antigens from *Staphylococcus aureus* strains associated with toxic-shock syndrome, *Science*, 211, 842, 1981.
13. **Vaessen, R. T. M. J., Kreike, J., and Groot, G. S. P.,** Protein transfer to nitrocellulose filters: a simple method for quantitation of single proteins in complex mixtures, *FEBS Lett.*, 124, 193, 1981.
14. **Newhall, W. J. V., Batteiger, B. E., and Jones, R. B.,** Analysis of the human serological response to proteins of *Chlamydia trachomatis*, *Infect. Immunol.*, 38, 1181, 1982.

6.3.1 POLYCLONAL ANTIBODIES AS PRIMARY REAGENT IN IMMUNOBLOTTING — USES AND LIMITATIONS

Jakob Ramlau

INTRODUCTION

The array of antibodies present in a good antiserum is truly bewildering in its complexity, and it is thus not without reason that the advent of "monoclonal technology" with its possibility of singling out only one well-characterized reagent, was greeted enthusiastically. However, classical polyclonal antibodies ("polyclonals" for short) will also continue to be used. Furthermore, they are involved in the serological response to an infectious microorganism, in autoimmune diseases, and in allergy.

The purpose of this chapter is to review the basal immunochemistry of polyclonals and in light of this, to comment upon the application of the polyclonals in immunoblotting. I intend to compare mono- and polyclonals with regard to virtues and disadvantages. Some special problems with polyclonals are discussed with practical "troubleshooting" in mind.

PRODUCTION AND PROPERTIES OF POLYCLONAL ANTIBODIES

The molecular nature of the antigenic determinant is a key question in immunology which has been studied for years. Some early results[1] led to the view that each protein possessed a limited number of rather circumscript structures, the determinants. These were due to a proper folding of the molecule, but this view seems to be giving way to a more flexible way of thinking: most of the solvent-accessible surface of the protein makes up a mosaic of overlapping determinants.[2,3]

The number of possible determinants on a molecule must therefore be quite high. It should also be noticed that the immunogen is proteolytically processed in immunogen-presenting cells.[4] This processing will probably make new determinants (previously hidden) accessible on the fragments.

Surprisingly small peptides, e.g., hexapeptides, can be immunogenic.[5] Thus, it is probably fair to say that both continuous (sequential, dependent only on primary structure) and discontinuous (conformational, owing their presence to folding of peptide chain(s), bringing otherwise separated residues into proximity due to both tertiary and quaternary structures), determinants exist together on protein molecules. The issue is made even more complex by the hypothesis that areas characterized by flexibility should be more immunogenic than other parts of the molecule.[6] Antisera obtained from a group of animals will thus exhibit a broad reactivity compared to the single specificity of a monoclonal antibody.

Users of antisera will be concerned with the strength of a given reagent often referred to as titer, a word with many definitions. This strength will be a sum of the products of concentration and affinity for each antibody specificity present. It must be remembered that there is more than one antibody reacting with each determinant on the complex protein antigen. This measure of strength is thus simple for a monoclonal antibody, but far too complex for strict analysis of a polyclonal.

The affinity of the antibody for the determinants increases with the time of immunization. This has led to the use of long-time immunization for the production of polyclonals.[7] It might thus be disconcerting that many immunization protocols aimed at production of monoclonals use large-dose, short-time immunization. However, the affinity of the ensuing monoclonal antibody can often be chosen by judicious screening in the cloning and subcloning procedure. There is thus no base for a general verdict about the relative avidities of monoclonals vs. polyclonals.

The concentration of specific antibodies in a polyclonal antiserum is seldom determined, but a good antiserum can contain 20 to 30% specific immunoglobulins (unpublished results). With a total immunoglobulin concentration of 6 to 7 mg/mℓ and a production of approximately 500 mℓ serum per rabbit per year, this compares favorably with 20 to 100 μg/mℓ present in most culture supernatants, while ascitic fluid is a more comparable competitor. Developments in tissue culture techniques might very well change this.

The "nonspecific" (the term should be taken to denote all immunoglobulins not reacting with the antigen of choice) immunoglobulins will normally make up a majority of the immunoglobulins present in a polyclonal antiserum. These nonspecific immunoglobulins can be divided into two groups: "background immunoglobulins" and "unspecific antibodies".

"Background immunoglobulins": This first group consists of the normal immunoglobulins which are also present in the sera from an unimmunized animal exposed to the same environmental immunogens. Many of these antibodies are directed against commensal microorganisms[8] and it is noteworthy that pathogen-free animals often have very low immunoglobulin concentrations.[9] Antibodies to foodstuffs have also been found.[34] These background antibodies can be problematical in serological and taxonomic work with blotting of microbial proteins as well as in immunochemical screening of expression libraries in, e.g., *Escherichia coli*. One could also envisage problems with natural antibodies to foodstuffs in the analysis of specific components in food and for allergy research.

"Unspecific antibodies": This term should be taken to denote the almost regular presence in polyclonals of unwanted specificities directed against minor impurities in the immunogen preparation. Less than 1% impurities in an immunogen can give rise to antibody responses out of proportion (results from my own experiments). The first approach to avoid this problem is, of course, to purify the immunogen as well as possible, though yield and purity are often inversely proportional. The rabbit is normally better at detecting impurities than we are at removing them. Absorption of the antibody to remove unwanted specificities is one of the worst problems in the preparation of good polyclonals.

The third kind of reactivity to be discussed is the concept of cross-reactivity. If the antibodies reacting with the immunizing antigen also react with another protein this will be a case of cross-reactivity. Such cross-reactivities can be demonstrated elegantly by precipitation-in-gel techniques.[11] These reactions of partial identity are often seen between functionally related proteins, e.g., the serum albumins, and can easily be explained by the sharing of some determinants. It is, however, worth mentioning that identical epitopes can be present on otherwise unrelated proteins.[12,13] For smaller immunogens, another kind of cross-reactivity can be relevant: different affinity for slightly different epitopes, i.e., "degree of fit". This kind of cross-reaction can only be investigated by competition experiments.

APPLICATION OF POLYCLONAL ANTIBODIES IN BLOTTING TECHNIQUES

Like all immunochemical techniques, immunoblotting can be used both for analysis of antigens in a mixture if a suitable antibody is present, or for the characterization of the reactivity of a given antibody with a mixture of antigens. The following discussion centers on problems related to the practical use of polyclonals.

Reactivity with the Immunogen

There is very little reason to believe that the binding to nitrocellulose in itself denatures protein antigens. There are some findings[14] that indicate a difference in reactivity with antibodies between free and absorbed antigen in enzyme-linked immunosorbent assays (ELISA) with monoclonal, but the broad reactivity of polyclonals makes it less probable that this will ever be a problem; however, the treatment of the sample before blotting might influence the reactivity. Some reports stress the native state of the blotted material,[15] but the majority of experiments are still performed with the denaturing sodium dodecyl sulfate-polyacrylamide gel electrophoresis (SDS-PAGE). Experiments with precipitation in gels indicated that the reactivity of most antigens was greatly diminished by the sample treatment,[16] often to a degree abolishing immunoprecipitability. This is in marked contrast to the often very good reactivity found by immunoblotting with polyclonal antibodies. Several reasons might exist:

1. The antigen might renature on transfer. The point has not yet been properly investigated. Early observations indicated that treatment with nonionic detergent could renature the antigen.[16]
2. Nonprecipitating antibodies were not observed in the precipitation in gels system.
3. The very high sensitivity of the detection systems in immunoblotting renders a reaction founded on rather few bound antibody molecules that are easily detectable.

Sample treatment involving boiling in SDS and reduction will probably destroy all tertiary and quaternary structures in the antigen. A number of observations indicate that some monoclonal antibodies fail to react with material so treated[17] as the determinant involved might have been discontinuous. The determinants that resist sample treatment are probably the continuous determinants depending only on primary structure. It is unknown how large a percentage the antibodies against these determinants make up compared to the antibodies to structural determinants. It might be noteworthy that the detection limit for a blotted band is higher than for the same protein dot-spotted in native form.[35] This might support the notion that only few and/or low avidity antibodies are involved in the reaction. It is often seen that antibodies elicited by immunization with small synthetic peptides react quite well with blotted bands. This is in accordance with the hypothesis that they make up single continuous determinants. However, one can almost always obtain a usuable reaction with polyclonal antibodies in blotting with antigens that are both reduced and SDS-denatured. However, if no reaction is obtained it might be tried to change sample treatment, with or without reduction, though such changes invalidate the determination of molecular weights. One can also change more radically into separation systems that do not rely on complete denaturation (see Chapter 8.5).

Titer and Avidity

As mentioned previously, a good polyclonal can contain 25% of specific antibodies, but many antibodies will contain less, especially after short immunizations with small doses of weak immunogens. Patients' antibodies must be taken as they are in serological studies. There is thus no general rule about what to expect. The reactivity observed after processing the blot is a product of amount and avidity of antibody. The antibody must be avid enough to hang on during the washings and secondary incubations. Most antibodies to protein immunogens will have binding constants between 10^6 and 10^8 M. Experiments with ELISA indicate that detection of antibodies binding to a coated antigen requires binding constants around 10^5 M.[18]

These points are of importance not only in specifying the quality of the wanted antibody, but even more in indicating a way to avoid unspecific binding. In all blotting systems one must perform a checkerboard titration with serial dilutions of both antigen and antibody to establish optimal conditions, conditions where there is a good signal-to-noise ratio (with strong staining of the wanted bands with low background staining and insignificant unspecific staining of other bands). Due to the low avidity and concentration of the unspecific antibodies this is often possible. This trick can be furthered by choosing incubation and washing conditions that only "strong" antibodies can withstand (see Chapter 7).

These kind of procedures should not be used uncritically in circumstances where weaker antibodies might be relevant: with monoclonals and some patient antibodies, to use some examples. The conditions favoring binding vary between epitope/antibody systems, but the basic pH/detergent system described in Chapter 1, which may be further augmented by high ionic strength, is rather efficient.

Thus far, our experience with good polyclonal antibodies indicates that they can routinely be diluted 1:1000 for incubations overnight in the system described in Chapter 1. This dilution refers to immunoglobulin concentrations comparable to neat serum, however with unknown, possibly weaker systems, titrations and/or changes must be made, always including the relevant controls.

Reactivity with Other Proteins

While too little reactivity with the antigen of choice is but rarely seen, just the opposite (including reactivity with other bands) is often seen. Much of the theoretical background for such reactions has already been discussed, and Chapter 7 outlines procedures to check all the possible causes of unwanted or unexpected reactions. The more obvious errors such as reactions with natural antibodies to microorganisms have also been mentioned. It must be stressed once again that any antibody must always be performance-tested in the same system in which it will be used. Antibodies functionally monospecific in precipitation-in-gel techniques might very well contain weak unspecific antibodies that turn up in blotting due to the much higher sensitivity, or nonprecipitating antibodies might be detected by the use of the solid-phase antigen.

The nonspecific reactions can easily be identified with control incubations with an immunoglobulin preparation or a serum from an nonimmunized animal. Such an experiment also includes nonimmune reactions: if a preparation of active Fc receptors (or protein A-like material from a microorganism) is blotted on nitrocellulose and incubated with any antibody (except Fab fragments) a positive reaction will be obtained. It is often seen that some proteins in the sample possess some kind of general stickiness, making them bind to any preparation of antibodies (see Figure 5, Chapter 9.1.2). In situations where high nonspecific binding is a problem it might worthwhile to include nonimmune immunoglobulin from the same species as used for production of secondary antibodies (if used) in the solution used for blocking. This will ensure that all molecules with an illegitimate affinity for immunoglobulins are saturated with an immunoglobulin that will not react with the visualizing reagent.

The unspecific reactions are more difficult. If we first consider a truly monospecific antibody that turns out to react with several bands, many innocent explanations can be given. The antigen in question might be present in many forms: precursors, intermediates more or less fully modified post-translatory, fragments either degraded in vivo or in vitro during extraction and fractionation or induced by sample treatment; inadequate reduction can lead to multimeric proteins migrating as a series of bands and even proteins with only internal thiol bridges will migrate differently as the partially reduced proteins have a smaller Stoke's radius. Fragments/subunits might even be shared between otherwise different proteins; the light chains of the immunoglobulin molecule is a typical example. Thus, all the tricks of the trade must be used to assess the influence of sample preparations and precautions taken to avoid proteolytic degradation. It might even be necessary to use peptide mapping as described by Cleveland[19] or other workers[20] to clarify the relationship between various bands. Even a truly monospecific antibody might react with other apparently unrelated proteins if they share a determinant that might be functional[21,22] or share a rather common group, such as carbohydrates.[23] Other studies, both with polyclonal[12,13] and monoclonal antibodies,[24] indicate that the specificity of the antibody can be quite relative and that increased sensitivity leads to detection of systems not detected by cruder methods. It is interesting to muse on these lines when one considers the number of proteins being found to possess homologies by computer comparisons of sequences — what might the immunochemical consequences be?

It is only fair to point out that the precipitation in gel methods does have a point in the comparison of antigens in all their forms; many reports indicate the possibility of studying partial identities between proteins[11] as applied in the investigations of complex degradation patterns.[25] The ''graphical'' relationship of the precipitation peaks can be used and interpreted in a way that is not possible with the blotting methods as they are used now. These reactions of partial identity due to shared determinants is an important advantage inherent in polyclonals. A monoclonal will either react or not, and the only kind of partial identity will be with respect to ''degree of fit''. The reactions of the polyclonals make it possible to use them against other antigens than those used for immunization. An antibody directed against human albumin can be counted upon to react with many determinants on albumins of many other species.

No simple universal procedure can be recommended for solving the problem of what constitutes unspecific/specific binding. In the case of multiple reactivities, one must consider the numerous possible heterogeneities of the antigen and nonspecific reactions, and finally verify/falsify the offending bands as being due to unspecificity by combinations of methods, dilutions of both antigen and antibody, and analysis of side fractions. If one band in the blotted pattern varies as does the wanted component in the fractionation, the identification is furthered while a number of contaminants can be identified in the same way. Absorption with these side fractions can often be helpful in producing a monospecific antibody from an otherwise oligospecific preparation. Another way of identifying the wanted component is often present in biological systems; intelligent manipulation or use of genetic variants can supply the researcher with an extract devoid of the activity of interest, but otherwise supposedly identical. These preparations can be used both as controls and for the absorptions needed to make the antibody monospecific. Antibodies specific for pregnancy-related proteins can thus be obtained by immunizing with sera from pregnant animals and absorbing the resulting antibody with sera from normal male animals.[26]

Many of the experiments needed for controls and their interpretations are discussed in Chapter 7.

Working with Patient's Antibodies

The titer of an antibody obtained by the unsystematic, haphazard way of natural infection or the unclarified processes in autoimmune diseases is not necessarily comparable to what skilled immunization of an experimental animal can yield. Furthermore, the immunogen in question might have been quite complex. The pattern obtained might therefore be difficult to interpret and care must be taken to obtain relevant controls as the natural antibodies present in a control group comparable to the patient in question might be different from the pool of healthy laboratory workers often used. These precautions might be especially relevant in serological work. If secondary antibodies are used to differentiate between isotypes in the serological response, one must remember that, at least in ELISA,[27] IgG can compete IgM away from the same epitope.

The antibodies in autoimmune diseases can be investigated with blotting techniques. An interesting report[28] demonstrates a number of the problems discussed earlier: nonspecific reactions and interesting reactions of the secondary antibody. Rheumatoid factors can cause many complications in ELISA,[29] and one should be aware of the risk in blotting as well.

Preparation of Antibodies

A few practical notes can be given on the preparation of polyclonal antibodies:

1. Choice of animal. There is probably no formal proof of the superiority of one species over another with regard to antibody production. One should choose an animal suitably taxonomically removed from the source of one's immunogen, but as no information will normally be present on the differences between one's immunogen and the corresponding substance in experimental animals, no rational choice can be made beforehand. A simple practical viewpoint is that the animal should be small enough to be easily handled, injected, and bled. The use of several smaller animals is advantageous as it equalizes the differences in response so often seen. Our experience with rabbits indicates that less than four to five animals is risky. It is logistically easy to use one species as all secondary reagents then become universal. It should be remembered that the often used "universal" protein A reagents do not react equally well with all IgG of all species.[30] Its reactivity with rabbit IgG is good, but mice and especially rat immunoglobulins can be difficult. The use of secondary antibodies circumvent this problem though it should be borne in mind that many secondary reagents can have a surprising cross-species reactivity. A rabbit antibody to human immunoglobulins can easily react with murine immunoglobulins and vice versa.
2. Immunization regimen. The protocol of Harboe and Ingild[7] with its recommendation of long-term, small-dose immunization can be recommended, good results are often obtained with doses as small as 25 to 50 μg per rabbit using Freund's incomplete adjuvant.
3. Fractionation of antisera. There are normally good reasons for purifying the immunoglobulin fraction from the antisera for precipitation in gel techniques and affinity chromatography. The salting-out/ion exchange method of Harboe and Ingild[7] is simple for larger volumes, while protein A columns are very fast for the odd small sample. However, the dilutions normally used in immunoblotting incubations and the numerous washes makes this unnecessary (though harmless) for sera intended for blotting. The stability of rabbit antisera and immunoglobulins is extremely good, they will keep for years at 4°C if preserved with 15 m*M* NaN_3. There are records showing that plasminogen, copurified with the immunoglobulins, can be troublesome for some purposes, the activated enzyme can be inhibited with aprotinin.[31]

As previously mentioned, all absorptions should be performed with solid-phase techniques. One special problem should be mentioned: a number of the antibodies present in a polyclonal might be directed against determinants on impurities hidden within the molecules. These determinants will not be accessible on the surface of the molecule and thus will not necessarily take part in precipitation in gel techniques. Nor will the resulting antibodies be removed by simple absorption on a column with intact impurities coupled to it (see Chapter 7).

Affinity purification should not be used uncritically. The elution of the bound antibodies might harm the antibodies or the best antibodies might never leave the column again. There is actually some reason to believe that the possible aggregation brought about by harsh elution might increase nonspecific binding (see Chapter 7). It must be remembered, however, that the use of "preparative blotting" for affinity purification of the wanted antibodies can be a very useful tool[32,33] (see Chapter 8.8).

CONCLUDING REMARKS

It is obvious that the use of polyclonals requires much insight if optimal results are to be obtained and the interpretation of results can be surprisingly difficult. However, the production of polyclonals is very simple and easy as the rabbits are doing most of the work compared to the more artificial situation with monoclonals. As far as our experience goes, it is easy to obtain 100 mℓ of good rabbit antibodies that diluted 1000 times will give 100 ℓ of solution ready for incubation. This means that even a small pool of antisera will be enough for years of experimentation; however, the care and thought required in assessing the quality of the reagent before its routine use is of the same kind as that required in the screening of monoclonals. This characterization can be carried out in an unhurried way compared with the pressing need for decisions in monoclonal screening. The most troublesome aspect of the production of polyclonals is definitely the absorptions. It is often valuable to immunize with a first generation of immunogen and use the resulting oligospecific antibody for the purification of a second generation high quality antigen.

The final advantage of the polyclonals is, in my opinion, their broadness of reactivities that makes them unsurpassed as a firm foundation for continued, more detailed work on fine structure with other tools, including monoclonals and antipeptide antibodies.

REFERENCES

1. **Atassi, M. Z. and Lee, C -L.,** The precise and entire antigenic structure of native lysozyme, *Biochem. J.*, 171, 429, 1978.
2. **Benjamin, D. C., Berzofsky, J. A., East, I. J., Gurd, F. R. N., Hannum, C., Leach, S. J., Margoliash, E., Michael, J. G., Miller, A., Prager, E. M., Reichlin, M., Sercarz, E. E., Smith-Gill, S. J., Todd, P. E., and Wilson, A. C.,** The antigenic structure of proteins: a reappraisal, *Ann. Rev. Immunol.*, 2, 67, 1984.
3. **Walter, G.,** Production and use of antibodies against synthetic peptides, *J. Immunol. Methods*, 88, 149, 1986.
4. **Sams, C. F., Hemelt, V. B., Pinkerton, F. D., Schroepfer, G. J., and Matthews, K. S.,** Exposure of antigenic sites during immunization, *J. Biol. Chem.*, 260, 1185, 1985.
5. **Young, C. R., Schmitz, H. E., and Atassi, M. Z.,** Antibodies with specificities to preselected protein regions evoked by free synthetic peptides representing protein antigenic sites or other surface locations: demonstration with myoglobin, *Mol. Immunol.*, 20, 567, 1983.
6. **Westhof, E., Altschuh, D., Moras, D., Bloomer, A. C., Mondragon, A., Klug, A., and Van Regen-Mortel, M. H. V.,** Correlation between segmental mobility and the location of antigenic determinants in proteins, *Nature (London)*, 311, 123, 1984.
7. **Harboe, N. M. G. and Ingild, A.,** Immunization, isolation of immunoglobulins and antibody titre determination, *Scand. J. Immunol.*, 17(Suppl. 10), 345, 1983.
8. **Bourne, F. J., Honour, J. W., and Pickup, J.,** Natural antibodies to *Escherichia coli* in the pig, *Immunology*, 20, 433, 1971.
9. **Eisen, H. N.,** in *Microbiology*, 2nd ed. Davis, B. D., Dulbecco, R., Eisen, H. N., Ginsberg, H. S., and Wood, W. B., Eds., Harper & Row, New York, 1973, 480.
10. **Johnstone, A. and Thorpe, R.,** *Immunochemistry in Practice*, Blackwell Scientific, Oxford, 1982, 74.
11. **Bock, E. and Axelsen, N. H.,** Partial antigenic identity, *Scand. J. Immunol.*, 17(Suppl. 10), 183, 1983.
12. **Sperling, R., Francus, T., and Siskind, G. W.,** Degeneracy of antibody specificity, *J. Immunol.*, 131, 882, 1983.
13. **Nigg, E. A., Walter, G., and Singer, S. J.,** On the nature of crossreactions observed with antibodies directed to defined epitopes, *Proc. Natl. Acad. Sci. U.S.A.*, 70, 5939, 1982.
14. **Dierks, S. E., Butler, J. E., and Richerson, H. B.,** Altered recognition of surface-adsorbed compared to antigen-bound antibodies in the ELISA, *Mol. Immunol.*, 23, 403, 1986.
15. **Reinhart, M. P. and Malamud, D.,** Protein transfer from isoelectric focusing gels: the native blot, *Anal. Biochem.*, 123, 229, 1982.
16. **Nielsen, C. S. and Bjerrum, O. J.,** Immunoelectrophoretic analysis of sodium dodecylsulphate-treated proteins, *Scand. J. Immunol.*, 4(Suppl. 2), 73, 1975.
17. **Thorpe, R., Bird, C. R., and Spitz, M.,** Immunoblotting with monoclonal antibodies: loss of immunoreactivity with human immunoglobulins arises from polypeptide chain separation, *J. Immunol. Methods*, 73, 259, 1984.
18. **Steward, M. W. and Lew, A. M.,** The importance of antibody affinity in the performance of immunoassays for antibody, *J. Immunol. Methods*, 78, 173, 1985.
19. **Cleveland, D. W., Fischer, S. G., Kirschner, M. W., and Laemmli, U. K.,** Peptide mapping by limited proteolysis in sodium dodecyl sulfate and analysis by gel electrophoresis, *J. Biol. Chem.*, 252, 1102, 1977.

20. **Walker, A. I. and Anderson, C. W.,** Partial proteolytic maps: Cleveland revisited, *Anal. Biochem.*, 146, 108, 1985.
21. **Staben, C. and Rabinowitz, J. C.,** Immunological crossreactivity of eukaryotic C_1-tetrahydrofolate synthase and prokaryotic 10-formyltetrahydrofolate synthetase, *Proc. Natl. Acad. Sci. U.S.A.*, 80, 6799, 1983.
22. **Katiyar, S. S. and Porter, J. W.,** Antibodies specific for NADPH-binding regions of enzymes possessing dehydrogenase activities, *Proc. Natl. Acad. Sci. U.S.A.*, 80, 1221, 1983.
23. **Triadou, N.,** Antigenic cross-reactions among human intestinal brush-border enzymes revealed by the immunoblotting method and rabbit anti-enzyme sera, *J. Immunol. Meth.*, 73, 283, 1984.
24. **Lane, D. and Koprowski, H.,** Molecular recognition and the future of monoclonal antibodies, *Nature (London)*, 296, 200, 1982.
25. **Bjerrum, O. J. and Bøg-Hansen, T. C.,** Analysis of partially degraded proteins by quantitative immunoelectrophoresis, *Scand. J. Immunol.*, 4(Suppl. 2), 89, 1975.
26. **Hau, J., Svendsen, P. E., Teisner, B., and Svehag, S.-E.,** Studies of pregnancy-associated murine serum proteins, *J. Reprod. Fert.*, 54, 239, 1978.
27. **Vejtorp, M.,** Solid phase IgM ELISA for detection of Rubella-specific IgM antibodies, *Acta Pathol. Microbiol. Scand. Sect. B*, 89, 123, 1981.
28. **Stricker, R. B., Abrams, D. I., Corash, L., and Shuman, M. A.,** Target platelet antigen in homosexual men with immune thrombocytopenia, *N. Engl. J. Med.*, 313, 1375, 1986.
29. **Vejtorp, M.,** The interference of IgM rheumatoid factor in enzyme-linked immunosorbent assays of Rubella IgM and IgG antibodies, *J. Virol. Methods*, 1, 1, 1980.
30. **Langone, J. J.,** Protein A of *Staphylococcus aureus* and related immunoglobulin receptors produced by *Streptococci* and *Pneumococci, Advances in Immunology*, Vol. 30, Academic Press, 1982, 157.
31. **Bjerrum, O. J., Ramlau, J., Clemmesen, I., Ingild, A., and Bøg-Hansen, T. C.,** An artefact in quantitative immunoelectrophoresis of spectrin caused by proteolytic activity in antibody preparations, *Scand. J. Immunol.*, 4(Suppl. 2), 81, 1975.
32. **Olmsted, J. B.,** Affinity purification of antibodies from diazotized paper blots of heterogeneous protein samples, *J. Biol. Chem.*, 256, 11955, 1981.
33. **Beall, J. A. and Mitchell, G. F.,** Identification of a particular antigen from a parasite cDNA library using antibodies affinity purified from selected portions of Western blots, *J. Immunol. Methods*, 86, 217, 1986.
34. **Harboe, N.,** personal communication.
35. **Bjerrum, O. J.,** personal communication.

6.3.2 MONOCLONAL ANTIBODIES AS FIRST OVERLAY IN IMMUNOBLOTTING — GENERAL REQUIREMENTS

Karl-Erik Johansson

INTRODUCTION

Monoclonal antibodies,[1] or simply monoclonals, are produced by immunizing a mouse, isolating antibody producing (plasma or B) cells from the spleen, and immortalizing them by fusion with a mouse myeloma cell line.[2-4] The hybrid cells or hybridomas are then grown in a selective medium, cloned by limited dilution, and screened for antibody production by, e.g., ELISA, (enzyme-linked immunosorbent assay). The antibody-producing hybridomas are further cultivated in vitro in suspension cultures or in vivo in mice as tumors. Monoclonals are highly specific and the supply is in principle unlimited. Furthermore, the antigen used for immunization does not have to be pure, instead clones which produce antibodies of a predefined specificity are selected. Many manuals are available about production of monoclonals.[5-11] Overlay is a term which has been recommended to be used for incubating a blot with a ligand.[12]

DIFFERENCES BETWEEN MONOCLONAL AND POLYCLONAL ANTIBODIES

Early immunochemical analysis methods were dependent on precipitating antibodies but modern ones, like immunoblotting and ELISA, are not. Monoclonals, which are not precipitating unless the antigen is composed of several identical subunits, are therefore very useful probes in immunoblotting experiments,[13] and immunoblotting can be used for characterization of monoclonals (see Chapter 8.1). Some inherent properties of the monoclonals must be considered for optimal design of the experiments and interpretation of the results.

In a conventional antiserum from an animal immunized with a pure antigen, there will be antibodies directed against the different epitopes of that particular antigen. Furthermore, there will also be a whole set of antibodies with different affinities.[4] Thus, a polyclonal antiserum is a very undefined product, even if it is monospecific, and the titers of individual antibodies are very sensitive to small variations in the properties of the antigen, the immunization route, and the animal. The fact that monoclonals are specific to a single epitope offers the possibility to design experiments with a unique specificity, but this high specificity might cause some problems which are discussed later. Techniques are available for determining their affinities.[14]

Monoclonals are by definition directed against one antigen only, since they originate from cells with identical genetic information. One should, however, be aware of the risk of cross-reactions with similar epitopes, which of course also applies for polyclonals. If a monospecific but polyclonal antiserum is used to characterize a particular antigen by immunoblotting after electrophoresis, and some faint bands are observed besides the main one, it is never known if the faint bands are due to a specific reaction or presence of antibodies directed against impurities in the original antigen preparation. If more than one band is found when immunoblotting is performed with monoclonals (see Figure 1), it must be due to the presence of the corresponding epitope in a modified form of the antigen, provided that the cloning was successful. The modification of the protein can be, for example, posttranslational, but proteolysis in the sample is also a possibility.

A large pool of antiserum has to be produced when several experiments are to be conducted with a polyclonal antiserum and comparisons are essential. It will never be possible to obtain a new serum pool with exactly the same properties when the first one has been consumed.

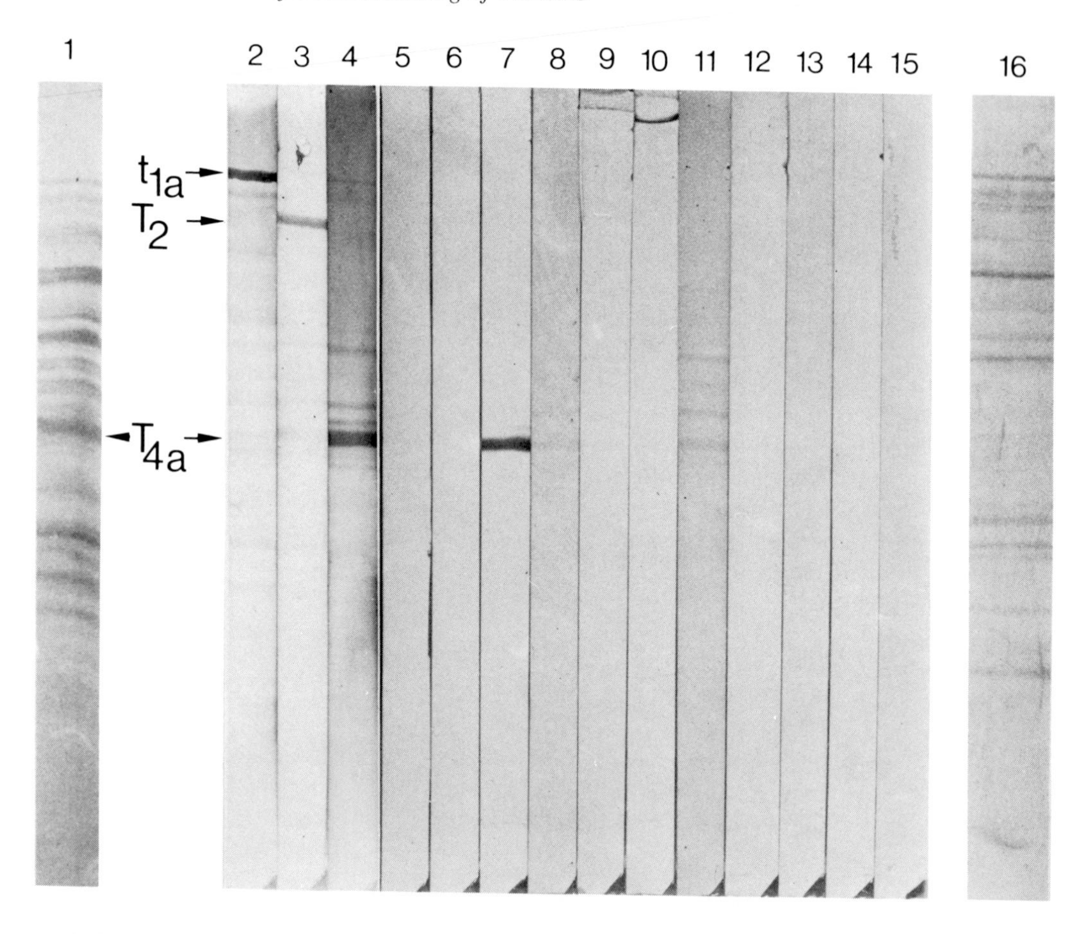

FIGURE 1. Screening of hybridoma supernatants, which were positive in ELISA, for production of monoclonal antibodies against the antigen T_{4a} of *Acholeplasma laidlawii*[26] by immunoblotting with horseradish peroxidase conjugated secondary antibodies. Strip 1, Amido Black stained nitrocellulose membrane after electrotransfer of SDS-PAGE separated *A. laidlawii* membrane proteins (Neville system, see Chapter 3). Strips 2 to 4, incubation of the nitrocellulose strips with monospecific primary antibodies against the *A. laidlawii* membrane protein antigens t_{1a}, T_2, and T_{4a},[27] respectively. Strips 5 to 15, incubation with different hybrid supernatants as primary antibodies. Strip 16, Coomassie blue stained gel after electrotransfer. (From Tigyi, G., Juntti, N., and Johansson, K.-E., *Protides Biol. Fluids*, 33, 575, 1985. Copyright Pergamon Journals, Ltd., Oxford. With permission.)

The supply of monoclonals, on the other hand, is in principle unlimited, provided that the hybridomas are stable antibody producers. Unfortunately, this is not always the case.

Furthermore, monoclonal antibodies can be intrinsically labeled by growing the hybridomas in the presence of a radiolabeled amino acid, for instance, ^{35}S-methionine,[5] which could facilitate detection of the antigen in any blotting system.

The concentration of monoclonals in the spent medium from a conventional suspension culture of the hybridomas is only about 10 μg/mℓ. The concentration of monoclonals in ascites fluid is 5 to 20 mg/mℓ, but in the ascites fluid there are also irrelevant antibodies from the mouse present, which constitute at least 10% of the total immunoglobulins. An interesting new technique to produce large amounts of monoclonals in vitro has been developed.[15] By growing the hybridomas in dialysis tubings containing medium of low protein concentration and changing the outside medium containing the fetal calf serum (FCS) regularly, it proved possible to get high concentration of monoclonals (up to 5 mg/mℓ), which are not contaminated with irrelevant immunoglobulins or with serum proteins. The concentration of immunoglobulins in a polyclonal serum is 10 to 20 mg/mℓ and the fraction of specific antibodies in an hyperimmune serum is at most a few percent. The above factors have to be considered when the dilution factor of monoclonals in a blotting experiment is to be estimated.

PROPERTIES OF MONOCLONAL ANTIBODIES

As described earlier, the advantages in using monoclonals in blotting experiments are obvious, but some drawbacks should also be discussed.

Stability

Monoclonals may be less stable than the average polyclonal antibody. If a small fraction of the antibodies in a polyclonal antiserum is less stable under certain conditions, it will in general not be observed, but in a monoclonal antiserum, all antibodies will have the same stability properties. They might, for instance, be less stable upon storage, sensitive to freezing, lyophilization, and to a nonphysiological pH. Thus, if a high pH (see Chapter 1) is used in the incubation buffer in the immunoassay, it should first be tested that the monoclonals are stable under such conditions. In the experiment shown in Figure 1, all antisera were diluted with phosphate-buffered saline (PBS, pH 7.4), but in the experiment shown in Figure 2, the antisera (including the monoclonals) were diluted in 50 m*M* Tris and 150 m*M* NaCl (pH 10.2), which gave less unspecific staining (see Chapter 1). The monoclonals used in Figure 1 were, however, also stable at pH 10.2.

Specificity

The specificity of monoclonals might be too narrow for certain purposes. For instance, if the monoclonals are to be used to detect the presence of microorganisms, it would be an advantage if they could be used to identify all strains within a species. However, the monoclonals could well be specific to an epitope, which is not common to all strains investigated. This could be an advantage for strain identification, but certainly not for species identification. Antigenic drift, which is well established for many viral antigens,[16] could also cause problems in detection with monoclonal antibodies, if they are specific to a drifting epitope. Monoclonals can, of course, also be utilized to study this antigenic drift.[17]

Monoclonals are sometimes used for identification of gene products in recombinant DNA experiments. The gene product from such an experiment could well be modified in such a way that the antigen has lost the corresponding epitope. For instance, the protein can be truncated or not correctly refolded. Discontinuous, or conformational epitopes, which are composed of different stretches of the polypeptide chain(s),[18] should be particularly sensitive to incorrect refolding. Thus, the epitopes of an antigen are not equally stable to denaturation and some might be difficult to renature after denaturation. This problem is not so serious when working with polyclonals; the chance is rather good that there is at least one epitope left which was not denatured or which reformed correctly. There is a risk, however, that the monoclonals are specific to an epitope, which is not easily renatured after SDS-PAGE (sodium dodecyl sulfate-polyacrylamide gel electrophoresis).[19] For some proteins, only 1/5 of the monoclonal antibodies reacts with SDS-denatured protein, and for other proteins more than 4/5 of the monoclonals also recognize the corresponding epitope after SDS-treatment.[20]

Monoclonals, which work well in ELISA, performed under native conditions, do not necessarily work in blotting systems where denaturing gels are analyzed. Renaturation can be facilitated by special treatment of the sample, the gel, or the nitrocellulose membrane.[19] It is always wise to try to produce a set of hybridomas which secrete antibodies against several epitopes of the same antigen. Such a panel of monoclonal antibodies can also be used for epitope mapping (see Chapter 8.2).

Sensitivity

Only one immunoglobulin molecule will bind to each antigen molecule when monoclonals

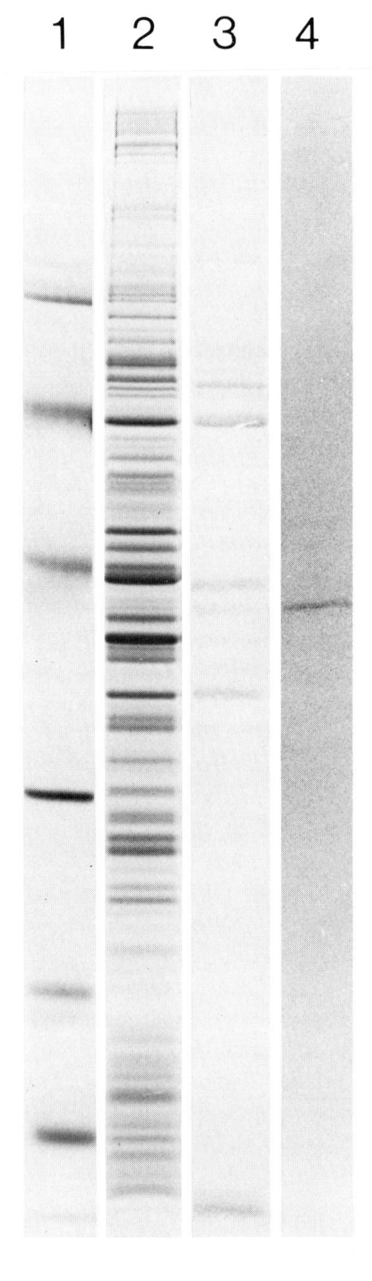

FIGURE 2. Double replica electroblotting[28] of *Mycoplasma hyorhinis* antigens. Lane 1, low M_r markers (Pharmacia, Sweden). Lane 2, Coomassie blue stained *M. hyorhinis* antigens after SDS-PAGE (Laemmli system, see Chapter 3). No. 3 and 4, immunoblots of the two nitrocellulose strips obtained from one lane in double replica electroblotting. A polyspecific (polyclonal) antiserum against whole cells of *M. hyorhinis* was used as primary antibodies with strip 3 and a monoclonal antiserum against one *M. hyorhinis* antigen was used with strip 4. (Taken from Johansson, and Bölske, manuscript in preparation.)

are used in immunoblotting. This will result in a lower sensitivity for a detection system based on monoclonals, particularly for high molecular weight proteins, which usually have many epitopes to which antibodies in a polyclonal serum can bind.

Reaction with Second Overlay

It should be kept in mind that monoclonals always constitute a particular subclass of the mouse immunoglobulins, which can be determined by immunodiffusion. SDS-PAGE can also be used, since the constituent polypeptide chains of different immunoglobulin subclasses migrate differently in such a gel.[21] If the detection system is based on a monoclonal antibody, it has to be directed against the subclass of the first overlay. The first overlay might, of course, be polyclonal, which will facilitate detection with a monoclonal second overlay, since the possibilities to make monoclonals in other species than the mouse are rather limited. If radiolabeled protein A from *Staphylococcus aureus,* which binds to the Fc part of most mammalian immunoglobulin molecules,[22] is used as second overlay, one must perform a test showing that the monoclonals are of a subclass that binds to protein A.[23] For instance, it seems as if mouse immunoglobulins of subclass IgG_1 has a low affinity for protein A.

EXPERIMENTAL APPLICATIONS

Screening for Monoclonals

Preliminary screening for secretion of relevant antibodies is most conveniently done by ELISA or dot immunobinding (see Chapter 2) techniques if the antigen is available in a reasonably pure form.[24] Antibodies from clones, which are positive in ELISA, can be further analyzed by electrophoresis and electroblotting, which is more specific and not dependent on a pure antigen (see Figure 1). SDS-PAGE in the Neville or the Laemmli systems (see Chapter 3) can be used, provided that the epitope will not be irreversibly denatured.[25] In such a case, a nondenaturing system (e.g., crossed immunoelectrophoresis or isoelectric focusing; see Chapter 8.5) must be used.

The antigen preparation can be very crude, provided it is known to which band in SDS-PAGE the monoclonals should be specific. The antigen sample should be applied over the whole width of the gel without using a slot former. The nitrocellulose membrane can then be cut in narrow strips (2 to 5 mm) without paying attention to sample lanes as shown in Figure 1. After blocking (see Chapter 6.2), which is performed prior to cutting, each strip is incubated with a small volume (1 to 3 mℓ) of supernatant culture (spent medium) diluted at least tenfold. It is useful to have a monospecific (but polyclonal) antiserum against the same antigen, which can be used as control to identify the antigen on the nitrocellulose membrane after electrophoresis and blotting. A second overlay must be used as a detection system (see Section 6.4) if the monoclonals are not labeled. In the experiment shown in Figure 1, preliminary screening was first performed on 480 clones by ELISA and 15 of them were positive. Only 1 clone of these 15 worked satisfactorily in immunoblotting on SDS gels (see strip no. 7 in Figure 1).

Identification of Antigens

Monoclonal antibodies can be used for the identification of the corresponding antigen and the experiments are in principle performed as when polyclonals are used. If monoclonals are not produced in the laboratory or cannot be obtained by a collaborative project, it is worthwhile to consult the international Hybridoma Data Bank[28] to check if the monoclonals are available elsewhere. Monoclonal antibodies have for diagnostic purposes been produced against many bacteria,[29] but have also been used in immunoblotting experiments for identification of antigens.[30]

Supernatant culture or ascites fluid can be used directly after centrifugation in blotting experiments after a dilution of 10- to 100-fold and at least 100- to 10,000-fold, respectively. Dilution must be done in a buffer, in which the monoclonals are stable for the time period necessary for incubation (2 to 24 hr). They should also be stable upon freezing and storage in the dilution buffer if the solution is to be reused.

Double Replica Electroblotting

When electrophoresis and blotting are used in the screening of hybridomas for antibody production, it is essential to produce as many strips as possible from one gel by making them as narrow as possible. Twice as many strips can be easily obtained by using the double replica electroblotting technique.[31] This is an ordinary liquid state electroblotting system, but two nitrocellulose membranes are used in the electroblotting package — one on each side of the gel. The proteins from the gel are then distributed on the two membranes by changing the polarity of the system in such a way that the time periods are continuously increasing. Another advantage is that cathodically migrating antigens are not lost during the electroblotting, but also trapped on the two membranes.[31] It also seems as if proteins are more efficiently eluted from the gel during a double replica electroblotting experiment. The double replica electroblotting system is also very useful when two antisera or the protein pattern and the blotting pattern are to compared, which is illustrated in Figure 2.

It is very convenient to use a computerized unit, which automatically changes the polarity of the system according to a predetermined schedule. The proteins will be evenly distributed on the two membranes if the electroelution time in each direction is short in the beginning of the electroblotting and increases during the course of the experiment.[31] Such an apparatus, which can be connected to any power supply, is available from PALM Sweden (Box 10066, S-750 10 Uppsala).

CONCLUDING REMARKS

A new era in immunochemistry began when Köhler and Milstein[1] in 1975 invented the technique to produce monoclonal antibodies with a predefined specificity. The blotting technique was introduced the same year by Southern[32] and O'Farrell[33] published his famous paper on 2D PAGE. The combination of these three techniques, i.e., immunoblotting of 2D gels with monoclonals, is a very powerful biochemical analytical method. The resolution capacity is up to 2000 components and the specificity of the immunological assay is the highest possible. These techniques will be used often in the future, when new monoclonals are produced and simplified techniques for 2D PAGE are developed.

ACKNOWLEDGMENTS

I thank Dr. Göran Bölske and Professor Bengt Hurvell for many valuable discussions and Carina Bohlin for typing the manuscript. Financial support was obtained from the board of the National Veterinary Institute and from the foundations of Magn. Bergvall, O. E. and Edla Johansson, and Helge Ax:son Johnson.

REFERENCES

1. **Köhler, G. and Milstein, C.,** Continuous cultures of fused cells secreting antibody of predefined specificity, *Nature (London),* 256, 495, 1975.
2. **Milstein, C.,** Monoclonal antibodies, *Sci. Am.,* 243, 56, 1980.
3. **Yelton, D. E. and Scharff, M. D.,** Monoclonal antibodies, *Am. Sci.,* 68, 510, 1980.
4. **Sevier, E. D., David, G. S., Martinis, J., Desmond, W. J., Bartholomew, R. M., and Wang, R. W.,** Monoclonal antibodies in clinical immunology, *Clin. Chem.,* 27, 1797, 1981.
5. **Hurrell, J. G. R., Ed.,** *Monoclonal Hybridoma Antibodies: Techniques and Application,* CRC Press, Boca Raton, Fla., 1982.
6. **Galfrè, G. and Milstein, C.,** Preparation of monoclonal antibodies: strategies and procedures, *Methods Enzymol.,* 73, 3, 1981.

7. **Köhler, G.,** The technique of hybridoma production, in *Immunological Methods*, Vol. 2, Lefkovits, I. and Pernis, B., Eds., Academic Press, New York, 1981, 285.
8. **Løvborg, U.,** *Monoclonal Antibodies. Production and Maintenance*, Heinemann Medical Books, London, 1982.
9. **Campbell, A. M.,** *Monoclonal Antibody Technology*, Elsevier, Amsterdam, 1984.
10. **Springer, T. A., Ed.,** *Hybridoma Technology in the Biosciences and Medicine*, Plenum Press, New York, 1985.
11. **Goding, J. W.,** *Monoclonal Antibodies: Principles and Practice. Production and Application of Monoclonal Antibodies in Cell Biology, Biochemistry and Immunology*, Academic Press, Orlando, Fla., 1985.
12. **Gershoni, J. M. and Palade, G. E.,** Protein blotting: principles and applications, *Anal. Biochem.*, 131, 1, 1983.
13. **Ogata, K., Arakawa, M., Kasahara, T., Shiori-Nakano, K., and Hiraoka, K.,** Detection of *Toxoplasma* membrane antigens transferred from SDS-polyacrylamide gel to nitrocellulose with monoclonal antibody and avidin-biotin, peroxidase anti-peroxidase and immunoperoxidase methods, *J. Immunol. Methods*, 65, 75, 1983.
14. **Mason, D. W. and Williams, A. F.,** The kinetics of antibody binding to membrane antigens in solution and at the cell surface, *Biochem. J.*, 187, 1, 1980.
15. **Sjögren-Jansson, E. and Jeansson, S.,** Large-scale production of monoclonal antibodies in dialysis tubing, *J. Immunol. Methods*, 84, 359, 1985.
16. **Norrby, E.,** Variability of antigen epitopes of monotypic viruses, in *Rapid Methods and Automation in Microbiology and Immunology*, Habermehl, K.-O., Ed., Springer-Verlag, Berlin, 1985, 83.
17. **Laver, W. G.,** The use of monoclonal antibodies to investigate antigenic drift in influenza virus, in *Monoclonal Hybridoma Antibodies: Techniques and Applications*, Hurrell, J. G. R., Ed., CRC Press, Boca Raton, Fla., 1982, 103.
18. **Van Regenmortel, M. H. V.,** Which structural features determine protein antigenicity?, *Trends Biochem. Sci.*, 11, 36, 1986.
19. **Bers, G. and Garfin, D.,** Protein and nucleic acid blotting and immunobiochemical detection, *BioTechniques*, 3, 276, 1985.
20. **Bjerrum, O. J., Selmer, J., Larsen, F., and Naaby-Hansen, S.,** Exploitation of antibodies in the study of cell membranes, in *Investigation and Exploitation of Antibody Combining Sites*, Reid, E., Cook, G. M. W., and Morre, D. J., Eds., Plenum Press, New York, 1985, 231.
21. **Köhler, G., Hengartner, H., and Schulman, M. J.,** Immunoglobulin production by lymphocyte hybridomas, *Eur. J. Immunol.*, 8, 82, 1978.
22. **Forsgren, A., Ghetie, V., Lindmark, R., and Sjöqvist, J.,** Protein A and its exploitation, in *Staphylococci and Staphylococcal Infections*, Vol. 2, *The Organism In Vivo and In Vitro*, Easmon, C. S. F. and Adlam, C., Eds., Academic Press, New York, 1983, 229.
23. **Lindmark, R., Thorén-Tolling, K., and Sjöquist, J.,** Binding of immunoglobulins to protein A and immunoglobulin levels in mammalian sera, *J. Immunol. Methods*, 62, 1, 1983.
24. **Stocker, J. W., Malavasi, F., and Trucco, M.,** Enzyme immunoassay for the detection of hybridoma products, in *Immunological Methods*, Vol. 2, Lefkovits, I. and Pernis, B., Eds., Academic Press, New York, 1981, 299.
25. **Tigyi, G., Juntti, N., and Johansson, K.-E.,** Screening for monoclonal antibodies against a mycoplasmal membrane protein by protein electroblotting, *Protides Biol. Fluids*, 33, 575, 1985.
26. **Johansson, K.-E.,** Membrane components of mycoplasmas, *Ann. Clin. Res.*, 14, 278, 1982.
27. **Johansson, K. -E.,** Characterization of the *Acholeplasma laidlawii* membrane by electroimmunochemical analysis methods, in *Electroimmunochemical Analysis of Membrane Proteins*, Bjerrum, O. J., Ed., Elsevier, Amsterdam, 1983, 321.
28. **Bussard, A., Krichevsky, M. I., and Blaine, L. D.,** An international hybridoma data bank: aims, structure, function, in *Monoclonal Antibodies Against Bacteria*, Vol. 1, Macario, A. J. L. and Conway de Macario, E., Eds., Academic Press, Orlando, Fla., 1985, 287.
29. **Macario, A. J. L. and Conway de Macario, E., Eds.,** *Monoclonal Antibodies Against Bacteria*, Vols. 1 and 2, Academic Press, Orlando, Fla., 1985.
30. **Jürs, M., Peters, H., Timmis, K. N., and Bitter-Suermann, D.,** Immunoblotting with monoclonal antibodies: a highly specific detection system for detection and identification of bacterial outer membrane proteins, in *Rapid Methods in Automation in Microbiology and Immunology*, Habermehl, K-O., Ed., Springer-Verlag, Berlin, 1985, 94.
31. **Johansson, K.-E.,** Double replica electroblotting: a method to produce two replicas from one gel, *J. Biochem. Biophys. Meth.*, 13, 197, 1986.
32. **Southern, E. M.,** Detection of specific sequences among DNA fragments separated by gel electrophoresis, *J. Mol. Biol.*, 98, 503, 1975.
33. **O'Farrell, P. H.,** High resolution two-dimensional electrophoresis of proteins, *J. Biol. Chem.*, 250, 4007, 1975.

6.4 IMMUNOVISUALIZATION WITH ANTI-ANTIBODIES:

6.4.1 IMMUNOVISUALIZATION WITH PEROXIDASE LABELED ANTI-ANTIBODIES, THE PEROXIDASE ANTIPEROXIDASE, AND BIOTIN-STREPTAVIDIN METHODS

Kenji Ogata

INTRODUCTION

Antigens transferred onto nitrocellulose filters can be visualized by various methods, including the indirect immunoperoxidase, the peroxidase antiperoxidase (PAP), and biotin-streptavidin method as described in this chapter. These are basic techniques for visualization of antigens on tissue sections by immunoenzymatic methods.[1-3]

The principle of the indirect immunoperoxidase method is shown in Figure 1A. An antigen bound to nitrocellulose is recognized by a specific antibody, which is then detected by a second horseradish peroxidase conjugated antibody.

As shown in Figure 1B, in the PAP method, an unlabeled second antibody is used for detection of the first antibody instead of the peroxidase conjugated second antibody utilized in the indirect immunoperoxidase method. The binding ability of the second antibody is also used to bind PAP to the immune complex of the primary antibody and the antigen on the nitrocellulose. The PAP method has higher sensitivity in comparison to the indirect immunoperoxidase method.[2,4]

Biotin and streptavidin have a high affinity interaction ($k_D = 10^{-15}\ M^{-1}$)[5] and this makes it possible to enhance the sensitivity of antigen detection systems. For immunoblotting, streptavidin is more useful than avidin, because streptavidin is a slightly acidic nonglycosylated protein from *Streptomyces avidinii,*[6] which gives a lower background and less non-specific binding to proteins on nitrocellulose than avidin, a basic glycosylated protein from egg white.[7] Therefore, the biotin-streptavidin method shown in Figure 1C is described in this chapter. As shown in Figure 1C, antigen is detected by rabbit primary antibody and followed by biotinylated goat antirabbit IgG antibody and peroxidase conjugated streptavidin.

Although antibodies derived from other animals can be used in antigen detection systems, to demonstrate the examples, rabbit and goat antibodies are utilized as primary and secondary antibodies, respectively. In the methods described in this chapter, nitrocellulose strips are covered with antibody solution on Parafilm. This method allows rapid manipulation and makes it possible to use a relatively small amount of antibody solution for the detection of antigens.

EQUIPMENT AND REAGENTS

Equipment

Glass plate (size should be greater than the nitrocellulose sheet); Parafilm (American Can Co., Greenwich, Conn.); Pasteur pipette; moist chamber; tray for washing nitrocellulose sheet (size should be greater than the sheet); forceps for nitrocellulose, razor blade; filter paper (Whatman No. 1).

Reagents

Phosphate buffered saline (PBS) (0.01 *M* phosphate, 0.15 *M* NaCl, pH 7.2)
 10× stock solution
  $NaH_2PO_4 \cdot H_2O$ 20.5 g

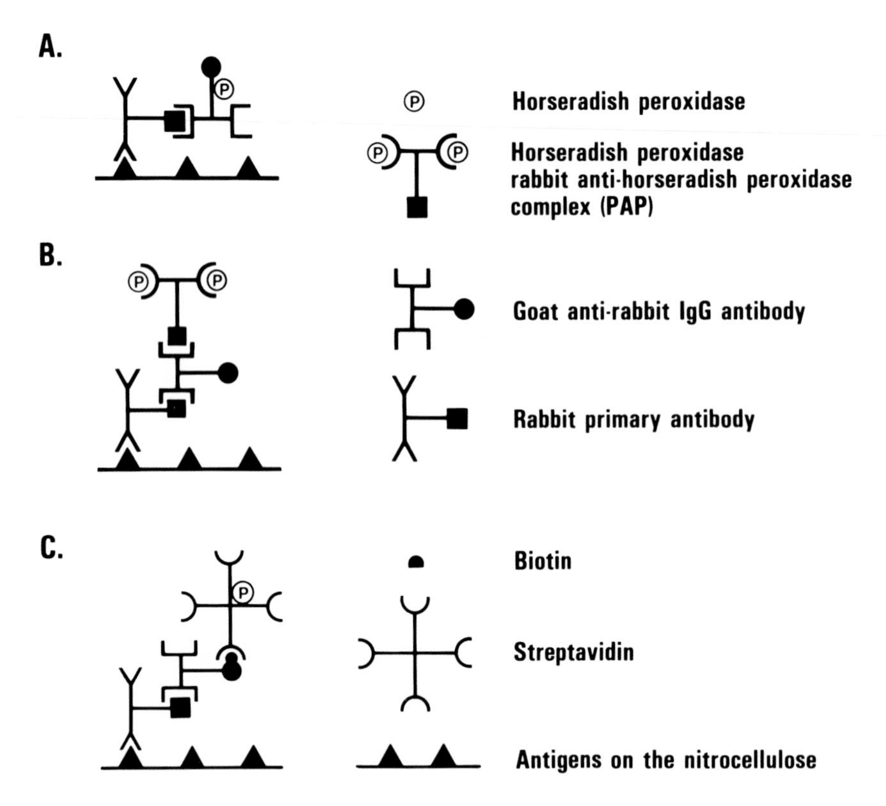

FIGURE 1. Principles of indirect immunoperoxidase (A), peroxidase antiperoxidase (PAP) (B), and biotin-streptavidin methods (C).

$Na_2HPO_4 \cdot 7H_2O$	179.9 g
NaCl	701.3 g
Thimerosal (Sigma, St. Louis)	8.0 g
Add distilled water to 8 ℓ	
Dilute stock 10 times with distilled water prior to use	
0.05% (v/v) Tween® 20 PBS: Washing buffer	
PBS	10 ℓ
Tween® 20	5 mℓ
3% (w/v) Bovine serum albumin ((BSA)-PBS): Blocking buffer	
PBS	1000 mℓ
BSA	30 g
Adjust to pH 7.2 with 0.1 *M* NaOH; make 50 mℓ aliquots and store at −20°C	
3% BSA-PBS-Tween® 20: Incubation buffer	
Washing buffer	1000 mℓ
BSA	30 g
Adjust to pH 7.2 with 0.1 *M* NaOH; make 50 mℓ aliquots and store at −20°C	
0.1% (w/v) Amido Black 10B	
Amido Black 10B	1 g
Acetic acid	70 mℓ
Distilled water	930 mℓ
7% (v/v) acetic acid	
Acetic acid	70 mℓ
Distilled water	930 mℓ
0.05 *M* Tris-HCl buffer, pH 7.4	
Tris base	6.05 g
Add to 750 mℓ distilled water; adjust to pH 7.4 with concentrated HCl; add distilled water to 1000 mℓ; make 40 mℓ aliquots and store at −20°C	
Substrate solution	
0.05 *M* Tris-HCl buffer, pH 7.4	40 mℓ

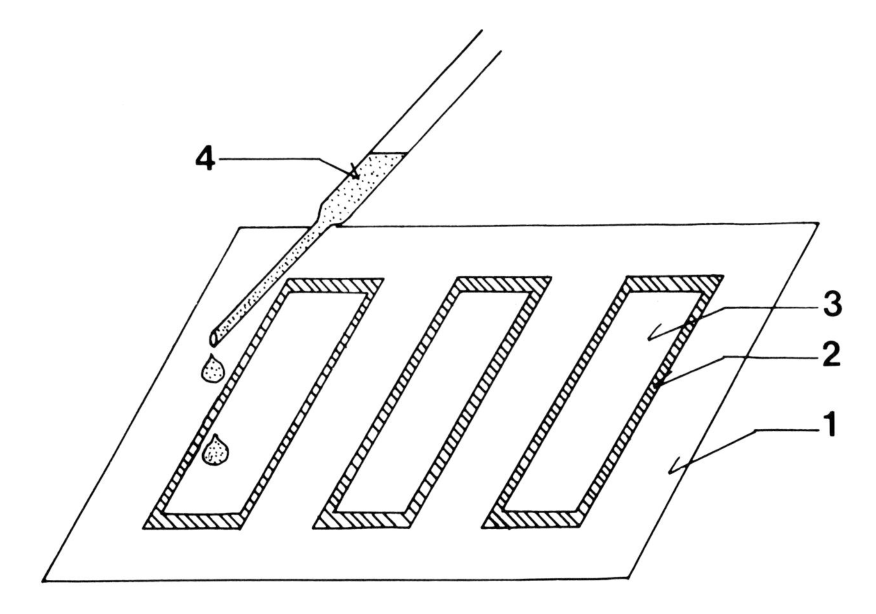

FIGURE 2. Procedure of an antibody treatment to nitrocellulose strips. (1) Glass plate, (2) Parafilm, (3) nitrocellulose strip, and (4) antibody solution.

3,3′-Diaminobenzidine tetrahydrochloride (Sigma)	10 mg
30% (w/v) H_2O_2	13 μℓ
Prepare just before use	

PROCEDURES

Preparation of Nitrocellulose for Immunological Detection of Antigens

A1. Electrophoretic transfer of antigens to a nitrocellulose sheet from sodium dodecyl sulfate (SDS)-polyacrylamide gels, see previous chapters.
A2. After transfer of antigens, wash nitrocellulose sheet with washing buffer for 1 hr at room temperature with shaking.
A3. On the clean filter paper, cut nitrocellulose (using sharp razor blade) into strips which correspond to the lanes of the SDS slab gel.
A4. Soak strips in 0.1% (v/v) Amido Black 10B for a few minutes to stain the proteins on the nitrocellulose and destain with 7% (v/v) acetic acid.
A5. Soak other strips in blocking buffer at room temperature for 1 hr with shaking to saturate remaining protein binding sites of the nitrocellulose.
A6. Wash strips 3 times with washing buffer for 10 min with shaking.

Indirect Immunoperoxidase Method

Reagents: rabbit primary antibody, peroxidase conjugated goat antirabbit IgG antibody (Tago).

B1. Place Parafilm on the glass plate, put nitrocellulose strips on Parafilm, and cover strips with rabbit primary antibody diluted in incubation buffer as shown in Figure 2 for 1 hr at room temperature in a moist chamber. (The volume of antibody solution should be approximately 100 μℓ/cm^2 of nitrocellulose membrane.
B2. Wash the strips 4 times with washing buffer for at least 30 min with shaking.
B3. Cover strips with horseradish peroxidase conjugated goat antirabbit IgG antibody diluted in incubation buffer on Parafilm for 1 hr at room temperature in a moist chamber.

B4. Repeat step 2.
B5. Soak strips in substrate solution for 5 min at room temperature to detect peroxidase activity.
B6. Wash strips with deionized water to stop the enzyme reaction.
B7. Dry strips on filter paper.

PAP Method

Reagents: rabbit primary antibody, goat antirabbit IgG antibody (Tago), and PAP complex (rabbit) (Miles Laboratories).

C1. Same as step 1, indirect immunoperoxidase method.
C2. Same as step 2, indirect immunoperoxidase method.
C3. Cover strips as above with unlabeled goat antirabbit IgG antibody diluted in incubation buffer on Parafilm for 1 hr at room temperature in moist chamber.
C4. Wash strips 4 times with washing buffer for at least 30 min with shaking.
C5. Cover strips with peroxidase antiperoxidase complex (rabbit) diluted in incubation buffer on Parafilm for 1 hr at room temperature in a moist chamber.
C6. Repeat step 4.
C7. Soak strips in substrate solution to 5 min at room temperature to detect peroxidase activity.
C8. Wash strips with deionized water to stop the enzyme reaction.
C9. Dry strips on the filter paper.

Biotin-Streptavidin Method

Reagents: rabbit primary antibody, biotinylated goat antirabbit IgG antibody (Vector Laboratory), and peroxidase conjugated streptavidin (Amersham).

D1. Same as step, indirect immunoperoxidase method.
D2. Same as step 2, indirect immunoperoxidase method.
D3. Cover strips with biotinylated goat antirabbit IgG antibody diluted in incubation buffer on Parafilm for 30 min at room temperature in a moist chamber.
D4. Repeat step 2.
D5. Cover the strips with peroxidase conjugated streptavidin diluted in washing buffer with 0.5% (w/v) BSA for 30 min at room temperature in a moist chamber.
D6. Repeat step 2.
D7. Soak strips in substrate solution for 5 min at room temperature to visualize peroxidase activity.
D8. Wash strips in deionized water to stop enzyme reaction.
D9. Dry strips on filter paper.

COMMENTS ON THE PROCEDURES

Point A1: Some antigens lose their antigenicity when boiled in SDS-sample buffer for SDS-PAGE. In such cases, it is recommended to mix antigens with SDS-sample buffer at room temperature overnight.

Point A2: After transfer of antigens onto nitrocellulose, washing the nitrocellulose with washing buffer may enhance sensitivity and reduce nonspecific binding of antibodies to proteins on the nitrocellulose. If it is hard to eliminate nonspecific reactions on nitrocellulose, 0.5% (v/v) Nonidet P40 in PBS can be used as the washing solution.

Points A5 and D5: Recently, 3 to 5% (w/v) nonfat dry milk has been used as a blocking reagent and also as a diluent for antibodies instead of BSA. This is also effective in reducing background and nonspecific binding of antibodies to proteins on the nitrocellulose. However, it has been reported that nonfat dry milk might block the binding of primary antibodies to certain antigens.[8] In addition, milk cannot be used as a diluent of peroxidase conjugated streptavidin, since it blocks the reaction of streptavidin to biotin.

Point B1: The method for covering nitrocellulose with antibody solution described here is advantageous in quick manipulation, and is particularly useful for narrow strips up to 2 to 3 cm in width. For wider nitrocellulose, incubation in a plastic bag is recommended to facilitate even distribution of antibody solution.

Point B2: Dilutions of antibodies must be determined by preliminary tests. Generally, 1:100 to 1:2000 dilutions have been used for polyclonal antibodies and 1 μg/mℓ for monoclonal antibodies.

Point B5: Since sodium azide inhibits peroxidase reactions, it cannot be used as a preservative; instead, use thimerosal.

Point C5: The optimum concentration of PAP should also be determined by preliminary testing. Usually, PAP diluted to 1:200 to 1:1000 is utilized.

Point D1: The present chapter outlines a three-step biotin-streptavidin method using rabbit primary antibody, biotinylated goat second antibody, and peroxidase conjugated streptavidin as an example of this system. However, since biotinylation[9] of antibody is relatively simple, as the following description shows, a two-step method using biotinylated primary antibody and peroxidases conjugated streptavidin may be considered as a more specific method.

Reagents: 0.1 *M* $NaHCO_3$, pH 8.3, (84 g $NaHCO_3$, add distilled water to 10 ℓ), dimethylsulfoxide (DMSO), *N*-hydroxysuccinimide biotin (Sigma).

Procedure: Dialyze antibody (at least DEAE purified grade) against 0.1 *M* $NaHCO_3$ and adjust its protein concentration to 1 mg/mℓ with 0.1 *M* $NaHCO_3$. Dissolve *N*-hydroxysuccinimide-biotin to 1 mg/mℓ in DMSO. Add 60 to 100 μℓ of this solution to 1 mℓ of antibody solution. Incubate for 4 hr at room temperature with rotating. Dialyze reaction mixture at least twice against ~1000 times the volume of PBS.

Point D5: When peroxidase conjugated avidin is used instead of peroxidase conjugated streptavidin, 3% (w/v) BSA in 0.1 *M* Tris-HCl buffer, pH 9.0, should be used as the diluent in order to prevent nonspecific binding of peroxidase conjugated avidin to proteins on the nitrocellulose.

Point D6: The optimum dilution of peroxidase conjugated streptavidin should be determined by preliminary testing. (Usually it can be used at 1:500 to 1:2000 dilution.)

EVALUATION AND EXAMPLES

The indirect immunoperoxidase method is simple but less sensitive than the other methods using, for example, the PAP, biotin-avidin, or biotin-streptavidin methods (Figure 5). Consequently, when it is unnecessary to use highly sensitive methods because sufficient amounts of certain antigens are blotted on the nitrocellulose, the indirect immunoperoxidase method is very useful.

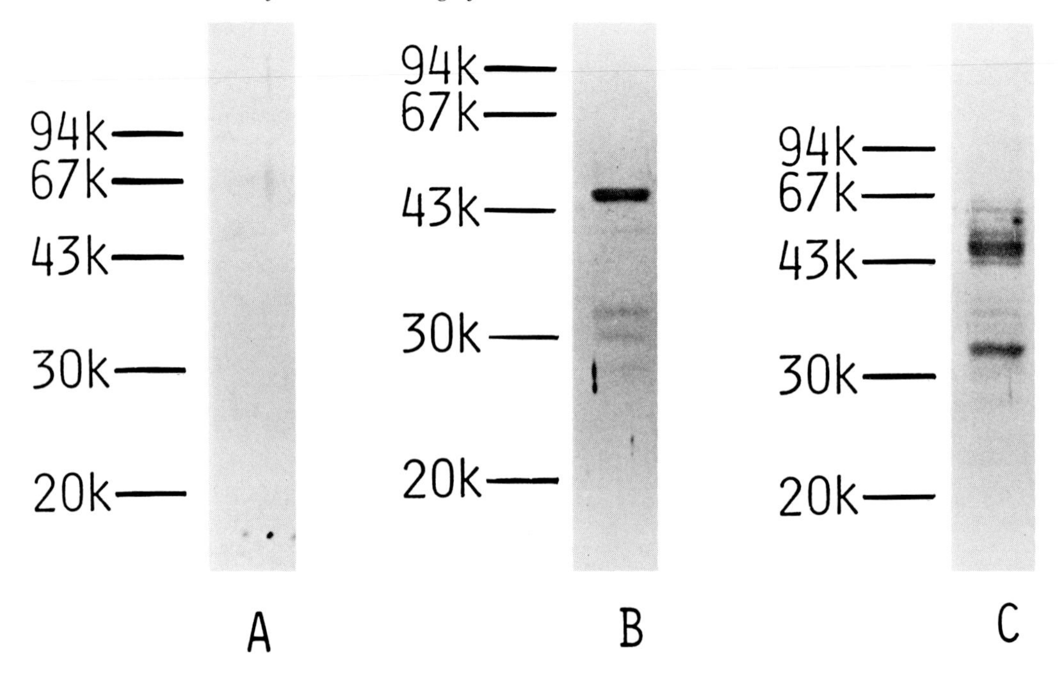

FIGURE 3. Use of PAP method to show appearance times of IgG antibody to Toxoplasma membrane antigens from rabbit infected with *Toxoplasma gondii*. An insoluble fraction of *Toxoplasma* was solubilized by 0.5% (v/v) Nonidet P40 and was used as antigen at a concentration of 90 μg/mℓ for immunoblotting. Rabbit anti-*Toxoplasma* antisera were taken 8 days (A), 18 days (B), and 28 days (C) after infection and diluted to 1:100 in incubation buffer. Goat antirabbit IgG serum and PAP were diluted to 1:400. 3′,3-Diaminobenzidine tetrahydrochloride was used as substrate. The proteins used as M_r markers were phosphorylase b (94k), BSA (67k), ovalbumin (43k), carbonic anhydrase (30k), soybean trypsin inhibitor (20k), and α-lactoalbumin (14.4k).

As PAP does not bind to other proteins, cross-reactions which occasionally have been observed between second or third antibody and antigens blotted on the nitrocellulose do not usually occur. Figure 3 shows an example of the PAP method. In this example, *Toxoplasma* membrane antigens solubilized with 0.5% (v/v) Nonidet P40 were transferred to nitrocellulose after SDS-PAGE. Rabbit sera taken 8 days (A), 18 days (B), and 28 days (C) after infection of *Toxoplasma gondii*[4] were used as a source of primary antibody. After experimental infection using *Toxoplasma* oocysts, the time of the appearance of IgG antibody against *Toxoplasma* membrane antigens could be clearly delineated by the PAP method. Figure 4 shows an example of the four-step PAP method. Antigens were the same as those in the previous example. This time, however, mouse monoclonal antibodies were used to detect *Toxoplasma* antigens. As the antigen was prepared from *Toxoplasma* grown in mouse abdominal cavity, it contained mouse immunoglobulin from mouse ascites as seen in lane D. In this case, nitrocellulose strips were treated with rabbit antimouse IgG antibody, goat antirabbit IgG antibody, and PAP, resulting in the reaction of antimouse IgG antibody with mouse immunoglobulins contained in the antigen solution.

The biotin-streptavidin method is advantageous in enhancing the sensitivity of immunological detection,[10,11] since the interaction of biotin with streptavidin is extremely strong and specific. The three-step biotin-streptavidin method described in this chapter is convenient, because a variety of biotinylated second antibodies against mouse, rabbit, goat, and human immunoglobulins are commercially available. Figure 5 shows an example of the three-step biotin-streptavidin method. In this study, 0.5% Nonidet P40 lysate of MOLT4 cells (human T cell lymphoma cell line) was used as the antigen. A cell lysate prepared from 5×10^5 cells was loaded to each lane. Proliferating cell nuclear antigen (PCNA),[12] one of the nuclear antigens recognized by autoantibodies from systemic lupus erythematosus (SLE) patients

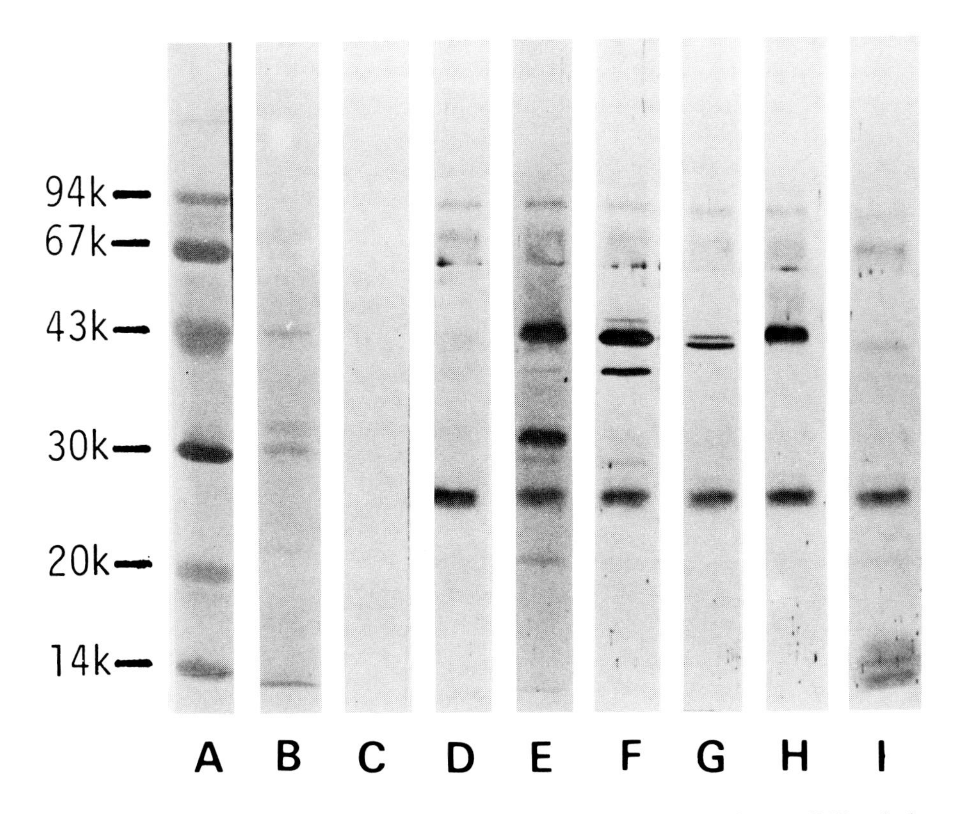

FIGURE 4. Detection of *Toxoplasma* membrane antigens on nitrocellulose strips by four-step PAP method. (A) Marker proteins stained by 0.1% (w/v) Amido Black 10B; (B) *Toxoplasma* membrane antigens stained by 0.1% (w/v) Amido Black 10B; (C) the strip was treated with goat antirabbit IgG antibody and PAP; (D) with rabbit antimouse IgG antibody, goat antirabbit IgG antibody, and PAP; (E) with mouse anti-*Toxoplasma* antiserum (1:100), rabbit antimouse IgG antibody, goat antirabbit antibody, and PAP; (F to I) with mouse monoclonal antibodies (1 μg/mℓ) against *Toxoplasma* membrane antigens, rabbit antimouse IgG antibody, goat antirabbit IgG antibody, and PAP. M_r marker and substrate were the same as those of Figure 3.

was specifically detected by human autoantibody (Figure 5, lane B), and mouse monoclonal antibodies (Figure 5, lanes D and E), but not by normal human serum (Figure 5, lane A) or normal mouse serum (Figure 5, lane C).

The two-step method mentioned in the Comments section (point D1) has the additional advantage of specific detection of antigen. Figure 6 shows an example of such a case. The antigen used in this study was the same as that in Figure 4. Mouse monoclonal antibody was used to detect 43,000 M_r antigen[13] in *Toxoplasma gondii.* A small amount of mouse immunoglobulin was contained in this antigen preparation, and although it could not be detected by the indirect immunoperoxidase method (Figure 6, lane A), it (arrow) was detected by more sensitive PAP method (Figure 6, lane B). The problem was solved by using biotinylated mouse monoclonal antibody (Figure 6, lane C) with its higher sensitivity than either of the two other methods, as demonstrated by comparing the 43,000 M_r bands obtained by each method.

ACKNOWLEDGMENT

This is publication 4388BCR from the Research Institute of Scripps Clinic, La Jolla, California.

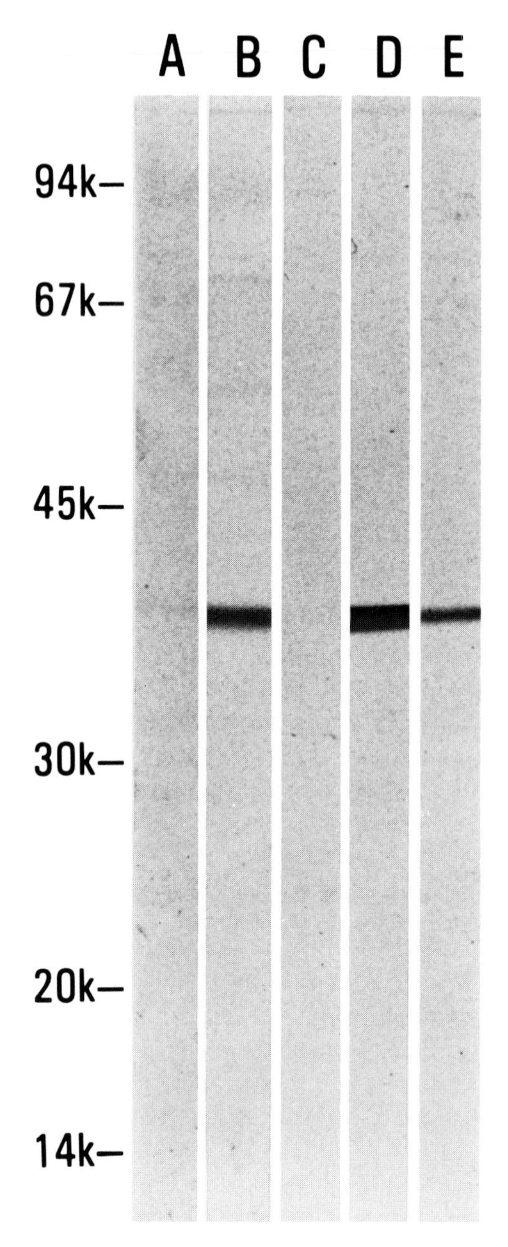

FIGURE 5. Detection of proliferating cell nuclear antigen on nitrocellulose by three-step biotin-streptavidin method. In each lane, 0.5% (w/v) Nonidet P40 lysate from 5×10^5 MOLT4 cells was loaded on SDS-polyacrylamide gel. After transferring the proteins, nitrocellulose was soaked in 3% (w/v) nonfat dry milk in PBS to saturate the remaining protein binding sites. As primary antibodies, normal human serum (1:100) (lane A), anti-PCNA autoantibody from systemic lupus erythematosus patients (10 μg/mℓ) (lane B), normal mouse serum (1:100) (lane C), mouse monoclonal anti-PCNA antibody (19A2) (1 μg/mℓ) (lane D), and mouse monoclonal anti-PCNA antibody (19F4) (1 μg/mℓ) (lane E) were applied to each nitrocellulose strip. Biotinylated goat antihuman IgG antibody (1:500) for lanes A and B, and biotinylated goat antimouse IgG antibody (1:500) for lanes C, D, and E were used to cover the strips as second antibody. As diluent of the primary and secondary antibodies, 3% (w/v) nonfat dry milk in washing buffer was used. Finally, peroxidase conjugated streptavidin (1:1000) in washing buffer with 0.5% (w/v) BSA was applied to the nitrocellulose. Substrate was the same as that of Figure 3.

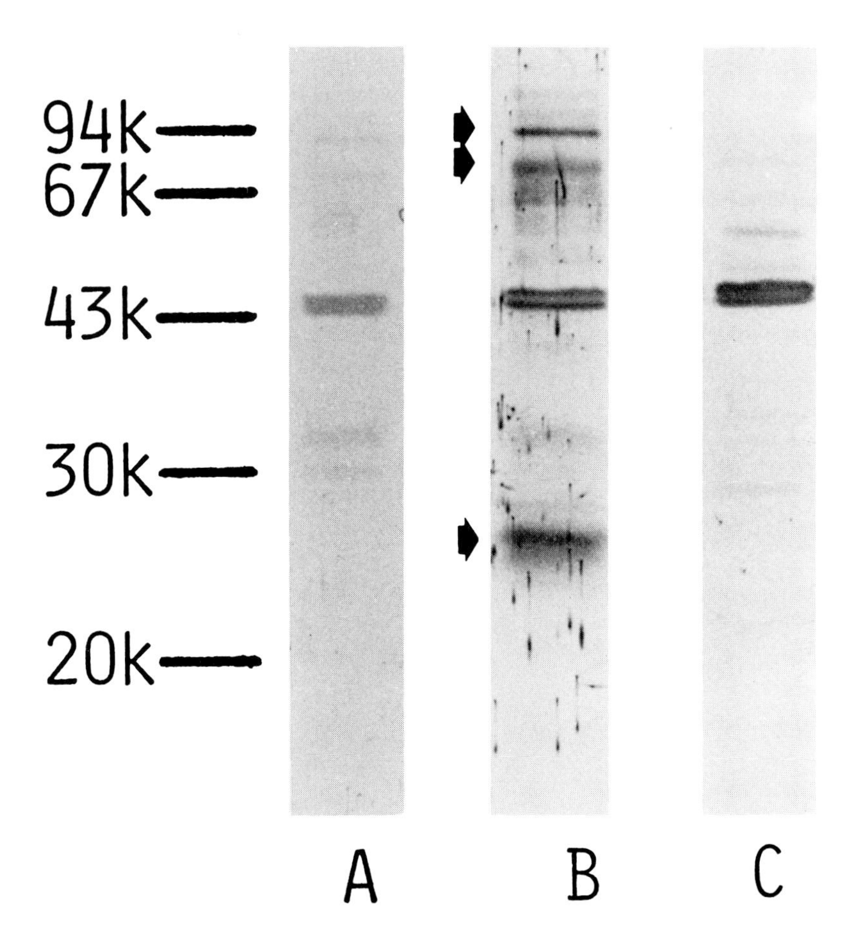

FIGURE 6. Comparison of indirect immunoperoxidase PAP and biotin-avidin methods. Antigen used in this experiment was the same as that of Figure 4. Mouse monoclonal anti-*Toxoplasma* antibody was used to detect 43,000 M_r antigen on the nitrocellulose. (Lane A) Indirect immunoperoxidase method, monoclonal antibody (1 μg/mℓ) bound to 43,000 M_r antigen was detected by peroxidase conjugated goat antimouse IgG antibody (1:400). (Lane B) PAP method, *Toxoplasma* antigen was detected by mouse monoclonal antibody (1 g/mℓ), rabbit antimouse IgG antibody (1:400), goat antirabbit IgG antibody (1:400), and PAP (rabbit) (1:400). (Lane C) Biotin-avidin method, the antigen was detected by biotinylated monoclonal antibody (1 μg/mℓ) and peroxidase conjugated avidin (1:1000). The diluent, substrate, and incubation time were the same as those mentioned in the text. Arrows indicate mouse IgG reacted with antimouse IgG antibody.

REFERENCES

1. **Graham, R. C., Jr. and Karnovsky, M. J.,** The early stages of absorption of injected horseradish peroxidase in the proximal tubules of mouse kidney: ultrastructural cytochemistry by a new technique, *J. Histochem. Cytochem.*, 14, 291, 1966.
2. **Sternberger, L. A., Hardy, P. H., Jr., Cuculis, J. J., and Meyer, H. G.,** The unlabeled antibody enzyme method of immunohistochemistry: preparation and properties of soluble antigen-antibody complex (horseradish peroxidase-antihorseradish peroxidase) and its use in identification of spirochetes, *J. Histochem. Cytochem.*, 18, 315, 1970.
3. **Hsu, S. M., Raine, L., and Fanger, H.,** A comparative study of the peroxidase-antiperoxidase method and an avidin-biotin complex mehtod for studying polypeptide hormones with radioimmunoassay antibodies, *Am. J. Clin. Pathol.*, 75, 734, 1981.

4. **Ogata, K., Arakawa, M., Kasahara, T., Shioiri-Nakano, K., and Hiraoka, K.,** Detection of toxoplasma membrane antigens transferred from SDS-polyacrylamide gel to nitrocellulose with monoclonal antibody and avidin biotin, peroxidase anti-peroxidase and immunoperoxidase methods, *J. Immunol. Methods,* 65, 75, 1983.
5. **Green, N. M. and Avidin, I.,** The use of [^{14}C]biotin for kinetic studies and for assay, *Biochem. J.,* 89, 585, 1963.
6. **Chaiet, L. and Wolf, F. J.,** The properties of streptavidin, a biotin-binding protein produced by *Streptomycetes, Arch. Biochem. Biophys.,* 106, 1, 1964.
7. **Gyorgy, P., Rose, C. S., Eakin, R. E., Snell, E. E., and Williams, R. J.,** Egg white injury as the result of nonabsorption or inactivation of biotin, *Science,* 93, 477, 1941.
8. **Spinola, S. M. and Cannon, J. G.,** Different blocking agents cause variation in the immunologic detection of proteins transferred to nitrocellulose membranes, *J. Immunol. Methods,* 81, 161, 1985.
9. **Guesdon, J. L., Ternynck, T., and Avrameas, S.,** The use of avidin-biotin interaction in immunoenzymatic techniques, *J. Histochem. Cytochem.,* 27, 1131, 1979.
10. **Lanzillo, J. J., Stevens, J., Tumas, J., and Fanburg, B. L.,** Avidin-biotin amplified immunoperoxidase staining of angiotensin-1-converting enzyme transferred to nitrocellulose after agarose isoelectric focusing, *Electrophoresis,* 4, 313, 1983.
11. **Brower, M. S., Brankel, C. L., and Garry, K.,** Immunodetection with streptavidin-acid phosphatase complex on western blots, *Anal. Biochem.,* 147, 382, 1985.
12. **Miyachi, K., Fritzler, M. J., and Tan, E. M.,** Autoantibody to a nuclear antigen in proliferating cells, *J. Immunol.,* 121, 2228, 1978.
13. **Ogata, K., Kasahara, T., Shioiri-Nakano, K., Igarashi, I., and Suzuki, M.,** Immunoenzymatic detection of three kinds of 43,000 molecular weight antigens by monoclonal antibodies in the insoluble fraction of *Toxoplasma gondii, Infect. Immunol.,* 43, 1047, 1984.

6.4.2 ALKALINE PHOSPHATASE LABELED REAGENTS

Karl-Johan Pluzek and Jakob Ramlau

INTRODUCTION

Although present in various mammalian organs and blood and in bacteria, alkaline phosphatase (AP) for conjugation purposes is most often derived from calf intestinal mucosa (M_r = 140,000) due to availability, high-specific activity, high stability, and the possibility of selectively inhibiting endogenous AP enzymes with levamisole without affecting the AP in the conjugate used, especially for immunohistochemistry.[1]

Since Engvall and colleagues in 1971 described AP-conjugates as the amplifying reagent for enzyme-linked immunosorbent assay (ELISA), using substrates which produced soluble reaction products,[2] the AP amplifying system has been applied to immunohistochemistry[3] and recently to immunoblotting.[4] These techniques employ substrates which yield insoluble reaction products, precipitating at the site of enzyme reaction. The substrates are adapted from assays detecting AP in biological systems, for example, detection of AP isoenzymes in serum after separation in gels.[5]

In immunoblotting AP conjugates have become popular due to excellent stability, high sensitivity, numerous substrates, minimal background staining, and minimal fading of most substrates. However, compared to peroxidase (see Chapter 6.4.1), conjugates based on AP are somewhat more expensive.

REAGENTS

Three AP amplifying systems are considered: antibody conjugates based on covalent linkage, complexes of AP and anti-AP, and biotin-avidin reagents based on covalent linkage between AP and avidin or on linkage of biotinylated AP with avidin.

Antibody Conjugates

Conjugates are the most widely used amplifying reagent system. Conjugates based on AP linked to the primary antibodies are not so frequently employed due to less sensitivity (see Table 3).

Conjugates based on secondary antibodies are produced by coupling AP to immunoglobulins directed against primary antibodies (e.g., mouse, rabbit, or human immunoglobulins). These antibodies should be purified to at least the immunoglobulin fraction, but affinity isolated antibodies are preferred due to less frequent nonspecific staining.

Excellent conjugates are obtained by the one-step glutaraldehyde coupling method,[2] where molecules with an M_r of approximately 1 million are produced, when coupling molecules like AP rich in primary amino groups. A more well-defined conjugate is obtained employing a heterobifunctional maleimide coupling reagent.[6] AP is treated with the reagent to introduce maleimide groups. The maleimide enzyme is then allowed to react with thiol groups of $F(ab')_2$ fragments of immunoglobulins. In this way, molecules with an M_r of approximately 200,000 are obtained.

Complexes of AP and Anti-AP Antibodies

Soluble complexes consisting of AP and monoclonal mouse anti-AP antibodies (APAAP) provide an alternative amplifying reagent, when primary mouse antibodies are applied. In a typical assay a secondary anti-mouse immunoglobulin constitutes a link between the mouse IgG from the complex and a mouse antibody directed against immobilized antigens. The linking antibody and the complex may be applied repeatedly once or twice to enhance the staining intensity (see Table 3). The use of such complexes for immunoblotting has been described by Cordell et al.[7]

Biotin-Avidin Reagents

The strong binding between biotin and avidin (or streptavidin) has been employed for various amplifying systems (see Chapter 6.4.1.)

Biotinylated secondary antibodies function as linking reagents between primary antibodies and the enzyme-avidin system. Two types of systems may be applied: an AP-conjugated avidin or a soluble complex consisting of biotinylated AP and avidin.[8]

Substrate Systems

AP is a phosphohydrolase which reacts with a phosphoester substrate at a pH optimum of about 9 to 10.[9] For immunoblotting as well as for immunohistochemistry the ideal reaction product of the reaction between enzyme and substrate is an insoluble, strongly colored, nonfading deposit.

The various substrates for AP can be grouped into the following types according to reaction mechanism:

1. Substituted naphthol phosphate ester derivatives, which after hydrolysis of phosphate form soluble, uncolored naphthol derivatives. These form insoluble azo dyes when reacting with coupling salts (diazonium compounds).[3]
2. Indoxyl phosphate esters, which after hydrolysis of phosphate form soluble, uncolored leuco-indigo derivatives, which in turn, after oxydation by tetrazolium salts, form insoluble, colored indigo derivatives. The reduced tetrazolium salts form insoluble colored products (diformazans), which precipitate on the same localization as the previously precipitated indigo stain.[10]
3. A third type, which to our knowledge has not yet been applied in immunoblotting, involves an initial dephosphorylation by AP of $NADP^+$ to NAD^+, which activates a NAD-dependent redox cycle, and this in turn generates a colored formazan dye from a tetrazolium salt.[11]

Substituted Naphthol Phosphate Esters

Some diffusion of the reaction product may take place, since the naphthol phosphate ester derivatives form a soluble product before the reaction with diazonium salts. Thus, the character of the coupling reaction, which differs for the various diazonium salts, is of importance for the diffusions. These characteristics are also responsible for the fact that with some of these substrates color appears on both sides of the nitrocellulose membrane.

Indoxyl Phosphate Esters

The reaction products of indoxyl phosphate ester substrates give rise to sharp spots or bands due to fast precipitation. Thus, the color appears only on one side of the blotting membrane. The reaction has been shown to be quantitative over a wide concentration range of proteins.[10]

In Table 1 examples of commonly used substrates and coupling salts are listed. It should be noted that some of the reagents can be considered toxic and should be handled thereafter. The developed stains are very stable and do not fade, if they are kept out of daylight.

Alkaline phosphatase is a metalloenzyme, which contains zinc and magnesium. Magnesium chloride or magnesium sulfate are often added to substrates used for immunoblotting to enhance the enzyme activity.[10,12]

Table 1
EXAMPLES OF SUBSTRATES, COUPLING SALTS, AND TETRAZOLIUM SALTS USED FOR THE REACTION WITH ALKALINE PHOSPHATASE

	Chemicals		Company
Substrate-phosphate derivative	Naphthol AS-MX phosphate (free acid)	3-Hydroxy-2-naphthoic acid 2,4-Dimethylanilide	Sigma N 4875
	Naphthol AS-MX phosphate (sodium salt)	3-Hydroxy-2-naphthoic acid 2,4-Dimethylanilide	Sigma N 5000
	Naphthol AS-GR phosphate (sodium salt)	3-Hydroxy-2-anthranoic acid 2-Methylanilide	Sigma N 3625
	Naphthol AS-BI phosphate (sodium salt)	6-bromo-2-hydroxy-3-naphthoic acid 2-Methoxyanilide	Sigma N 2250
	Naphthol AS-TR phosphate (sodium salt)	3-Hydroxy-2-naphthoic acid 4′-Chloro-2′-methylanilide	Sigma N 6125
	1-Naphthyl phosphate (sodium salt)		Merck 6815
	5-Bromo-4-chloro-3-indolyl phosphate (*p*-toluidine salt)		Sigma B 8503
Coupling salt	Fast Red TR salt	5-Chloro-2-methyl-benzenediazonium chloride hemi-zinc chloride	Sigma F 1500
	Fast Red ITR salt	*N*′,*N*′-Diethyl-4-methoxymethanilamide-diazonium salt	Sigma F 1375
	Fast Red Violet LB salt	5-Chloro-4-benzamido-2-methyl-benzene-diazonium chloride, hemi-zinc chloride salt	Sigma F 3381
	Fast Violet B salt	6-Benzamido-4-methoxy-*m*-toluidine, diazonium chloride	Sigma F 1631
	Fast Garnet GBC salt	4′-Amino-2,3′-dimethylazobenzene diazonium salt	Sigma F 6504
	Fast Blue BB salt	Diazotized 4′-amino-2′,5′-diethoxybenzanilide, zinc chloride salt	Sigma F 0250
	Fast Blue RR salt	Diazotized 4-benzoylamino-2,5-dimethoxyaniline, zinc chloride salt	Sigma F 0500
	Variamine Blue B salt	*N*-*p*-Methoxy-phenyl:*p*-phenylenediamine, diazonium salt	Merck 8507
	New Fuchsin		Merck 4041
Tetrazolium salt	Nitro blue tetrazolium	2,2′-di-*p*-Nitrophenyl-5,5′-diphenyl-3,3′[3,3′-dimethoxy-4,4′-diphenylene] ditetrazolium chloride	Sigma N 6876
	Tetranitro blue Tetrazolium	2,2′,5,5′-Tetra-*p*-nitrophenyl-5,5′-diphenyl-3,3′[3,3′-dimethoxy-4,4′-diphenylene] ditetrazolium chloride	Sigma T 4000
	Tetrazolium red	2,3,5-Triphenyltetrazolium chloride	Sigma T 8877
	Tetrazolium violet	2,5-Diphenyl-3-α-naphthyl-tetrazolium chloride	Sigma T 0138
	Tetrazolium blue chloride	3,3′-3,3′-Dimethoxy-(1,1′-biphenyl)-4,4′-diyl-bis 2,5-diphenyl-2H-tetrazolium dichloride	Sigma T 4375

EVALUATION

Comparison of Various Substrate Systems

We have compared several substrate and buffer systems employing a modified dot immunobinding assay.

Dot Immunobinding Assay[13]

The following standard conditions were employed: all incubations were performed at room temperature on a rocking table (IKA-VIBRAX-VXR, Janke & Kunkel, Staufen I, Br., West Germany. Nitrocellulose (Schleicher & Schuell, BA 85 0.45 μm) was cut in size to cover the 96 wells of a microtiter plate (immuno plate I, code no. 2-39454 NUNC, Denmark). The membrane was incubated overnight in 20 μg/mℓ of human IgG (Code No. 737-043 AB KABI, Sweden), 50 m*M* Tris-HCl, and 100 m*M* NaCl, pH 7.2 (Tris-buffered saline or TBS). Blocking was performed with 0.5% (v/v) Tween® 20 in TBS for 30 min. After washing (5 min) in 0.1% (v/v) Tween® 20 in TBS (washing buffer), the membrane was incubated for 2 hr with rabbit antihuman IgG (Dakopatts) diluted 1:8000 in washing buffer. After three further washings (5 min), the membrane was incubated for 2 hr with AP-conjugated swine antirabbit immunoglobulins (Dakopatts) diluted 1:1000 in washing buffer. After a final washing cycle, excess liquid was shaken off and the membrane placed on the microtiter plate covering the wells. Prior to this, substrate systems with varying compositions had been pipetted into the wells. The membrane was covered with two layers of Parafilm M (American Can Company, Greenwich, Conn.) and a piece of foam rubber (2 mm). Finally, another (empty) microtiter plate was placed upside down at the top, so that the three small holes at the edge of both microtiter plates were opposite each other. Three wooden pegs penetrated each set of oppositely placed holes to keep the two plates in a fixed position. The plates were forced together with rubber bands to keep them tight. The plates were then inverted, and the upper plate was tapped with the fingers to collect all the liquid on the membrane. Incubation with substrate was carried out for 20 min. During this incubation a heavy load (1 kg) was placed on the plates to secure close contact between the two plates. The plates were again inverted, dismantled, and the membrane was washed in distilled water and dried by pressing between filter papers.

Results

The goal has been to compare substrate systems under identical conditions. Therefore, we have not attempted to optimize the reaction conditions for all the individual combinations. The variety of developed colors in dot immunobinding obtained by combining a number of substrates and coupling salts are shown in Plates 1 and 2. The colors vary due to both coupling salts and substrates (see color insert, Plate 1). None of the substrates can be selected to be the most suitable, because they are dependent on the coupling salt. Of the coupling salts, Fast Violet B, Fast Garnet GBC, Fast Blue BB or RR, and Variamine Blue B gave the most intense colors.

We have only studied one indoxyl phosphate ester, 5-bromo-4-chloro-3-indolyl phosphate (BCIP). In order to find the most suitable tetrazolium salt to be combined with BCIP, we compared five different salts as shown in Plate 2. The best combinations were BCIP combined with Nitro blue tetrazolium salt (NBT) or tetranitro blue tetrazolium salt (TNBT).[8,10,14] Other combinations developed faint blue colors. Later experiments (not shown) indicated that NBT and TNBT are of equal sensitivity, which is approximately four times better than the three other tetrazolium salts.

Naphthol phosphate derivatives are available as disodium salts or as free acids. In experiments (not shown) the disodium salt or the free acid of naphthol AS-MX phosphate together with Fast Red TR or Fast Blue BB or RR have been applied using 0.1 *M* Tris-HCl at pH 7.0 to 10.0. Tris can be used in the whole range, because very low buffer capacity is needed for the reaction. No significant difference in staining intensities was observed in the pH range.

In negative control experiments, where the AP-conjugate was omitted, Fast Red TR did not show any nonspecific staining. Fast Blue BB or RR gave a faint yellow-brown color; the higher the pH, the stronger the color. This held for both naphthol forms. The effect may be due to decomposition of coupling salt under alkaline conditions.[14] The background staining can be lowered by decreasing the concentration of the coupling salt (see Appendix).

Various buffers have been reported to be suitable for the enzyme reaction, barbital, Tris, propandiol, carbonate, and borate buffer at a pH range of 7.5 to 10.0.[4,10,12,15] Phosphate buffers must be avoided as they inhibit the enzymatic reactions. In the experiments shown in Plates 1 and 2, 0.1 *M* Tris-HCl, pH 9.0 was used. In other analogous experiments (not shown), identical colors and intensities were obtained with 0.1 *M* ethanolamine, pH 9.0. The effect of changing pH or buffer was further investigated by applying the combinations naphthol AS-MX phosphate with Fast Red TR or hexazotized New Fuchsin and BCIP with NBT at pH 7.0 to 10.0 using the following buffers, all containing 1 m*M* $MgCl_2$: 0.1 *M* Tris-HCl (pK_a 8.0), 0.1 *M* propandiol (2-amino-2-methyl-1,3-propandiol) (pK_a 8.8), 0.1 *M* ethanolamine (pK_a 9.5). No obvious pH optimum was observed. pH 7.0 gave the least intense staining, though not always obvious. Tris and propandiol buffers gave the most reproducible and intense staining. Ethanolamine buffer was unstable below pH 9.5 on storage. All three buffers gave identical colors for a given combination of substrates except for naphthol AS-MX phosphate with hexazotized New Fuchsin in ethanolamine buffer, where a violet instead of red color was obtained.

In conclusion, all three buffers may be used with good results at the pH range 7.5 to 9.5. For ethanolamine, pH should be adjusted prior to use. Although AP have a pH optimum at 9 to 10, a reaction time of 20 min in the employed pH range should be sufficient.

Detection Limits for Immunoblotting with Selected Substrate Combinations

On the basis of the above-mentioned results concerning color intensities, we selected combinations of substrates and coupling salts, which could be suitable for immunoblotting.

Experimental Set-Up

Three samples of mouse monoclonal antibodies (different in purity, pI, and concentrations) from tissue culture supernatants were separated by agarose gel electrophoresis and bound to nitrocellulose by capillary blotting (see Chapter 4.1). The membrane was placed on the surface of the agarose, filter papers were added, and then a pressure was applied on the gel. The blocking and incubations followed the descriptions in the Evaluation section. After blocking, the membrane was incubated overnight with rabbit antimouse immunoglobulins (Dakopatts) diluted 1:500 in washing buffer. After washing three times for 5 min, the membrane was incubated for 2 hr with a soluble complex of AP and mouse anti-AP (Dakopatts) diluted 1:500 in washing buffer. After washing, the selected substrate-coupling salt combinations in various buffers were applied for 10 min, and the membrane was washed in distilled water and dried by pressing between filter papers.

In order to compare the sensitivites under the various conditions, we applied a dot immunobinding assay simultaneously (see Chapter 2) as follows: 3 μℓ of a series of dilutions of purified polyclonal mouse IgG were pipetted onto a nitrocellulose membrane. Successive dilutions in TBS 1 + 1 from 100 μg/mℓ to 0.05 μg/mℓ were applied, which equals 300 to 0.15 ng protein. The membrane was left to air dry for 30 min. The blocking steps and subsequent incubations were identical to those described earlier.

Results

In Table 2, the applied substrate combinations are listed together with an evaluation of colors and detection limits. With optimization of incubation times, concentrations, and antibody amplification systems, the sensitivity can be increased. In color Plate 3, examples of the immunoblotting experiments are shown.

Table 2
DETECTION LIMITS FOR VARIOUS SUBSTRATE COMBINATIONS IN DOT IMMUNOBINDING ASSAY

Substrate phosphate derivative	Coupling salt	Employed buffer systems[a]	Color of stain	Detection limit (pg/mm²)
5-Bromo-4-chloro-3-indolyl	Nitro blue tetrazolium	1	Bluish, gray	40
	(oxidizing agent)	2	Bluish, gray	40
		3	Bluish, gray	40
Naphthol AS-MX (free acid)	Hexazotized New	4	Red	80
	Fuchsin	5	Red (violet background)	80
		6	Red	160
Naphthol AS-MX (sodium salt)	Fast Red TR	7	Red	320
Naphthol AS-MX (free acid)	Fast Red TR	7	Red	320
Naphthol AS-MX (free acid)	Fast Blue RR	7	Blue	40
Naphthol AS-MX (free acid)	Variamine Blue B	4	Blue	80
Naphthol AS-GR (free acid)	Variamine Blue B	4	Green	80
1-Naphthyl (sodium salt)	Variamine Blue B	4	Brown	160
1-Naphthyl (sodium salt)	Fast Garnet GBC	4	Gray	160
1-Naphthyl (sodium salt)	Fast Violet B	4	Brown	160

Note: Detection limits (pg protein/mm² in dots of 8 mm²) determined by dot immunobinding with immobilized mouse IgG followed by rabbit antimouse immunoglobulins and soluble complexes of AP and mouse anti-AP. Substrate incubation time, 10 min. Concentrations of substrates and coupling salts were as described in Plates 1 and 2.

[a] (1) 0.2 *M* Tris-HCl pH 9.5, (2) 0.1 *M* ethanolamine pH 9.5, (3) 0.1 *M* propandiol pH 9.5, (4) 0.2 *M* Tris-HCl pH 9.0, (5) 0.1 *M* ethanolamine pH 9.0, (6) 0.1 *M* propandiol pH 9.0, (7) 0.2 *M* Tris-HCl pH 8.0.

Differences in color intensities and detection limits were minimal between free acid and disodium salt of naphthol AS-MX phosphate with Fast Red TR as a coupling salt. Immediately after substrate incubation the free acid form gave a yellowish background staining. Only a weak background staining remained after drying, which did not disturb the evaluation. When Variamine Blue B was applied, a brown background staining appeared during substrate incubation, but it partly disappeared upon drying (see color Plate 3, e and f). Analogously, Fast Garnet GBC and Fast Violet B immediately after termination of substrate incubation gave a greenish and dark yellowish background, respectively. The green color disappeared, while the yellow color, originating from Fast Violet B, remained, although more pale. In general, the background staining can be lowered by decreasing the concentration of coupling salt and using a lower pH. Simultaneously, an improvement of sensitivity has been observed (see color Plate 3, b and c and Appendix). The three buffers employed Tris-HCl, ethanolamine, and propandiol gave almost identical staining.

The combination of 5-bromo-4-chloro-3-indolyl phosphate (BCIP) and Nitro blue tetrazolium (NBT) turned out to give the highest sensitivity (40 pg protein per square millimeter) combined with sharp bands and with low background staining (see Plate 3a and Table 2). The suitability of this combination has previously been reported.[8,10,14] Naphthol AS-MX phosphate together with Fast Blue RR exposed an equally low detection limit,[14] and also with hexazotized New Fuchsin, this substrate showed relatively low detection limits (80 pg protein per square millimeter).[7,14,15] However, for preparation of hexazotized New Fuchsin, the rather hazardous chemicals sodium nitrite and 2 *M* hydrochloric acid are used (see Appendix).

If very low detection limits are not required, naphthol AS-MX phosphate, naphthol AS-GR phosphate, or 1-naphthyl phosphate together with the coupling salt Variamine Blue B can be employed. 1-Naphthyl phosphate together with Fast Garnet GBC is also a good

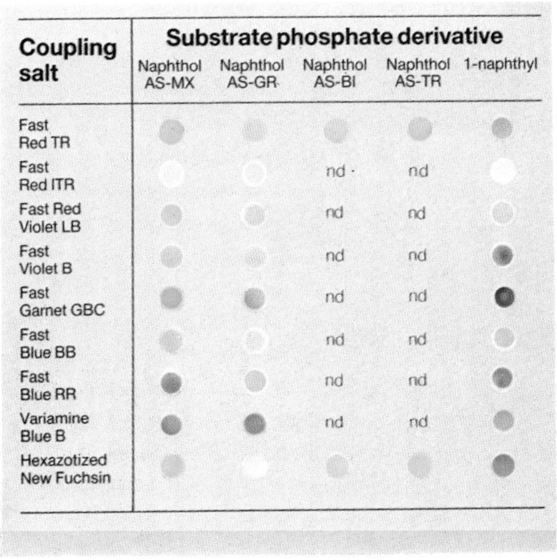

Coupling salt	Substrate phosphate derivative				
	Naphthol AS-MX	Naphthol AS-GR	Naphthol AS-BI	Naphthol AS-TR	1-naphthyl
Fast Red TR					
Fast Red ITR			nd	nd	
Fast Red Violet LB			nd	nd	
Fast Violet B			nd	nd	
Fast Garnet GBC			nd	nd	
Fast Blue BB			nd	nd	
Fast Blue RR			nd	nd	
Variamine Blue B			nd	nd	
Hexazotized New Fuchsin					

Plate 1. Comparison in *dot immunobinding* of combinations of naphthol phosphate ester substrates and coupling salts. The dots have been cut out and combined. The concentrations of naphthol phosphate derivatives and coupling salts were 0.2 to 0.4 and 1 mg/mℓ, respectively, in 0.1 *M* Tris-HCl, pH 9.0, with 1 m*M* $MgCl_2$. When hexazotized New Fuchsin was employed as the coupling salt, the concentrations were 0.6 and 0.16 mg/mℓ, respectively. Incubation (20 min) was applied for all combinations. The antigen-antibody system was as follows: human IgG — rabbit anti-human IgG — AP-conjugated swine anti-rabbit immunoglobulins. For details of names and supplier of substrates and coupling salts, refer to Table 1. "nd" indicates not determined.

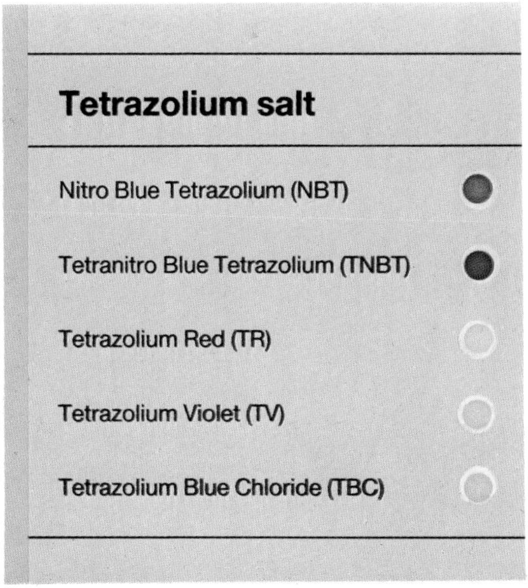

Tetrazolium salt	
Nitro Blue Tetrazolium (NBT)	
Tetranitro Blue Tetrazolium (TNBT)	
Tetrazolium Red (TR)	
Tetrazolium Violet (TV)	
Tetrazolium Blue Chloride (TBC)	

Plate 2. Comparison in *dot immunobinding* of combinations of the indoxyl phosphate ester substrate, 5-bromo-4-chloro-3-indolyl phosphate (BCIP), with various tetrazolium salts. The concentrations of the indoxyl phosphate ester and the tetrazolium salts were kept at 0.06 and 0.1 mg/mℓ, respectively. The experimental conditions and other designations were as for Plate 1.

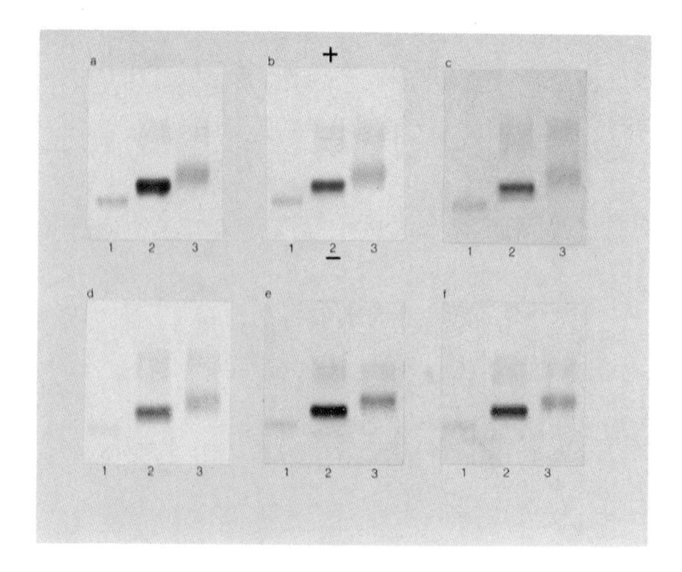

Plate 3. Immunoblotting of three mouse monoclonal antibodies (1 to 3) from tissue culture supernatants separated by agarose gel electrophoresis. Sample volume 4 $\mu\ell$. Immobilized proteins were detected by rabbit anti-mouse IgG and soluble complexes of AP and mouse anti-AP. Various substrate combinations were applied for 10 min. Concentrations of substrates and coupling salts as described in Plates 1 and 2, except for combinations b, where a concentration of 0.5 mg/mℓ for both reagents was used. (a) 5-Bromo-4-chloro-3-indolyl phosphate and Nitro Blue Tetrazolium salt in 0.1 *M* propandiol, pH 9.5, (b) naphthol AS-MX phosphate and Fast Blue RR in 0.2 *M* Tris-HCl, pH 8.0, (c) naphthol AS-MX phosphate and Fast Blue RR in 0.2 *M* Tris-HCl, pH 9.5, (d) naphthol AS-MX phosphate and hexazotized New Fuchsin in 0.1 *M* propandiol, pH 9.0, (e) naphthol AS-GR phosphate and Variamine Blue B in 0.2 *M* Tris-HCl, pH 9.0, (f) 1-naphthyl phosphate and Variamine Blue B in 0.2 *M* Tris-HCl, pH 9.0.

combination. Naphthol AS-MX phosphate together with Fast Red TR gave low background staining and have been used by many laboratories with good results.[4,16,17] However, we do not find this combination suitable for immunoblotting, due to low sensitivity (320 pg/mm^2). Many of the substrate combinations have been tested in a dot immunobinding technique by De Jong et al.,[14] who found essentially similar relative sensitivites as given here.

Various combinations of naphthol phosphate derivatives and coupling salts in various buffer systems have been systematically compared for use in immunohistochemistry.[3,14,15,18-21] The results regarding buffer systems and intensities of stains are consistent with our observations. However, due to different requirements for substrates used in immunohistochemistry (e.g., contrast between colors of substrate and counterstaining substance, size and homogeneity of precipitates visualized by light microscopy, and nonspecific binding to tissues) some coupling salts found to be less suitable for immunoblotting may be adapted to immunohistochemistry (e.g., Fast Red TR).

Comparison of Antibody Amplification Systems

Various amplification systems were compared using essentially the same dot immunobinding assay as described above. After the application of mouse IgG, sequential incubations of different primary and secondary antibodies/conjugates with intermediate washing steps were performed as outlined in Table 3. We have selected BCIP together with NBT for substrate combinations in these experiments.

Results

The lowest detection limit (less than 20 pg protein per square millimeter) was obtained with an indirect technique (IV), where AP-conjugated swine antirabbit immunoglobulins are linked via rabbit antimouse immunoglobulins to mouse IgG on the nitrocellulose membrane. An AP-conjugated avidin system (VI) also exhibited low detection limits.

We applied another biotin-avidin system linked to the mouse IgG via a biotinylated rabbit antimouse immunoglobulin (V). Avidin was mixed with biotinylated AP in optimal proportions (assessed by an ELISA technique) and applied in a suitable dilution.[8] The sensitivity of the biotin-avidin complex system (V) was slightly less compared to the AP-avidin system (VI). However, in our set-up, the former method gave the lowest background staining of the two. As little as 0.5 pg of target DNA was detected using an optimized AP-avidin system analogous to VI by a hybridization reaction on nitrocellulose.[22]

The unlabeled bridge technique (I), where the complex AP-mouse anti-AP (APAAP) is linked to mouse IgG via rabbit antimouse immunoglobulins, also showed a low detection limit and low background staining. An equal improvement was obtained by repeating the incubations of the two employed reagents (II) or increasing substrate incubation time from 20 to 40 min (10 min incubation time for this modification did not give an acceptable sensitivity). In general, extension of substrate incubation time improved the sensitivity,[10] but often with a simultaneous increase in background staining. For very long incubation times (hours) the color development should proceed protected from light.[8] Substitution of rabbit antimouse immunoglobulins with the analogous AP-conjugated product (III) did not change the sensitivity in spite of an assumed increased number of AP molecules bound at the same localization. This observation could be explained by the assumption that once the stain has covered the enzyme, the reaction stops, irrespective of the number of reactive AP molecules present beneath.

Direct techniques were also evaluated where AP, conjugated to the $F(ab')_2$ fragment of rabbit antimouse immunoglobulins (VII) or to the whole antibody (VIII), was bound to the mouse IgG. Although the AP-$F(ab')_2$ fragment was used in a much higher concentration than the AP conjugated whole antibody, the latter gave the highest sensitivity. This may be due to the presence of more AP molecules on an extended area, when the conjugated whole

Table 3
COMPARISON OF ALKALINE PHOSPHATASE (AP) AMPLIFICATION SYSTEMS USING A DOT IMMUNOBINDING ASSAY

Incubation time (on rocking table)	Amplification system							
	I Unlabeled bridge	II Unlabeled bridge repeated steps	III Labeled bridge	IV Indirect AP-antibody	V Biotin-AP avidin complex	VI AP-avidin	VII Direct AP-Fab′ antibody	VIII Direct AP-antibody
	Mouse IgG 0.15—100 ng in 3 μℓ							
Overnight	RbaMoIg[a]	RbaMoIg[a]	AP-RbaMoIg[b]	RbaMoIg[a]	B-RbaMoIg[c]	B-RbaMoIg[c]	AP-Rb(Fab′)aMoIg[d]	AP-RbaMoIg[b]
2 hr	APAAP[e]	APAAP[e]	APAAP[e]	AP-SwaRbIg[f]	B-AP-avidin complex[g]	AP-avidin[h]		
20 min		RbaMoIg[a]						
20 min		APAAP[e]						
20 min					Substrate BCIP + NBT[i]			
Detection limit (pg/mm²)[j]	40	20	40	≤20	20	≤20	150	80

[a] Rabbit antimouse immunoglobulins (Dakopatts Z 259 lot 104) 1:500.
[b] AP-conjugated affinity isolated rabbit antimouse immunoglobulins (Dakopatts D 314 lot 065) 1:500.
[c] Biotinylated affinity isolated rabbit antimouse immunoglobulins (Dakopatts E 354 lot 115) 1:500.
[d] AP-conjugated Fab′ fragments of rabbit antimouse immunoglobulins (prepared in the author's laboratory).
[e] AP-anti-AP complex (Dakopatts D 651 lot 0384) 1:500.
[f] AP-conjugated affinity isolated swine antirabbit immunoglobulins (Dakopatts D 306 lot 055) 1:500.
[g] Biotinylated AP-avidin complex (prepared in the author's laboratory).
[h] AP-conjugated avidin (prepared in the author's laboratory).
[i] 5-Bromo-4-chloro-3-indolyl phosphate 0.06 mg/mℓ, nitro blue tetrazolium salt 0.1 mg/mℓ in 0.2 *M* Tris-HCl, pH 9.5, with 3.5 m*M* $MgCl_2$. For comparison, 10 and 40 min incubation times were also used for some of the modifications.
[j] The stated values have been obtained from experiments performed on the same day. They indicate picogram protein/mm² in dots of 8 mm.[2]

antibody is applied (due to a much higher molecular weight). These methods gave the lowest background staining of the examined modifications, but also the lowest sensitivity.

Addition of Magnesium Ions

It has been shown that magnesium ions enhance the enzyme activity of alkaline phosphatase from calf intestinal mucosa in biochemical reactions.[9] Normally, our substrate buffers contained 1 to 3.5 m*M* $MgCl_2$, adapted from ELISA substrate buffer. The effect of magnesium on the AP reactivity in immunoblotting was evaluated by omitting it from systems II and IV. We observed no difference regarding sensitivity or intensity. In another assay identical to system I (not shown) we used 0 to 15 m*M* of $MgCl_2$ in 0.2 *M* Tris-HCl buffer, pH 9.5. By incubation with the substrate combinations BCIP + NBT, BCIP + TNBT, or naphthol AS-MX phosphate + Fast Blue RR for 10 min, we did not observe any significant effect upon addition of $MgCl_2$. De Jong et al.[14] reported that maximum AP activity is obtained at a $MgCl_2$ concentration of 10 m*M* in dot immunobinding assay. The conflicting results may be explained by a difference between AP conjugates regarding the degree of saturation with zinc and magnesium ions during storage before preparation of working dilution.[23,24] Another explanation may be that the substrate incubation time chosen by us gives the enzyme, in spite of lower reactivity, sufficient time to develop the maximal amount of reaction product.

CONCLUSION

The aim of this chapter has been to investigate the use of alkaline phosphatase (AP)-labeled reagents in immunoblotting procedures.

Among the various immunochemical detection systems tested (see Table 3), the indirect technique (IV) (antigen/primary antibody/secondary, AP-conjugated antibody) was the most sensitive (less that 20 pg protein per square millimeter dot detected). If the AP-labeled primary antibody is available, the simpler and faster direct technique (VIII) can be performed with a small decrease in sensitivity (approximately four times). Methods using biotinylated antibodies and AP-avidin reagents (V, VI) were also very sensitive, but did not exceed the indirect technique with regard to sensitivity and simplicity.

Although some of the methods could be selected to be the most suitable, improvements for other methods might be obtained by varying reagent concentrations, incubation times, or temperature. The quality of the antibodies/conjugates also has a great impact on the sensitivites.

With regard to the staining reaction, a few recipes have been chosen after optimization from the multitude of substrates and coupling salts. The 5-bromo-4-chloro-3-indolyl phosphate/Nitro blue tetrazolium stain is very sensitive and gives sharp, well-delineated bands/spots. However, the bluish-gray color is not bright and aesthetically pleasant, while the naphthol phosphate/Fast Blue RR system yields a clear blue reaction product and is practically as sensitive. Both stains are very simple to use, while the best red stain, naphthol phosphate/hexazotized New Fuchsin, requires diazotization of the fuchsin reagent immediately before use with attendant inconvenience.

Though alkaline phosphatase has a basic pH-optimum (9 to 10), it works well enough in the pH-range 7.5 to 9.5. We recommend a Tris buffer, pH 9.5, but other buffers may be used as well. For some coupling salts (e.g., Fast Blue RR) a lower pH, 8.0, is recommended, due to higher background staining at higher pH. Magnesium ions should be added to the buffer at 4 m*M*, although there is reason to believe that the enzyme in the conjugate is nearly saturated with Mg^{2+}. The addition is a simple safeguard.

Thus, we find that in applying AP conjugates in the classical indirect technique, only a small panel of secondary antibodies is required, and by chosing appropriate staining reagents, sensitivities below 100 pg/mm^2 dot and practically no background staining is obtained.

APPENDIX
Recipes for Some Alkaline Phosphatase Staining Solutions

Reagents

Chemicals

Magnesium chloride, methanol, acetone, dimethylformamide, sodium nitrite, and New Fuchsin were from Merck, Schuchardt, West Germany. 5-Bromo-4-chloro-3-indolyl phosphate, nitro blue tetrazolium salt, naphthol phosphate esters, coupling salts (see Table 1), levamisole, and Tris 27-9 (Tris (hydroxymethyl) aminomethane) were obtained from Sigma, St. Louis.

Solutions

Tris buffer 0.2 *M*, $MgCl_2$ 4 m*M*, pH 9.5 or 8.0

Tris	24.2 g
$MgCl_2$, $6H_2O$	0.81 g
Titrated with 5 *M* HCl to pH 9.5 or 8.0	
Distilled water	ad 1000 mℓ

Indoxyl phosphate substrate stock solution 4 mg/mℓ

Methanol	10.0 mℓ
Acetone	5.0 mℓ
5-Bromo-4-chloro-3-indolyl phosphate (BCIP)	60.0 mg
Store aliquoted at −20°C; stable for months	

Naphthol phosphate substrate stock solution 5 mg/mℓ

Tris buffer, 0.2 *M*, $MgCl_2$ 4 m*M*, pH 8.0	50 mℓ
Naphthol AS-MX phosphate, disodium salt	250 mg

Store at 4°C; stable for at least 6 months; alternatively, other naphthol phosphate derivatives listed in Table 1 can be used; if the free acid form is employed, it should be dissolved in a small amount of dimethylformamide before mixing with the buffer

Staining Solutions

Indoxyl Phosphate Staining Solution (Freshly Prepared)

Nitro blue tetrazolium (crystalline)	5 mg
Indoxyl phosphate substrate, stock solution 4 mg/mℓ	0.75 mℓ
Tris buffer, 0.2 *M*, $MgCl_2$ 4 m*M*, pH 9.5	ad 50 mℓ

The incubation is carried out for 15 to 20 min on a rocking table at room temperature. Use at high concentration does not increase the sensitivity, but increases the background staining. Nonspecific precipitation of the solution does not occur.[14] Alternatively, Tetranitro blue tetrazolium salt listed in Table 1 can be substituted for nitro blue tetrazolium salt.

Naphthol Phosphate Staining Solution (Freshly Prepared)

Fast Blue RR salt (crystalline)	25 mg
Naphthol phosphate substrate stock solution 5 mg/mℓ	5 mℓ
Tris buffer, 0.2 *M*, $MgCl_2$ 4 m*M*, pH 8.0	ad 50 mℓ

The incubation is carried out as above, but protected against light. With increased concentration of coupling salt, the background staining also increases. With fixed concentration of coupling salt, the background staining decreases with increasing concentration of naphthol phosphate substrate.

Alternatively, other coupling salts listed in Table 1 can be substituted for Fast Blue RR salt. When practical grades are used, the dye contents (15 to 25%) should be considered.

If hexazotized New Fuchsin is chosen as coupling salt, the staining solution is prepared in a fume cupboard as follows:

New Fuchsin 40 mg/mℓ in 2 *M* HCl (freshly prepared) 0.2 mℓ
Sodium nitrite 40 mg/mℓ in distilled water (freshly prepared) 0.4 mℓ
Mix for 1 min and add to a solution of Naphthol phosphate substrate
stock solution 5 mg/mℓ 6 mℓ
and Tris buffer, 0.2 *M*, $MgCl_2$ 4 m*M*, pH 9.5 44 mℓ

Various combinations of substrate systems showed different degrees of turbidity due to decomposition of the diazonium salts at alkaline pH. The turbidity varies with pH and type of buffers and increases upon prolonged incubation. Fast Blue RR from different sources has shown variable stability.[12] Fast Blue RR is readily soluble in 0.2 *M* Tris-HCl, pH 8.0, and the staining solution is stable at least for the period required for staining. The higher the pH, the less soluble the coupling salt, leading to more background staining, probably due to nonspecific binding of decomposed diazonium salt (see Figure 3b and 3c). The background staining at high pH is less pronounced with red colored coupling salts such as New Fuchsin. It is advisable to filter turbid solutions, although it is not imperative.

Comments

When samples containing endogenous AP are applied, isoenzymes, except for intestinal, stomach, and placenta AP, can be inhibited by 1 m*M* levamisole in the substrate solution. The concentration fo levamisole will not inhibit the conjugate AP when this is derived from calf intestinal mucosa.[20,25] Alternatively, inactivation of endogenous AP with 10 m*M* EDTA prior to AP-conjugate incubation has been reported.[26]

Other buffer systems can be used, for example, propandiol. Ethanolamine is not recommended in conjunction with New Fuchsin. Avoid phosphate buffer as it inhibits the enzymatic reaction.

The substrate incubation is performed at room temperature for 15 to 20 min. We have not found it necessary to incubate for the longer times or higher temperatures sometimes reported.[10,27]

REFERENCES

1. **Goldstein, D. J., Rogers, C. E., and Harris, H.,** Expression of alkaline phosphatase loci in mammalian tissues, *Proc. Natl. Acad. Sci. U.S.A.*, 77, 2857, 1980.
2. **Engvall, E., Jonsson, K., and Perlmann, P.,** Enzyme-linked immunosorbent Assay. II. Quantitative assay of protein antigen, immunoglobulin G, by means of enzyme-labelled antigen and antibody-coated tubes, *Biochem. Biophys. Acta*, 251, 427, 1971.
3. **Mason, D. Y. and Sammons, R.,** Alkaline phosphatase and peroxidase for double immunoenzymatic labelling of cellular constituents, *J. Clin. Pathol.*, 31, 454, 1978.
4. **O'Conner, C. G. and Ashman, L. K.,** Application of the nitrocellulose transfer technique and alkaline phosphatase conjugated anti-immunoglobulin for determination of the specificity of monoclonal antibodies to protein mixtures, *J. Immunol. Methods*, 54, 267, 1982.
5. **Lee, L. M. Y. and Kenny, M. A.,** Electrophoretic method for assessing the normal and pathological distribution of alakaline phosphatase isoenzymes in serum, *Clin. Chem.*, 21, 1128, 1975.
6. **Ishikawa, E., Imagawa, M., Hashida, S., Yoshitake, S., Hamaguchi, Y., and Ueno, T.,** Enzyme-labeling of antibodies and their fragments for enzyme immunoassay and immunohistochemical staining, *J. Immunoassay*, 4, 209, 1983.
7. **Cordell, J. L., Falini, B., Erber, W. N., Ghosh, A. K., Abdulaziz, Z., MacDonald, S., Pulford, K. A. F., Stein, H., and Mason, D. Y.,** Immunoenzymatic labeling of monoclonal antibodies using immune complexes of alkaline phosphatase and monoclonal anti-alkaline phosphatase (APAAP complexes), *J. Histochem. Cytochem.*, 32, 219, 1984.
8. **Leary, J. J., Brigati, D. J., and Ward, D. C.,** Rapid and sensitive colorimetric method for visualizing biotin-labeled DNA probes hybridized to DNA or RNA immobilized on nitrocellulose: bio-blots, *Proc. Natl. Acad. Sci. U.S.A.*, 80, 4045, 1983.

9. **Morton, R. K.,** Some properties of alkaline phosphatase of cow's milk and calf intestinal mucosa, *Biochem. J.,* 60, 573, 1955.
10. **Blake, M. S., Johnston, K. H., Russel-Jones, G. J., and Gotschlich, E. C.,** A rapid, sensitive method for detection of alkaline phosphatase-conjugated anti-antibody on Western blots, *Anal. Biochem.,* 136, 175, 1984.
11. **Stanley, C. J., Johannsson, A., and Self, C. H.,** Enzyme Amplification Can Enhance Both the Speed and the Sensitivity of Immunoassays, *J. Immunol. Methods,* 83, 89, 1985.
12. **Turner, B. M.,** The use of alkaline-phosphatase-conjugated second antibody for the visualization of electrophoretically separated proteins recognized by monoclonal antibodies, *J. Immunol. Methods,* 63, 1, 1983.
13. **Bennett, F. C. and Yeoman, L. C.,** An improved procedure for the dot immunobinding analysis of hybridoma supernatants, *J. Immunol. Methods,* 61, 201, 1983.
14. **De Jong, A. S. H., van Kessel-van Vark, M., and Raap, A. K.,** Sensitivity of various visualization methods for peroxidase and alkaline phosphatase activity in immunoenzyme histochemistry, *Histochem. J.,* 17, 1119, 1985.
15. **Feller, A. C., Parwaresch, M. R., Wacker, H.-H., Radzun, H.-J., and Lennert, K.,** Combined immunohistochemical staining for surface IgD and T-lymphocyte subsets with monoclonal antibodies in human tonsils, *Histochem. J.,* 15, 557, 1983.
16. **Burnie, J. P., Matthews, R. C., Fox, A., and Tobaqchali, S.,** Use of immunoblotting to identify antigenic differences between the yeast and mycelial phases of Candida albicans, *J. Clin. Pathol.,* 38, 701, 1985.
17. **Sidberry, H., Kaufman, B., Wright, D. C., and Sadoff, J.,** Immunoenzymatic analysis by monoclonal antibodies of bacterial lipopolysaccharides after transfer to nitrocellulose, *J. Immunol. Methods,* 76, 299, 1985.
18. **Makler, M. T., Mlecko, C. R., and Pesce, A. J.,** Immunoenzymometric assays for simultaneous quantitation and distribution of erythrocyte antigen, *Clin. Chem.,* 27, 1609, 1981.
19. **Frickhofen, N., Bross, K. J., Heit, W., and Heimpel, H.,** Modified immunocytochemical slide technique for demonstrating surface antigens on viable cells, *J. Clin. Pathol.,* 38, 671, 1985.
20. **Yam, L. T., English, M. C., Janckila, A. J., Ziesmer, S., and Li, C.-Y.,** Immunocytochemical characterization of human blood cells, *Am. J. Clin. Pathol.,* 80, 314, 1983.
21. **Li, C.-Y., Ziesmer, S. C., Yam, L. T., English, M. C., and Janckila, A. J.,** Practical immunocytochemical identification of human blood cells, *Am. J. Clin. Pathol.,* 81, 204, 1984.
22. **Forster, A. C., McInnes, J. L., Skingle, D. C., and Symons, R. H.,** Non-radioactive hybridization probes prepared by the chemical labelling of DNA and RNA with a novel reagent, photobiotin, *Nucl. Acids Res.,* 13, 745, 1985.
23. **Van Belle, H.,** Alkaline phosphatase. II. Conditions affecting determination of total activity in serum, *Clin. Chem.,* 22, 977, 1976.
24. **Portmann, P., Schaller, H., Leva, G., Venetz, W., and Müller, T.,** The calf intestinal alkaline phosphatase. II. Reaction between the metal content and the enzyme activity, *Helv. Chim. Acta,* 66, 871, 1983.
25. **Ponder, B. A. and Wilkinson, M. M.,** Inhibition of endogenous tissue alkaline phosphatase with the use of alkaline phosphatase conjugates in immunohistochemistry, *J. Histochem. Cytochem.,* 29, 981, 1981.
26. **Dao, M. L.,** An improved method of antigen detection on nitrocellulose: *in situ* staining of alkaline phosphatase conjugated antibody, *J. Immunol. Methods,* 82, 225, 1985.
27. **Jalkanen, M. and Jalkanen, S.,** Immunological detection of proteins after isoelectric focusing in thin layer agarose gel: a specific application for the characterization of immunoglobulin diversity, *J. Clin. Lab. Immunol.,* 10, 225, 1983.

6.4.3 RADIOIMMUNODETECTION — IMMUNOVISUALIZATION BY MEANS OF ^{14}C-LABELED ANTI-ANTIBODIES AND SODIUM SALICYLATE FLUOROGRAPHY ON NITROCELLULOSE

Niels H. H. Heegaard

INTRODUCTION

Film detection methods in immunoblotting experiments were introduced by Burnette[1] in 1981, and have been widely used since. They alternate with the enzymatic procedures (Chapters 6.4.1 and 6.4.2) for immunovisualization on blots. At present about 50% of published immunoblotting experiments are performed with film detection (475 randomly selected references). In about 92% of these experiments ^{125}I is the label employed, either conjugated with protein A (50%), with primary or secondary antibodies (35%), or conjugated with other ligands such as lectins, hormones, or directly with the transferred molecules. Rarely, other isotopes, including ^{131}I, ^{35}S, ^{32}P, ^{14}C, or ^{3}H, have been employed.

The 50-50 distribution of enzymatic vs. radioactive imaging systems illustrates that none is superior in all respects. The use of ^{125}I, for example, implies radiation hazards and a short shelf life. Isotopes emitting weak β-radiation such as ^{3}H and ^{14}C have not been employed much. Therefore in this study, the usability on nitrocellulose blots of the long-lived, less hazardous isotope ^{14}C in connection with the water-soluble fluorophore sodium salicylate[2,3] is evaluated. Anti-antibodies are labeled with the isotope and human erythrocyte membrane proteins, human serum albumin, and corresponding rabbit antibodies are model systems.

REAGENTS AND PROCEDURES

Chemicals and Materials

Sodium salicylate, boric acid, sodium tetraborate, and sodium azide were from Merck, and Amersham delivered $K^{14}CNO$, 52 mCi/mmol. Bovine serum albumin, CRG-7 was obtained from Armour and human albumin (Reinst) from Behringwerke. X-Ray films were Kodak® X-Omat L and X-Omat AR-5. Dirty X-ray cassettes were cleared in Folipur from Siemens. Glass plates, 1-mm thick, were used for mounting nitrocellulose strips which had a pore size of 0.22 μm and came from Millipore (type GSWP 293 25).

Antibodies

Rabbit antibodies against albumin and human erythrocyte membrane proteins and swine antibodies against rabbit IgG and peroxidase-conjugated swine antirabbit IgG antibodies were delivered by Dakopatts, Glostrup, Denmark. The swine antirabbit IgG antibodies were labeled with ^{14}C by carbamylation with $K^{14}CNO$ to a specific activity of 62×10^6 dpm/mg.[3]

Solutions

^{14}C-Anti-Antibodies Solution[3]	
Phosphate-buffered saline, pH 7.4	50 mℓ
BSA	50 mg
^{14}C-swine antirabbit IgG antibodies	250 μℓ
Sodium azide, 4.5 m*M*	150 μℓ of 1.5 *M* stock
Store at 4°C; can be reused and is durable for more than 6 months	
Fluorography	
Sodium salicylate, 0.7 *M*	11.2 g
Distilled water	ad 100 mℓ
Store at 4°C; can be reused until the solution turns yellowish	

Procedures

Electroblotting

Electroblotting in a buffer tank was performed exactly as described in the Appendix to Chapter 1.

Radiolabeling of Blots and Fluorography

After electroblotting, strips were incubated with primary antibodies, diluted 1:2000 if not otherwise mentioned, as described in the Appendix to Chapter 1. Then, the strips incubated in the ^{14}C-labeled anti-antibodies solution for 24 hr at room temperature were washed twice for 30 min in 0.1 *M* NaCl and finally immersed in 0.7 *M* sodium salicylate for 15 min at room temperature. Both the anti-antibody and the sodium salicylate solutions were recovered and reused. The strips were then placed on a glass plate and dried in a stream of hot air. After drying, careful handling is essential as the nitrocellulose becomes brittle at this point. The glass plate was placed in an X-ray cassette with the strips against the film at −80°C for exposure (typically 3 to 5 days). X-Omat AR-5 films should be handled in complete darkness. When unloading, it is important that the cassette is dry as otherwise problems with nitrocellulose sticking to the film might occur. Peroxidase staining with 3-amino-9-ethyl carbazole was performed as described in the Appendix to Chapter 1.

RESULTS

By spotting ^{14}C-labeled probes onto nitrocellulose it was found that sodium salicylate did enhance detection of ^{14}C on nitrocellulose and that the optimal soaking condition was 0.7 *M* sodium salicylate for 15 min. Less concentrations gave less blackening of films and higher concentrations caused blurring of the image. The enhancing effect of sodium salicylate was most pronounced at higher levels of radioactivity. At low levels the sensitivity was approximately doubled.

Employing these conditions the minimum detectable radioactivity was evaluated by spotting the ^{14}C-labeled anti-antibodies in volumes of 5 μℓ onto the nitrocellulose giving dots of approximately 0.126 cm^2. On Kodak® X-Omat L films the minimal detectable amount after 48 hr of exposure was found to be 160 dpm/cm^2 (0.07 nCi/mm^2), while this could be visualized after 24 hr of exposure by the AR-5 film.

The lowest amount of antigen detectable in dot immunobinding experiments with albumin-anti-albumin probed with ^{14}C- and peroxidase-labeled anti-antibodies was for both found to be about 5 ng protein applied in a volume of 5 μℓ, i.e, corresponding to 40 ng/cm^2.

The application of ^{14}C-labeled anti-antibodies in a complete immunoblotting experiment is shown in Figure 1. Two series of experiments varying the antigen load (Figure 1A) and amount of primary antibody (Figure 1B) were performed. With a fixed amount of primary antibody (0.5 μℓ/mℓ), the spectrin bands (sp) are still visible at about 150 ng total protein load for SDS-PAGE (sodium dodecyl sulfate-polyacrylamide gel electrophoresis). Assuming a transfer efficiency for these high M_r species of 50% and a relative protein content of 20% of applied protein for each of the two spectrin bands down to approximately 15 ng spectrin is detected here. For the band 3 protein (b3) the figure is 37.5 ng. A protein load between 2 and 5 μg for SDS-PAGE seems to give the clearest image with preservation of most bands in this system. When the amount of antigen is fixed at 20 μg (Figure 1B) some bands are still detected at a primary antibody concentration of 0.001 μg/mℓ. The optimal concentration, however, is between 0.05 and 0.8 μg/mℓ. Since these values are valid for a concentrated antibody preparation, the figures for a crude antiserum would be about five times higher. The labeled anti-antibody solution was kept at 4°C and could be used after more than 6 months without changes in binding activity.

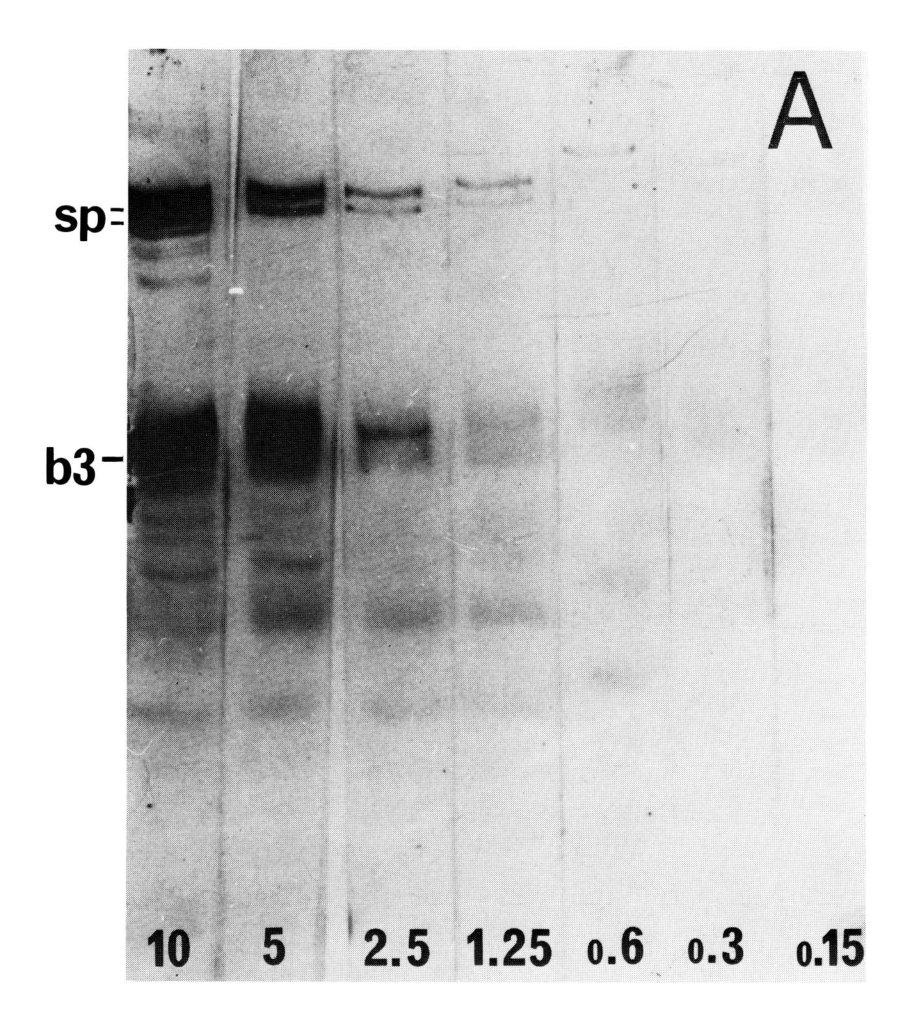

FIGURE 1. Fluorographs of SDS-PAGE-separated human erythrocyte membrane proteins immobilized on nitrocellulose, incubated with rabbit antihuman erythrocyte membrane proteins antibodies, and ^{14}C-labeled swine antirabbit IgG antibodies. (A) Variation of antigen load. Amount of analyzed protein (μg) is stated on the figure. (B) Variation of amount of primary antibody. Concentrations (μg/mℓ) are stated on the figure. Antigen load: 20 μg. SDS-PAGE of human erythrocyte membrane proteins was performed according to Fairbanks et al.;[17] see also Chapter 7. Incubation with primary antibody, A, 1:2000 in incubation buffer and B, as given above with 0.1% (w/v) bovine serum albumin included in the incubation buffers and otherwise as described in the Appendix to Chapter 1. Further processing of strips was as described in the Reagents and Procedures section. Exposure to X-Omat AR-5 film at −80°C for 68 hr. Designations: sp, spectrin bands; b3: band 3 protein. Bar = 1 cm.

DISCUSSION

The general advantages of radioactive imaging are high sensitivity, sharply defined patterns with low background staining, no conservation problems, easy densitometric scanning of films (and therefore good possibilities for quantitative measurements), exposure to films might be performed several times, general protein staining can be performed after autoradiography, and the whole radiographic procedure is easy to work with. The disadvantages are that the procedure is time consuming, that the coloration process cannot be surveyed while taking place as in enzymostaining, and that it implies use of compounds with health

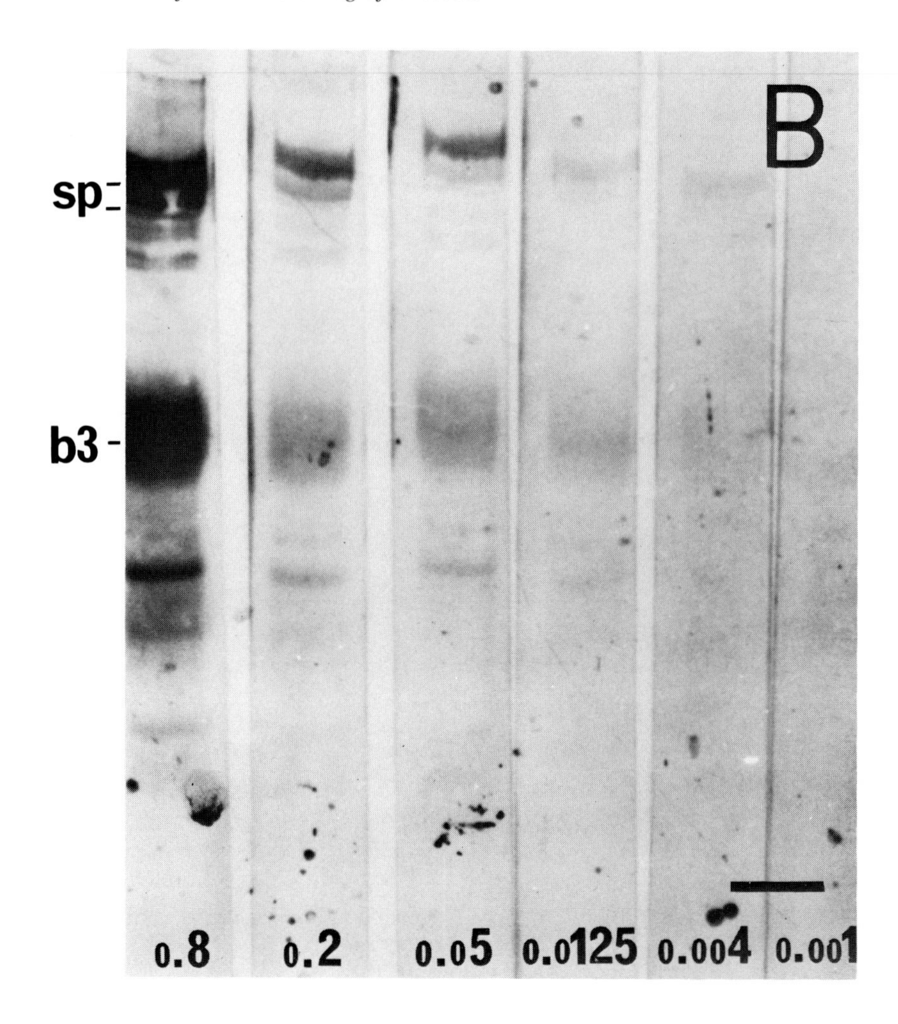

FIGURE 1B.

hazards and disposal problems. On the other hand, the enzyme substrates might be toxic or suspected carcinogens[4] and the staining might fade (see Chapter 7).

By employment of weak β-emitters such as ^{3}H and ^{14}C, the hazards of radioimmunodetection can be partially circumvented and direct autoradiography using these isotopes is aided by the low self-absorption of the thin blotting membrane compared with gels.[5] These isotopes are also more long-lived than ^{125}I, which has a half-life of 60 days. The half-life of ^{14}C (5700 years) ensures a durability which makes it likely that the use of ^{14}C-labeled anti-antibody solutions is cheaper in the long run than the enzymatic detection systems.

The commercial availability of ^{14}C-labeled ligands is not comparable with the supply of ^{125}I-labeled compounds, but ^{14}C is easily incorporated into proteins by reductive methylation[6] or by carbamylation.[7] The latter method was used here since it is a mild procedure known to have a minimal effect on the reactivity of an antibody.[8] Iodination procedures are more efficient but may disturb the molecular reactivity to a greater extent. The labeling efficiency probably explains the superior sensitivity of ^{125}I-detection in radioimmunoblotting.[9]

The weakest β-emitter, tritium, can also be employed for fluorography on blots with sodium salicylate,[10] or with commercially available scintillants such as Enhance®.[11] The label is easily introduced,[12] but the efficiency of fluorography is more important here because of the very low energy of radiation.

The results show that fluorography with sodium salicylate and ^{14}C-labeled anti-antibodies is a feasible method with a sensitivity in our model systems comparable to peroxidase detection with carbazole staining. Optimal dilutions and amounts of antigens and antibodies also correspond with the figures known from enzymatic staining (see Chapter 1).

Roberts[5] compared different fluorographical methods for detection of ^{35}S, ^{14}C, and ^{3}H on nitrocellulose. The best values for ^{14}C-detection were obtained with diphenyloxazole in toluene. The detection limit for a 24-hr exposure was 100 dpm/cm^2 comparable with the results of this study, but since presensitized film which is up to ten times more sensitive than unexposed film[13,14] was used, it appears that sodium salicylate fluorography ranks among the best methods for detection on ^{14}C-labeled blots. Other valuable features of sodium salicylate are that it is cheap, nontoxic, and water-soluble. The fluorographed blots can thus be stained with general protein stains after washing in distilled water. A disadvantage is that the sodium salicylate-impregnated nitrocellulose becomes rather brittle upon drying and crystallization, making some kind of support (e.g., glass plates) indispensable. The strips are thus mounted before drying and insertion into the X-ray cassette where the glass plate also ensures a close contact between film emulsion and nitrocellulose. Otherwise, sodium salicylate has no deleterious effects in contrast with, for example, the fluorophore methyl anthranilate which is optimal for fluorography on thin-layer chromatograms,[15] but destroys nitrocellulose completely.

A general disadvantage of fluorography vs. autoradiography for immunodetection on blots is that a not-well-defined amplifying step is introduced into the procedure. The relationship between amounts of radiation, fluorophoric output, and effect on films is not simply linear;[14,16] note, for example, the above-mentioned preferential amplification of high-dose ^{14}C dots over lower doses which were enhanced less. This may make attempts to quantify fluorograms less reliable than quantification on autoradiograms. The sensitivity of different approaches can only be compared with caution as affinities of ligand preparations and of different antibody specificities, recovery of blotted species, and conditions for transfer may vary.

In summary, the presented method shares characteristics with other radioactive detection systems, but merits from safety and simplicity in daily use, from the favorable economics caused by the reusability and stability of the preparation, and from the minimal influence of the labeling on the molecular characteristics. With further optimization of sensitivity the method constitutes a usable alternative to the other radioimmunodetection methods on blots.

ACKNOWLEDGMENTS

The technical skills of Mrs. Jane Tullberg are greatly appreciated. I thank Dr. Ole J. Bjerrum for helpful discussions and the Weimanns Foundation and Harboefonden for financial support.

REFERENCES

1. **Burnette, W. N.,** Western blotting: electrophoretic transfer of proteins from sodium dodecyl sulfate-polyacrylamide gels to unmodified nitrocellulose and radiographic detection with antibody and radioiodinated protein A, *Anal. Biochem.*, 112, 115, 1981.
2. **Chamberlain, J. P.,** Fluorographic detection of radioactivity in polyacrylamide gels with the water-soluble fluor, sodium salicylate, *Anal. Biochem.*, 98, 132, 1979.
3. **Heegaard, N. H. H., Hebsgaard, K. P., and Bjerrum, O. J.,** Sodium salicylate for fluorographical detection of immunoprecipitated proteins in agarose gels, *Electrophoresis*, 5, 263, 1984.

4. **Ogata, K., Arakawa, M., Kasahara, T., Shioiri-Nakano, K., and Hiroka, K. -I.,** Detection of *Toxoplasma* membrane antigens transferred from polyacrylamide gel to nitrocellulose with monoclonal antibody and avidin-biotin, peroxidase anti-peroxidase and immunoperoxidase methods, *J. Immunol. Methods*, 65, 75, 1983.
5. **Roberts, P. L.,** Comparison of fluorographic methods for detecting radioactivity in polyacrylamide gels or on nitrocellulose filters, *Anal. Biochem.*, 147, 521, 1985.
6. **Jentoft, N. and Dearborn, D. G.,** Labelling of proteins by reductive methylation using sodium cyanoborohydride, *J. Biol. Chem.*, 254, 4359, 1979.
7. **Bjerrum, O. J. and Løwenstein, H.,** Immunoelectrophoresis utilizing carbamylated antibodies, *Scand. J. Immunol.*, 17(Suppl. 10), 225, 1983.
8. **Bjerrum, O. J., Ingild, A., and Løwenstein, H.,** Carbamylation of antibodies; investigation of the number of converted amino groups, electrophoretic mobility and immunoprecipitation titre, *Immunochemistry*, 11, 797, 1974.
9. **Towbin, H. and Gordon, J.,** Immunoblotting and dot immunobinding — current status and outlook, *J. Immunol. Methods*, 71, 313, 1984.
10. **Naaby-Hansen, S. and Bjerrum, O. J.,** Auto- and isoantigens of the human spermatozoa detected by immunoblotting with human sera after SDS-PAGE, *J. Reprod. Immunol.*, 7, 41, 1985.
11. **Hoch, S. O.,** DNA-binding domains of fibronectin probed using western blots, *Biochem. Biophys. Res. Commun.*, 106, 1353, 1982.
12. **Gahmberg, C. G. and Andersson, L. C.,** Selective radioactive labelling of cell surface sialoglycoproteins by periodatetritiated borohydride, *J. Biol. Chem.*, 252, 5888, 1977.
13. **Bonner, W. M.,** Use of fluorography for sensitive detection on polyacrylamide gel electrophoresis and related techniques, *Methods Enzymol.*, 96, 215, 1983.
14. **Laskey, R. A. and Mills, A. D.,** Quantitative film detection of ^{3}H and ^{14}C in polyacrylamide gels by fluorography, *Eur. J. Biochem.*, 56, 335, 1975.
15. **Bochner, B. R. and Ames, B. N.,** Sensitive fluorographic detection of ^{3}H and ^{14}C on chromatograms using methyl anthranilate as a scintillant, *Anal. Biochem.*, 131, 510, 1983.
16. **Pulleyblank, D. E., Shure, M., and Vinograd, J.,** The quantitation of fluorescence by photography, *Nucl. Acids Res.*, 4, 1409, 1977.
17. **Fairbanks, G., Steck, T. L., and Wallach, D. F. H.,** Electrophoretic analysis of the major polypeptides of the human erythrocyte membrane, *Biochemistry*, 10, 2606, 1971.

6.4.4. IMMUNOGOLD AND IMMUNOGOLD/SILVER STAINING FOR DETECTION OF ANTIGENS

Marc Moeremans, Guy Daneels, Marc De Raeymaeker, and Jan De Mey

INTRODUCTION

Most visualization systems in immunoblotting are based on immunocytochemical techniques. We developed immunogold staining (IGS) for specific detection of antigen-antibody interactions on transfer membranes. We have enhanced the signal by a silver precipitation step (immunogold/silver staining, IGSS), which makes of it one of the most sensitive nonradioactive detection methods.

Antigens can be detected either directly or indirectly. In the direct method the primary antibody is conjugated to 20 nm colloidal gold particles. For the preparation of these specific gold probes the reader is referred to References 3 and 4. In general, the immunodetection step is the same as for the indirect method, which will be explained in detail in the following.

The basic procedure consists in incubating a saturated blot with a primary antibody. The interaction between the primary antibody and the antigen on the blot is visualized by incubation in an immunogold reagent, i.e., secondary antibodies linked to colloidal gold particles.[5,6] The detection sensitivity which is of intermediate sensitivity can be intensified at least ten times by a silver enhancement step.[6-11] In addition, this immunodetection can be combined with overall protein staining on the same membrane.[12]

This visualization technique is not limited to antigen-antibody interactions, but is more generally applicable in other overlay assays provided that a suitable gold probe can be prepared.

IMMUNOGOLD AND IMMUNOGOLD/SILVER STAINING

Equipment and Reagents

Apparatus

Rocking table, Petri dishes, incubator at 37°C, tweezers, pH meter, gloves, apparatus for heat-sealing household plastic bags (e.g., Krups), electrophoresis, and blotting apparatus.

Chemicals

Tris(hydroxymethyl)-aminomethane, sodium chloride, sodium azide, trisodium citrate, and citric acid were obtained from Merck, AuroProbe B1 GAR kit, containing AuroProbe B1 GAR (Goat antirabbit IgG linked to 20 nm colloidal gold particles), normal goat serum, bovine serum albumin, gelatin (Bactogelatin, Difco), and IntenSE (Initiator and Enhancer powders and fixing solution) for silver enhancement, were obtained from Janssen Life Sciences Products, Beerse, Belgium. For the original light-sensitive silver lactate method: silver lactate and hydroquinone from Janssen Chimica and fixing solution from Agfa-Gevaert. The Vectastain ABC kit (avidin: biotinylated horseradish peroxidase complex) was obtained from Vector.

Solutions

Tris buffered saline (TBS, pH 8.2)

20 m*M* Tris	2.42 g
150 m*M* sodium chloride	8.77 g
Distilled water	ad 900 mℓ
Adjust the pH with 1 N HCl	

20 m*M* sodium azide	1.3 g
Bidistilled water	ad 1000 mℓ
Blocking solution (nitrocellulose)	
5% (w/v) BSA	5 g
TBS, pH 8.2	ad 100 mℓ
Blocking solution (charged modified nylon)	
10% (w/v) BSA	10 g
TBS, pH 8.2	ad 100 mℓ
Washing buffer	
0.1% (w/v) BSA	1 g
TBS, pH 8.2	ad 1000 mℓ
TBS-0.1% BSA supplemented with gelatin	
0.4% (w/v) gelatin	0.4 g
TBS-0.1% (w/v) BSA	ad 100 mℓ
Heat to 50°C to dissolve the gelatin	
Citrate buffer (2 M)	
Trisodium citrate dihydrate	23.5 g
Citric acid monohydrate	25.5 g
Bidistilled water	ad 100 mℓ
Silver lactate	
5.5 m*M* silver lactate	0.11 g
Bidistilled water	ad 15 mℓ
Prepare just before use, protect from light	
Hydroquinone	
77 m*M* hydroquinone	0.85 g
Bidistilled water	ad 15 mℓ
Prepare just before use, protect from light	
Fixing solution	
Fixing solution	10 mℓ
Bidistilled water	ad 100 mℓ
Initiator (Janssen Life Sciences Products)	
Initiator	3.9 g
Bidistilled water	ad 50 mℓ
Enhancer (Janssen Life Sciences Products)	
Enhancer	2.35 g
Bidistilled water	ad 50 mℓ

Procedure

Washing and Blocking

After transfer to the membrane, which can be either nitrocellulose, unmodified, or charge-modified nylon-based, excessive binding sites are quenched by incubating in blocking solution. Due to the higher binding capacity of charge-modified membranes, their blocking conditions are different. Blocking for 45 min at 37°C with 5% BSA is sufficient for nitrocellulose membranes; however, in the case of charge-modified membranes, incubation must be carried out with 10% BSA at 37°C overnight.

After quenching excessive binding sites, the blot is washed at room temperature with 100 mℓ washing buffer in a Petri dish for 2 × 5 min.

Incubation with Primary Antibody

The blot is subsequently incubated in the primary antibody solution. The antibody is diluted to a concentration of 1 to 2 μg/mℓ in washing buffer supplemented with 1% normal serum (prepared from the same species as the secondary antibody). To reduce the amount of antibody one can prepare tailored plastic bags which, after addition of the incubation fluid, are heat sealed. In this case a volume of 15 to 20 mℓ is sufficient for a 10 × 15 cm membrane. A typical incubation time, at room temperature, is 2 hr. To ensure that the whole membrane is flooded by the antibody solution the plastic bag must be placed horizontally on a tilting apparatus. A negative control consists in replacing the primary antibody solution by washing buffer +1% normal serum (prepared from the same species as the secondary antibody).

Immunogold Staining

After incubation with the primary antibody, the membrane is washed with washing buffer at room temperature for 3 × 10 min. The gold-labeled secondary antibody is diluted in washing buffer supplemented with 0.4% (w/v) gelatin. To reduce the amount of gold reagent used, the incubation can be performed in tailored plastic bags. About 20 mℓ gold reagent is sufficient for a 10 × 15 cm blot. The incubation time depends on the dilution used. If the gold probe is diluted 25 times, an incubation of 2 hr at room temperature is sufficient; if diluted 100 times, incubation should be carried out overnight. After incubation with the gold probe, the membrane is washed with TBS-0.1% BSA buffer at room temperature for 2 × 5 min and air dried. If the signal needs to be silver enhanced, the membrane is processed as described later.

Immunogold/Silver Staining

Light-Sensitive Enhancement[6,7,10,11]

After incubation in the gold probe and washing in washing buffer for 2 × 5 min, the membrane is washed extensively with distilled water for 2 × 5 min to remove chloride ions. Before incubation in the developer the membrane is rinsed in 10x diluted citrate buffer for 2 min. The developer is prepared freshly immediately before use by mixing (in a vessel protected from light!) 60 mℓ bidistilled water, 10 mℓ 2 *M* citrate buffer, 15 mℓ 77 m*M* hydroquinone, and 15 mℓ 5.5 m*M* silver lactate. The silver lactate and hydroquinone are prepared just before use in a vessel protected from light. After rinsing in citrate buffer, the membrane is incubated in the developer in a Petri dish wrapped in aluminum foil. The silver enhancement can be checked visually but one must avoid exposure to strong daylight. After the reaction has been completed, the membrane is transferred to the fixing solution for 5 min, washed extensively in excess distilled water, and air dried. If the intensity diminishes by incubation in the fixing solution, reduce the incubation time of fixation. After washing in excess water (3 × 5 min), the membrane is air dried.

Comments on the Procedure

The incubation time in the immunogold reagent is influenced by the concentration of the reagent. A dilution of 1:100 (OD_{520} = 0.05) or even 1:200 (OD_{520} = 0.025) can be used overnight. However, when a dilution of 1:25 (OD_{520} = 0.20) is used, optimal results are obtained within 2 hr.

The IGS reagent can be used several times, certainly in the case where it is diluted 1:25. Of course, by reusing the reagent, longer incubation times may become necessary. Dot blot tests have shown that a 100-fold diluted reagent can be used at least five times.

The enhancement reagents are extremely sensitive to the purity of the water used. Low quality water results in the formation of precipitates which lower the reactivity of the enhancement reagent and in most cases produce a background. We therefore advise the use of bidistilled water throughout the procedure.

Controls are important. It is advisable to incubate one strip without primary antibody or if possible with an unrelated primary antibody. A positive control, i.e., the use of an antibody of known reactivity, is helpful to judge the experimental conditions.

It is essential to add gelatin to the immunogold reagent. In our hands this prevented the nonspecific binding between the immunogold reagent and tropomyosin as demonstrated by the negative control. The source of the gelatin is equally important. If other brands of gelatin are used than the one supplied with the kit it is necessary to include a negative control.

Because both metallic gold and silver catalyze the reduction of silver ions, the reaction which is initially catalytic soon becomes autocatalytic. Antigen spot tests show that high concentrations of antigen (100 and 10 ng/mm^2) result in the same signal intensity as judged with the naked eye. Therefore we advise experimenters to monitor the enhancement from time to time. Long incubations can result in the visualization of bands which are of minor importance compared to the main signal.

As already explained in the procedure, the incubation in fixing solution must be monitored. If the signal intensity diminishes before the 5 min have passed, the fixation time can be shortened. This is especially the case for weak signals. As indicated in the procedure it is necessary to remove the fixative by thoroughly washing with distilled water.

A light-insensitive developer offers the attractive possibility to monitor the development continuously.

BSA is expensive. Due to the addition of 20 m*M* NaN_3 to the blocking solutions, they can be reused several times.

Evaluation

To assess the utility of the IGS and IGSS different nonradioactive visualization systems were compared. (Figure 1). The experimental conditions (sample, electrophoresis, blotting, primary antibody) were identical except for the secondary antibody (other source) and the marker itself. These experiments showed that the sensitivity of the IGS without enhancement is comparable with the indirect peroxidase. In addition, the formation of the signal can be monitored continuously during the incubation in the immunogold reagent. Although the IGS itself is a simple one-step method where no additional reagents have to be used, its sensitivity is rather low. The main advantage of this technique lies in the possibility of enhancing this signal with silver.

The silver enhancement makes it one of the most sensitive nonradioactive methods as compared with the avidin:biotinylated horseradish peroxidase complex and the peroxidase antiperoxidase method. Initially, the silver enhancement was perhaps not so attractive because of its light sensitivity. However, it is now more user-friendly by the development of a light-insensitive system. The IGSS results in a high contrasted dark-brown to black and stable signal with practically no background. Enzymatic methods usually require the need of toxic or carcinogenic substances, they result in higher background, and the signal tends to fade with time.

Characterization of an antibody to chicken gizzard light meromyosin is shown in Figure 2. To determine the specificity of the antibody a total cell extract of chicken lung epithelial cells and of chicken heart fibroblasts was prepared. Purified reference proteins were included. After electrophoresis one gel unit was silver stained, the other two were transferred to Zetaprobe. One blot unit was stained by FerriDye, the other one was processed by the IGSS after incubation with the primary antibody.

DOUBLE STAINING[12]

Equipment and Reagents

Apparatus

See the IGS/IGSS Section earlier in the chapter.

Chemicals

Disodium hydrogen phosphate dihydrate, sodium dihydrogen phosphate monohydrate, and sodium chloride were obtained from Merck; Tween® 20 from Bio-rad; AuroProbe B1 GAR kit; and AuroDye® from Janssen Life Sciences Products.

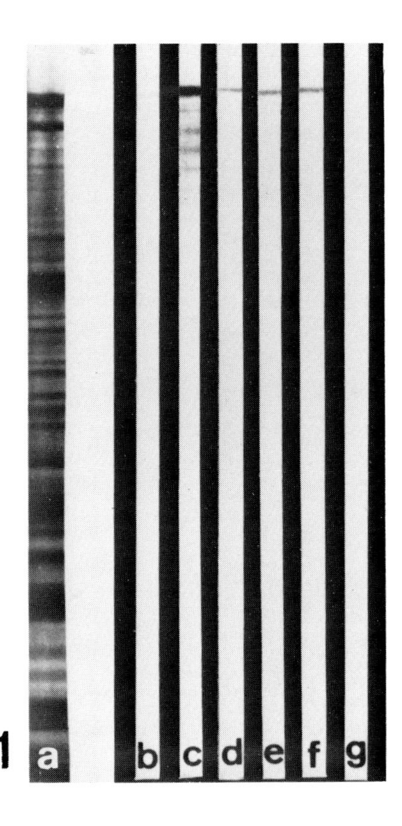

FIGURE 1. Comparison of various detection methods on strips of identical nitrocellulose transfers. Strip (a) shows overall protein pattern of chicken lung epithelial cell proteins as stained by colloidal gold, i.e., AuroDye®. Strips (b) to (f) were incubated with an affinity purified rabbit antifilamin antibody (directed against chicken gizzard filamin[8]) at a concentration of 1 μg/mℓ. Strip (g) is a control of the immunogold/silver staining where the primary antibody was omitted. Visualization method: (b) and (c) Immunogold and immunogold/silver staining using AuroProbe BL GAR, (d) indirect peroxidase using GAR-HRP, (e) avidin:biotinylated horseradish peroxidase complex method (according to the instructions of the manufacturer), and (f) peroxidase antiperoxidase.[16] For the enzymatic methods, the strips were developed with 4-chloro-1-naphthol. (From Moeremans, M. et al., *J. Immunol. Meth.*, 74, 353, 1984. With permission.)

Solutions

Phosphate buffered saline (PBS, pH 7.6)	
1.9 m*M* sodium dihydrogen phosphate monohydrate	0.262 g
8.1 m*M* disodium hydrogen phosphate dihydrate	1.442 g
150 m*M* sodium chloride	8.76 g
20 m*M* sodium azide	1.3 g
Bidistilled water	ad 1000 mℓ
Blocking solution	
0.3% (v/v) Tween® 20	1.5 mℓ
PBS, pH 7.6	ad 500 mℓ
Washing buffer	
0.05% (v/v) Tween® 20	250 μℓ
PBS, pH 7.6	500 mℓ

Citrate buffer, silver lactate, hydroquinone, fixing solution, initiator, enhancer all as in the IGS/IGSS section.

Procedure

Washing and Blocking

After transfer the membrane is washed extensively in blocking solution for 30 min at

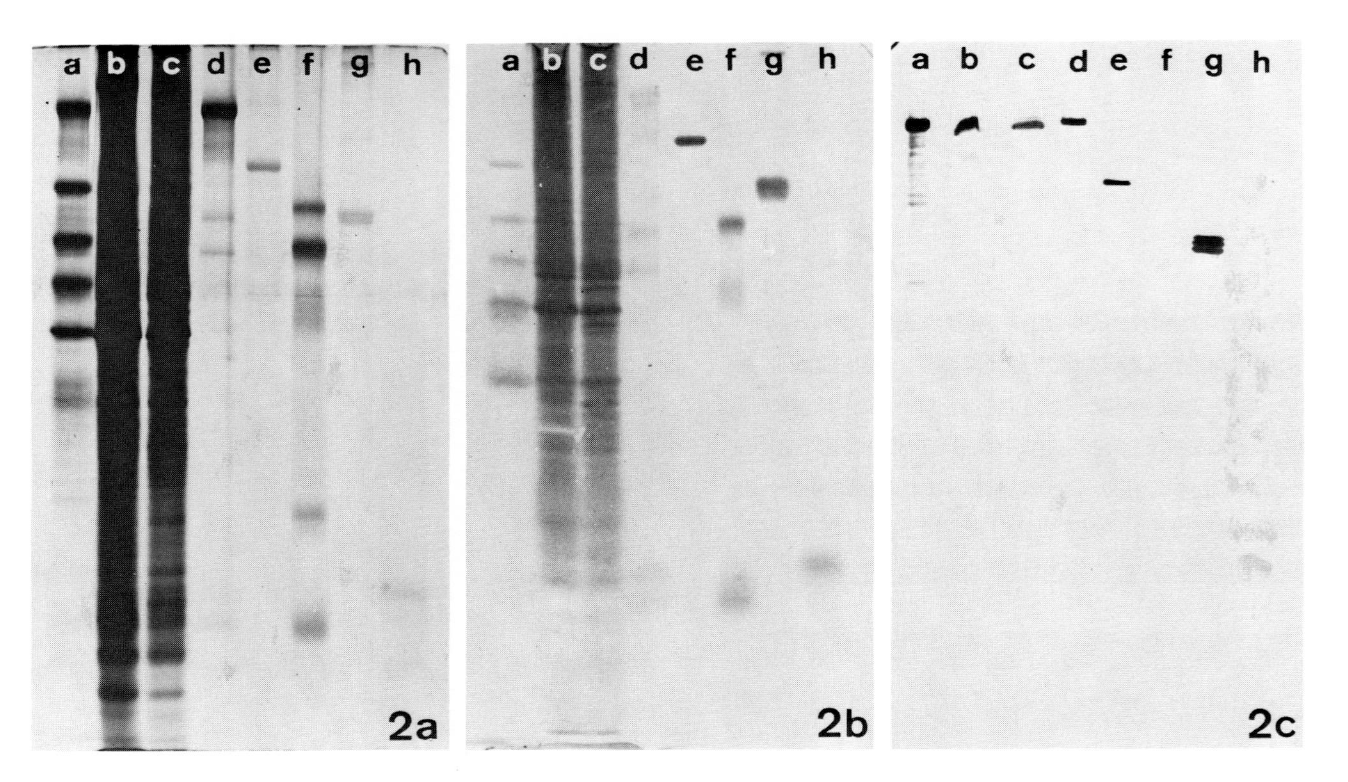

FIGURE 2. Characterization of an antibody to chicken gizzard light meromyosin.[9] Panel a: silver staining of SDS-polyacrylamide gel,[14] Panel b: visualization of total protein pattern on Zetaprobe by colloidal iron, i.e., FerriDye®, Panel c: immunodetection of antigen after incubation with antibody. Strip (a) Purified reference proteins at 0.5 μg/band. (1) chicken gizzard myosin heavy chain; (2) chicken gizzard α-actinin; (3) BSA; (4) rat brain tubulin; (5) chicken gizzard actin; (6) pig stomach tropomyosin. Strip (b) Chicken heart fibroblast.[9] Strip (c) Chicken lung epithelial cells.[9] Strip (d) Chicken gizzard myosin heavy chain (0.25 μg). Strip (e) Chicken gizzard myosin rods (0.25 μg). Strip (f) Chicken gizzard myosin subfragment 1 (0.25 μg). Strip (g) Chicken gizzard light meromyosin (0.25 μg). Strip (h) Chicken gizzard L_{20} (0.25 μg). (From Langanger, G. et al., *J. Cell Biol.*, 102, 200, 1986. With permission.)

37°C and three times for 15 min at room temperature. About 100 mℓ is sufficient for a blot of 15 × 10 cm. This extensive washing is necessary to remove adherent polyacrylamide gel particles and residual sodium dodecyl sulfate (SDS).

After quenching excessive binding sites with blocking buffer, the blot is washed at room temperature with washing buffer in a Petri dish for 2 × 5 min.

Incubation with Primary Antibody

Instead of using Tris-BSA buffers (see the IGS/IGSS Section earlier in this chapter), the antibody is diluted in PBS-0.05% Tween® 20. No normal serum is used.

Immunogold Staining

After incubation in primary antibody, the membrane is washed with PBS-0.05% Tween® buffer at room temperature for 3 × 10 min. The gold-labeled secondary antibody is diluted (100x) to an OD of 0.05 at 520 nm in washing buffer. Gelatin cannot be used here. The incubation can be performed in tailored plastic bags to reduce the amount of gold reagent used. The incubation time depends on the dilution of the gold probe. If diluted 25 times, an incubation of 2 hr at room temperature is sufficient; if diluted 100 times, incubation should be carried out overnight. After incubation in the gold probe the membrane is washed with PBS-0.05% Tween® buffer at room temperature for 3 × 5 min and processed as described earlier. No TBS-0.1% BSA wash.

Overall Protein Staining

After silver enhancement and washing in excess water, the membrane is incubated in PBS-0.3% (v/v) Tween® 20 at room temperature for 5 min. Subsequently the blot is incubated in the colloidal gold stain (Chapter 6.1.3).

Comments on the Procedure

The use of Tween® 20 as a blocking agent can give other results than when using BSA.[15] It is therefore essential to include a negative control. In addition, due to the subsequent protein staining, it is not possible to include gelatin in the immunogold reagent which may result in nonspecific binding of the gold probe. Normally, this is not a problem because nonspecific interactions are shown by the negative control and can be easily distinguished from the positive signal.

It is sometimes difficult to recognize an immunodetected band or spot after overall protein staining, so it is advisable to take a photograph after immunodetection before performing the overall protein staining.

The same comments as for the IGS/IGSS are applicable in this system.

Evaluation

As explained in Chapter 6.1.3, the overall protein staining with colloidal metal on a duplicate blot facilitates the correlation of an immunodetected band with the overall protein pattern. The preparation of a duplicate blot is difficult and time consuming, especially in 2D separations. Therefore, to demonstrate the utility of this method, a total cell extract was run on a 2D gel, electrotransferred to nitrocellulose, and processed for the double staining. Figure 3a shows the immunodetection of tubulin (arrow) with a rabbit antitubulin antibody on a 2D blot of a total cell extract of lung epithelial cells. Figure 3b shows the sequential overall protein staining.

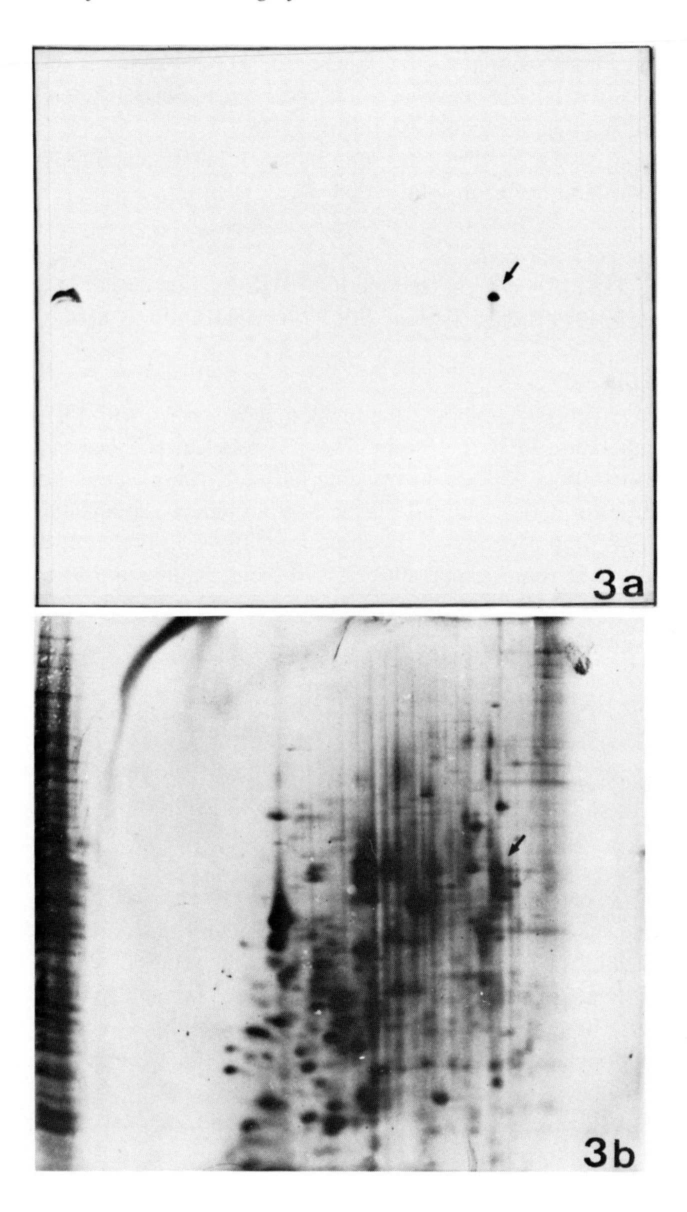

FIGURE 3. A whole cell extract of rat hepatoma cells was prepared for 2D electrophoresis according to Garrels.[17] After non-equilibrium pH gradient electrophoresis (NEPHGE) in the first dimension, proteins were separated according to their M_r with SDS-electrophoresis on a 12% gel[18] and electrotransferred to nitrocellulose.[19] (a) Immunodetection (indirect IGS/IGSS) of tubulin (arrow) with rabbit antitubulin and AuroProbe BL GAR. (b) Sequential overall protein staining of the same blot with AuroDye®. (From Daneels, G. et al., *J. Immunol. Meth.*, 89, 89, 1986. With permission.)

REFERENCES

1. **Gershoni, J. M. and Palade, G. E.,** Protein blotting: principles and applications, *Anal. Biochem.*, 131, 1, 1983.
2. **Towbin, H. and Gordon, J.,** Immunoblotting and dot immunobinding — current status and outlook, *J. Immunol. Methods*, 72, 313, 1984.
3. **De Mey, J.,** The preparation and use of gold probes, in *Immunocytochemistry. Practical Applications in Pathology and Biology*, 2nd ed., Polak, J. and Van Noorden, S., Eds., Wright, Bristol, England, in press, 1986.
4. **Roth, J.,** The colloidal gold marker system for light and electron microscopic cytochemistry, *Techniques in Immunocytochemistry*, Vol. 2, Bullock, G. R. and Petrusz, P., Eds., Academic Press, London, 1983, 217.
5. **Surek, B. and Latzko, E.,** Visualization of antigenic proteins blotted onto nitrocellulose using the immunogold-staining (IGS) method *Biochem. Biophys. Res. Commun.*, 121, 284, 1984.
6. **Moeremans, M., Daneels, G., Van Dijck, A., Langanger, G., and De Mey, J.,** Sensitive visualization of antigen-antibody reactions in dot and blot immune overlay assays with immunogold and immunogold/silver staining, *J. Immunol. Methods*, 74, 353, 1984.
7. **Holgate, C. S., Jackson, P., Cowen, P. N., and Bird, C. C.,** Immunogold-silver staining: new methods of immunostaining with enhanced sensitivity, *J. Histochem. Cytochem.*, 31, 938, 1983.
8. **Langanger, G., De Mey, J., Moeremans, M., Daneels, G., De Brabander, M., and Small, J. V.,** Ultrastructural localization of alpha-actinin and filamin in cultured cells with the immunogold staining (IGS) method, *J. Cell Biol.*, 99, 1324, 1984.
9. **Langanger, G., Moeremans, M., Daneels, G., Sobieszek, A., De Brabander, M., and De Mey, J.,** The molecular organization of myosin in stress fibers of cultured cells, *J. Cell Biol.*, 102, 200, 1986.
10. **Danscher, G.,** Histochemical demonstration of heavy metals, *Histochemistry*, 71, 1, 1981.
11. **Danscher, G. and Nörgaard, R. J. O.,** Light microscopic visualization of colloidal gold on resin-embedded tissue, *J. Histochem. Cytochem.*, 31, 1394, 1983.
12. **Daneels, G., Moeremans, M., De Raeymaeker, M., and De Mey, J.,** Sequential immunostaining (gold/silver) and complete protein staining (AuroDye) on western blots, *J. Immunol. Methods*, 89, 89, 1986.
13. **Laemmli, U. K.,** Cleavage of structural proteins during the assembly of the head of bacteriophage T4, *Nature (London)*, 227, 680, 1970.
14. **Morissey, J. H.,** Silver stain for proteins in polyacrylamide gels: a modified procedure with enhanced uniform sensitivity, *Anal. Biochem.*, 117, 307, 1981.
15. **Spinola, S. M. and Cannon, J. G.,** Different blocking agents cause variation in the immunologic detection of proteins transferred to nitrocellulose membranes, *J. Immunol. Methods*, 81, 161, 1985.
16. **De Blas, A. L. and Cherwinski, H. M.,** Detection of antigens on nitrocellulose paper immunoblots with monoclonal antibodies, *Anal. Biochem.*, 133, 214, 1983.
17. **Garrels, J. I.,** Two-dimensional gel electrophoresis and computer analysis of proteins synthesized by clonal cell lines, *J. Biol. Chem.*, 254, 7961, 1979.
18. **Blattler, D. P., Garner, F., Van Slijke, K., and Bradley, A.,** Quantitative electrophoresis in polyacrylamide gels of 2-40%, *J. Chromatogr.*, 64, 147, 1972.
19. **Towbin, H., Staehelin, T., and Gordon, J.,** Electrophoretic transfer of proteins from polyacrylamide gels to nitrocellulose sheets: procedure and some applications, *Proc. Natl. Acad. Sci. U.S.A.*, 76, 4350, 1979.

6.4.5 GENERAL PROTEIN VISUALIZATION BY DERIVATIZATION AND IMMUNOSTAINING WITH ANTIHAPTEN ANTIBODIES

Jason M. Kittler, Natalie T. Meisler, and John W. Thanassi

INTRODUCTION

Several methods are available to detect proteins transferred from gels to immobilizing matrices such as nitrocellulose. Earlier chapters describe the use of general protein stains such as Amido Black (Chapter 1) and India ink (Chapter 6.1.2) and techniques using silver and colloidal metal staining (Chapters 6.1.1 and 6.1.3). This chapter outlines an alternate approach for monitoring protein transfer. Transferred proteins are derivatized with a hapten and antihapten antibodies are then used to visualize the derivatized proteins immunochemically. Although this chapter describes one such method in detail, the principles can be used to construct conceptually similar detection systems.

The derivatization reaction described in this chapter involves incubation of the blot with pyridoxal 5′-phosphate, reduction with sodium borohydride, and detection with anti-phosphopyridoxyl antibodies. Subsequent immunovisualization is performed with anti-antibodies and horseradish peroxidase staining. This detection method is one of the few general protein staining techniques available for detecting proteins on both nitrocellulose and Zetabind membranes.

EQUIPMENT AND REAGENTS

Sodium dodecyl sulfate-polyacrylamide gel electrophoresis (SDS-PAGE) is performed on a Hoefer SE600 apparatus, San Francisco. Electroblotting is performed on a Bio-Rad Trans-Blot cell, Richmond, Calif. Nitrocellulose (BA80, pore size 0.15 μm) is a product of Schleicher & Schuell, Keene, N.H. Calf serum is heat-inactivated by heating to 56°C for 30 min.

Pyridoxal 5′-phosphate, silver nitrate, sodium borohydride, dinitrofluorobenzene (DNFB), BSA, and 3,3′-diaminobenzidine were purchased from Sigma Chemical Company, St. Louis. Zetabind (Zetaprobe blotting membrane) is purchased from Bio-Rad. Rabbit antiserum against dinitrophenol (DNP) and peroxidase antiperoxidase are from Miles Laboratories, Naperville, Ill. Goat antimouse $F(ab')_2$ antibodies, biotin-conjugated $F(ab')_2$ fragment of goat antimouse IgG (H and L chains) antibodies, and horseradish peroxidase-conjugated avidin are purchased from Cappel Laboratories, Malvern, Pa. Mouse monoclonal antiphosphopyridoxyl antibodies are from a previous study.[1] All other chemicals and reagents are reagent grade or the highest quality available.

PROCEDURES

Detection of Pyridoxal 5′-Phosphate-Derivatized Proteins with Anti-phosphopyridoxyl Antibodies

Rat liver cytosolic preparations are obtained by homogenizing liver in 3 vol (v/w) of 0.25 *M* sucrose and centrifugation at 100,000 × *g* for 1 hr at 4°C.[1] Cytosolic extracts are subjected to SDS-PAGE by the method of Haas and Kennett[2] using a 7.5% running gel and a 5% stacking gel.[3] Gels are 1.5-mm thick and contain the indicated amounts of protein per well. The histone H1 fractions of rat liver chromatin are fractionated by the method of Chiu et al.[4] and prepared for electrophoresis by a modification of the proceudre of Glass et al.[5] Histone samples are applied to the gel in equal volumes of a solution containing 0.1% (w/

v) SDS, 0.1% (v/v) 2-mercaptoethanol, and 8 *M* urea in 0.1 *M* sodium phosphate buffer (pH 7.0) with bromophenol blue as the tracking dye. Gels are run at ambient temperature for approximately 19 hr at 70 mA according to the procedure of Prince and Campbell.[6] Protein concentrations are determined by the method of Lowry et al.[7] Silver staining of SDS-PAGE gels is by the method of Wray et al.[8]

Following SDS-PAGE the polyacrylamide gels are soaked 3 × 20 min in the electroelution buffer of Towbin et al.[9] The pH of the electroelution buffer is adjusted to 8.8 with saturated sodium hydroxide for chromatin and histone samples.[10] Electroblotting of nonhistone samples is performed at 200 mA for 4 hr at room temperature. Electroblotting of histone samples is performed at 400 mA for 8 hr at 4°C. Electroblotting to Zetabind membrane[11] is identical to that for nitrocellulose.

Monoclonal antibodies exhibiting various specificities for pyridoxal 5′-phosphate were prepared from an antigen preparation derivatized by reaction with pyridoxal 5′-phosphate and sodium borohydride.[1]

Derivatization with Pyridoxal 5′-Phosphate

After electroelution of proteins from gels to nitrocellulose filter or Zetabind, a positively charged nylon membrane, the blots are washed in 10 m*M* phosphate-buffered saline (PBS), pH 8.0, for 5 min at room temperature. The matrix-immobilized proteins are derivatized by incubation with shaking at room temperature for 20 min in a 0.3 m*M* pyridoxal 5′-phosphate solution. The pyridoxal 5′-phosphate solution is prepared just before use in PBS, the pH is adjusted to 8.0 with saturated sodium hydroxide, and the solution is briefly gassed with nitrogen (5 to 10 min) in a brown or foil-wrapped bottle. The blots are then washed briefly (5 min) in PBS, pH 8, and incubated in a freshly prepared 63 m*M* solution of sodium borohydride in PBS, pH 8, for 20 min at room temperature with shaking. Excess sodium borohydride is removed by five washings in PBS, pH 7.3, 5 min each at room temperature.

Blocking Reactions

The blocking of nitrocellulose blots involves incubation at 37°C with shaking for 1 hr in a solution of 1% (w/v) bovine serum albumin (BSA) and 2.5% (v/v) dialyzed human plasma in 10 m*M* PBS, pH 7.3. The human plasma is previously dialyzed for 24 hr at 4°C against 5 m*M* hydroxylamine in PBS, pH 7.3, followed by dialysis for 24 hr, 4°C against PBS, pH 7.3. This treatment removes 90 to 95% of the endogenous pyridoxal 5′-phosphate.[1] The blocking procedure for Zetabind blots requires incubation at 45°C for 12 hr in PBS, pH 7.3, containing 10% (w/v) BSA and 2.5% (v/v) hydroxylamine-dialyzed human plasma. Quenched blots are then washed three times, 5 min each wash, in PBS, pH 7.3.

Immunolocalization and Visualization Procedures

Mouse monoclonal antiphosphopyridoxyl antibody is from a 50% saturation ammonium sulfate precipitation of antibody-containing ascites fluid from mice inoculated with hybridoma E6(2)2.[1] The antibody is reconstituted to its original volume of ascites fluid and then diluted 1:100,000 in 1% (w/v) BSA in PBS, pH 7.3. The blots are incubated in this solution, with shaking, at 37°C for 90 min followed by an overnight incubation at 4°C. Washing of the blots is performed at room temperature in PBS, pH 7.3, containing 0.05% (v/v) Tween® 20, with three washes of buffer, 10 min per wash.

Two methods are used for immunoblot detection involving either horseradish peroxidase-conjugated anti-antibody or biotin-conjugated anti-antibody followed by horseradish peroxidase-conjugated avidin. For the first method, horseradish peroxidase-conjugated goat antimouse $F(ab')_2$ antibody is diluted 1:10,000 in 1% BSA, PBS, pH 7.3 (Figure 1). The second method (Figures 1 and 2) uses a 1:10,000 dilution of biotin-conjugated $F(ab')_2$ fragment of goat antimouse IgG (H and L chains) in 1% (w/v) BSA, PBS, pH 7.3. Blots

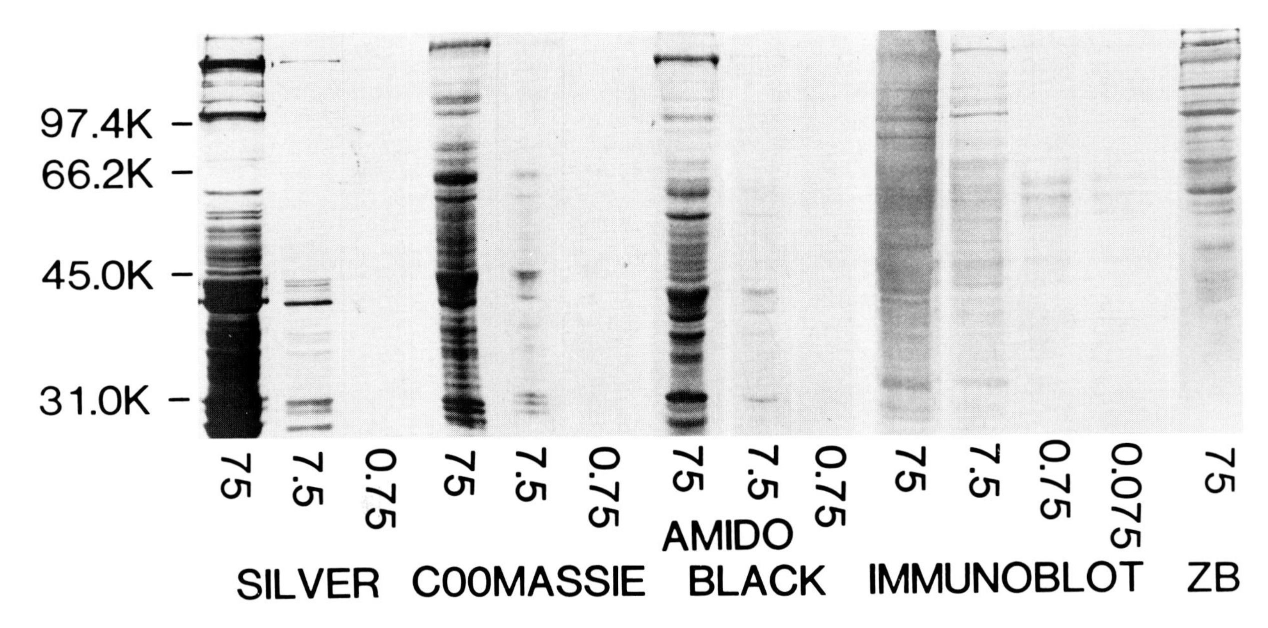

FIGURE 1. Protein visualization in gels with silver and Coomassie and after blotting with Amido Black and immunostaining of pyridoxal 5′-phosphate-derivatized proteins. Rat liver cytosolic extracts of samples containing the indicated micrograms of protein are subjected to SDS-PAGE. Gels are stained with silver stain[8] or Coomassie stain. Following electroelution, nitrocellulose blots are stained directly with Amido Black, and with immunostaining after derivatization of the proteins with pyridoxal 5′-phosphate and sodium borohydride. The latter is developed with monoclonal antiphosphopyridoxyl antibodies, biotin-conjugated anti-antibody, and peroxidase-conjugated avidin using 3,3′-diaminobenzidine staining. The lane marked ZB uses Zetabind membrane instead of nitrocellulose as the immobilizing matrix and employs peroxidase-labeled anti-antibody system of staining with the same substrate. Protein determinations are by the method of Lowry et al.[7] The positions of molecular weight markers are indicated. (From Kittler, J. M. et al., *Anal. Biochem.*, 137, 210, 1984. With permission.)

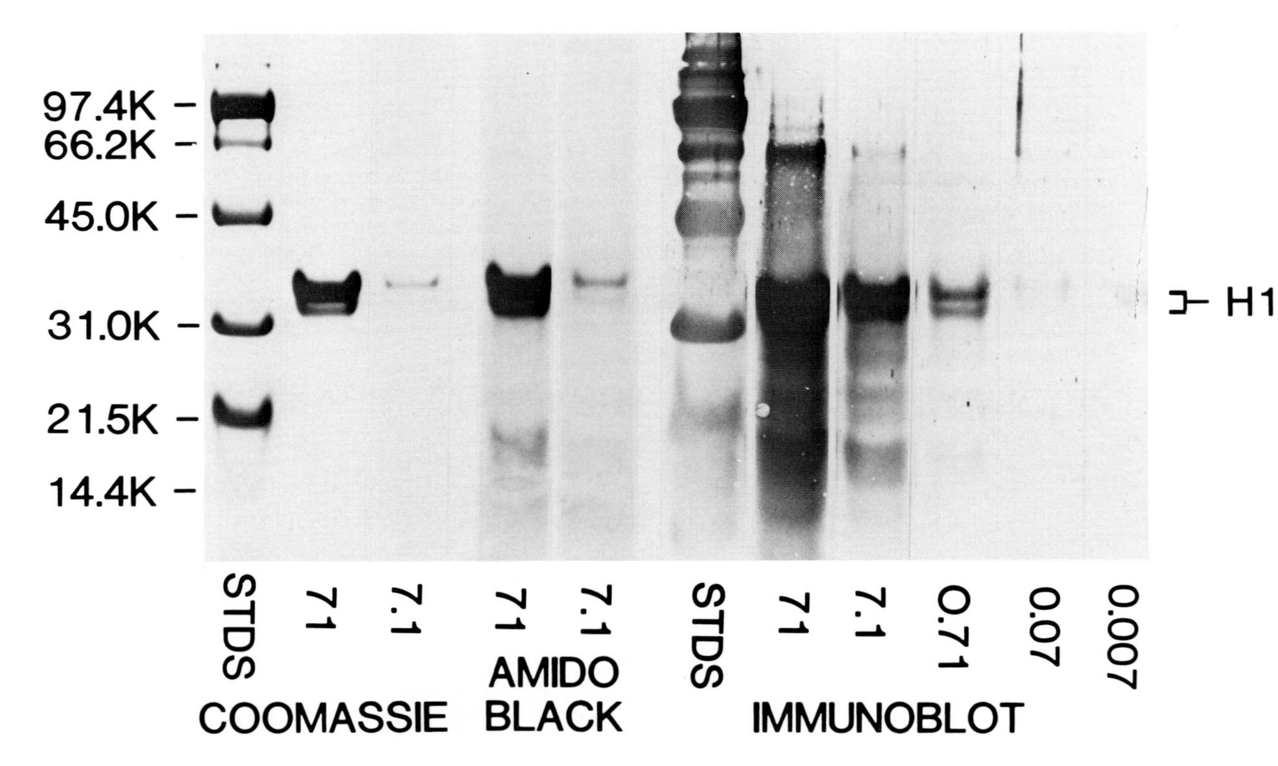

FIGURE 2. Protein visualization of the histone H1 fraction of rat liver chromatin in gels with Coomassie, and after blotting with Amido Black and immunostaining following pyridoxal 5′-phosphate derivatization. Histone H1 is prepared by the fractionation of chromatin by the method of Chiu et al.[4] and samples containing the indicated micrograms of protein are analyzed by SDS-PAGE. Electroelution is performed in 25 m*M* Tris, 192 m*M* glycine, 20% methanol (pH adjusted to 8.8 with saturated sodium hydroxide) at 400 mA for 8 hr at 4°C using a Bio-Rad Trans-Blot cell. Immunochemical detection of matrix-bound proteins with monoclonal antiphosphopyridoxyl antibodies is performed by the peroxidase-labeled second antibody staining system using 3,3′-diaminobenzidine as the substrate. The positions of molecular weight markers are indicated at the left edge of the figure in the lane marked Stds (standards). H1 designates the position of the histone H1 fraction. (From Kittler, J. M. et al., *Anal. Biochem.*, 137, 210, 1984. With permission.)

are incubated in the appropriate second antibody solution for 1 hr at 37°C, followed by an overnight incubation at 4°C if desired. For blots treated with the biotin-avidin system, the second antibody solution is decanted and the blots washed three times, 10 min per wash in 0.05% Tween® 20 in PBS, pH 7.3. Horseradish peroxidase-conjugated avidin diluted 1:10,000 in 1% BSA, PBS, pH 7.3, is added and incubated for 90 min at 37°C, then overnight at 4°C. Prior to staining, the blots are washed three times, 10 min per wash, with 0.05% Tween® 20 in PBS, pH 7.3. Blots are visualized by staining in a solution containing 30 mg of 3,3′-diaminobenzidine in 100 mℓ of 50 m*M* Tris-HCl buffer, pH 7.6, containing 17 μℓ of 30% (w/w) hydrogen peroxide. This solution is gassed with nitrogen for several minutes and filtered before use through one layer of Whatman No. 1 filter paper. Staining is performed at room temperature until the desired color intensity is obtained — usually within 5 to 10 min. The staining solution is removed by washing with water. Since nitrocellulose and Zetabind are fragile when dry, immunoblots are routinely stored in water prior to photographing to keep the blots more flexible. The 3,3′-diaminobenzidine staining procedure provides a permanent record of the protein pattern on the immobilizing matrix that is not subject to fading.

Immunodetection of Proteins by Other General Methods

Wojtkowiak et al.[12] developed a method for the visualization of polypeptides on nitrocellulose membranes in which the proteins are reacted with 2,4-dinitrofluorobenzene (DNFB) and detected with antiserum to dinitrophenol (DNP). Folowing transfer of the proteins from the polyacrylamide gels to the nitrocellulose sheets,[9,13] the sheets are incubated for 10 min at room temperature in DNFB at a concentration of 0.00001 to 0.1% (w/v) in 50 m*M* $NaHCO_3$ in 50% (v/v) dimethylsulfoxide (DMSO).[12] Excess DNFB is removed by several washes in 50 m*M* $NaHCO_3$ in 50% DMSO. Blots are then washed in 10 m*M* PBS, pH 7.2, five times, 5 min per wash; incubated in 3% (w/v) BSA, 10% (v/v) heat-inactivated calf serum in PBS, pH 7.2, at 40°C for 60 min; and then incubated overnight at 4°C in rabbit antiserum to DNP diluted 1:200 in 3% (w/v) BSA, 10% (v/v) calf serum in PBS, pH 7.2. The blots are then washed four times, 5 min per wash, in PBS, pH 7.2, and incubated with goat antirabbit IgG diluted 1:40 in 3% (w/v) BSA, 10% (v/v) calf serum in PBS, pH 7.2, for 30 min at room temperature. The blots are washed as before, and then incubated with a 1:100 dilution of peroxidase-antiperoxidase (from rabbit) in 3% (w/v) BSA, 10% (v/v) calf serum in PBS, pH 7.2, for 30 min at room temperature. Following three washes, 5 min per wash in PBS, the blots are stained with the 3,3′-diaminobenzidine staining solution as described earlier for phosphopyridoxyl proteins.

A third general detection method has been developed by Grøndahl-Hansen et al.[14] Following SDS-PAGE and electroblotting, the proteins on the blots are reacted with 2,4,6-trinitrobenzene sulfonic acid. Staining of the trinitrophenyl-derivatized proteins is performed with a monoclonal antitrinitrophenyl antibody and a peroxidase-conjugated anti-antibody. Like the other methods described to date, trinitrophenyl derivatization occurs at primary amino groups, resulting in a sensitivity that varies among proteins with different lysine contents. This method is both rapid and sensitive.

COMMENTS ON THE PROCEDURE

The usefulness of pyridoxal 5′-phosphate as a protein-labeling reagent and the availability of monoclonal antibodies against pyridoxal 5′-phosphate have been employed in the development of a general immunochemical method for detecting proteins on blots. At neutral pH, pyridoxal 5′-phosphate will form a Schiff base with primary amino groups such as the ϵ-amino group of lysine side chains. To covalently attach the pyridoxal 5′-phosphate, the Schiff's base is reduced with sodium borohydride, producing the 5′-phosphopyridoxyl group against which monoclonal antibodies are produced. Some of the proteins that have been derivatized with pyridoxal 5′-phosphate as a reagent for many kinds of biochemical analyses include lactic dehydrogenase,[15] reovirus transcriptase,[16] α-thrombin,[17] and ribulose-1,5 bisphosphate carboxylase/oxygenase.[18]

Figure 1 shows the results when SDS-PAGE separated rat liver cytosolic extracts are electroblotted and subjected to derivatization with pyridoxal 5′-phosphate and subsequent immunodetection. The proteins on the polyacrylamide gels are visualized by staining with silver[8] or Coomassie brilliant blue. Nitrocellulose blots are stained with Amido Black. Phosphopyridoxyl proteins on the nitrocellulose immunoblots are detected by monoclonal antiphosphopyridoxyl antibodies which are subsequently visualized by the avidin/biotin method. The Zetabind blot is detected in the same way, but the visualization is performed with the horseradish peroxidase-conjugated goat antimouse antibody. The results indicate that the immunoblot procedure developed for detecting phosphopyridoxyl-proteins using antiphosphopyridoxyl antibodies is at least as sensitive as the silver staining method used on polyacrylamide gels and approximately 100-fold greater in sensitivity than Amido Black staining of nitrocellulose blots. The Zetabind blot lane is included to demonstrate that this method is applicable for detecting proteins transferred to Zetabind membrane.

Figure 2 shows the results when the H1 histone fraction of rat liver chromatin is subjected to SDS-gel electrophoresis and nitrocellulose blotting. Figure 2 demonstrates that immunochemical detection of phosphopyridoxyl-derivatized histone H1 is 100 times or more sensitive than either Coomassie or Amido Black staining procedures. The method is capable therefore of detecting nanogram amounts of fractionated histones on nitrocellulose.

The actual sensitivity of the phosphopyridoxyl/antiphosphopyridoxyl immunoblot staining techniques is apparent from Figures 1 and 2. On nitrocellulose, a mixture of proteins such as rat liver cytosol can be detected to as low as 75 ng total protein. When a single protein such as histone H1 is analyzed by this method, as little as 7 ng can be detected. For Zetabind, a sensitivity at least equal to that of a silver-stained gel is achieved. It should be noted that conditions for optimal sensitivity with Zetabind blots have not been worked out and further analysis may greatly improve the sensitivity of this system.

The washing and incubation conditions described were chosen for this system for several reasons. Incubations for 90 min at 37°C followed by overnight incubations at 4°C allow for decreased backgrounds and second and third antibodies or proteins can be used at high dilutions (1:10,000). Shorter incubation times give slightly less optimal results.[19] The biotin-conjugated second antibody and avidin-peroxidase method gives increased sensitivity over the peroxidase-conjugated anti-antibody method.[20] The peroxidase anti-peroxidase method employed by Wojtkowiak et al.[12] results in 100-fold sensitivity over Amido Black staining of nitrocellulose transfers, but utilizes a 1:100 dilution of the peroxidase-antiperoxidase complex. For peroxidase-antiperoxidase staining, the first and third antibody reagents must be from the same species, restricting the types of antibodies or immune sera one can develop. Wojtkowiak et al.[12] use rabbit antiserum to DNP and peroxidase-antiperoxidase complex from rabbits, with a goat antirabbit linker antibody. Alternately, a biotin-conjugated second antibody can be prepared using any affinity-purified anti-antibody of choice.

The use of 3,3′-diaminobenzidine as the precipitating substrate for the enzyme horseradish peroxidase produces a permanent record of the stained immunoblot. Caution should be taken when handling 3,3′-diaminobenzidine since it is a suspected carcinogen.

EVALUATION

The methods described for detecting proteins on blots by using antihapten antibodies involve derivatization of proteins with one of the following: pyridoxal 5′-phosphate, trinitrobenzene sulfonic acid, or dinitrofluorobenzene. In all cases, derivatization occurs at primary amino groups, such as the ϵ-amino groups of lysine side chains. Although the methods described show increased sensitivity (10- to 100-fold) over Amido Black stain for staining proteins on nitrocellulose and share the properties of simplicity and general applicability, there are major differences among the methods. The DNFB/anti-DNP procedure of Wojtkowiak et al.[12] uses an organic solvent, DMSO, and antisera from different sources which may introduce biological variability.[21] The phosphopyridoxyl/antiphosphopyridoxyl method uses mild aqueous reaction conditions, monoclonal antibodies which eliminate the inherent biological variability present among different antisera, and requires second and third antibodies at low concentrations (1:10,000 dilutions). In addition, it is applicable for use on Zetabind membrane, or any other matrix that does not react with pyridoxal 5′-phosphate, providing, to our knowledge, the first general method for detecting proteins transferred to this matrix. The trinitrobenzene sulfonic acid method is both rapid and sensitive.[14]

All methods described depend on amino group availability and reactivity. Thus, proteins that are lysine-rich should stain better than lysine-poor proteins. This may, in part, explain the sensitivity of the histone H1 staining (Figure 2). On a molar basis, larger M_r proteins containing an average percentage of lysine residues would stain with greater intensity than smaller M_r proteins containing the same percentage of lysines. The smaller M_r proteins would contain fewer possible derivatization sites.

In theory, any general derivatization agent for proteins can be used provided that antibodies or antisera exist that specifically bind the hapten. However, consideration must be given to the conditions that will be needed to derivatize the proteins, and the derivatizing agent must not react with the immobilizing matrix in a nonspecific manner. In developing a general immunodetection method the following questions must be asked. Will the derivatization depend on the availability of certain components of a protein? In the methods available at present, primary amino groups are used for derivatization. Will it be proportional to the M_r and stain on a molar basis? Assuming an average content of derivatizing groups, staining should be proportional to the M_r. Will it stain proteins equally, or will it prefer certain subclasses of proteins over others? For the methods described, lysine-rich proteins would be expected to stain with greater intensity. Can it be used for quantitation of proteins, or merely for visualization purposes? Quantitation of the same protein in different samples is possible using a scanning densitometer. Comparisons of different proteins with dissimilar derivatization determinants would not be as straightforward, for the reasons described earlier. All the advantages and disadvantages of a proposed detection procedure must be considered along with its general applicability and ease of performance in the development of immunovisualization procedures for proteins on blots.

ACKNOWLEDGMENT

Work from the authors' laboratories was supported by U.S.P.H.S. grants AM25490 and CA35878.

REFERENCES

1. **Viceps-Madore, D., Cidlowski, J. A., Kittler, J. M., and Thanassi, J. W.,** Preparation, characterization, and use of monclonal antibodies to vitamin B-6, *J. Biol. Chem.*, 258, 2689, 1983.
2. **Haas, J. B. and Kennett, R. H.,** Characterization of hybridoma immunoglubulins by sodium dodecyl sulfate-polyacrylamide gel electrophoresis, in *Monoclonal Antibodies,* Kennett, R. H., McKearn, T. J., and Bechtol, K. B., Eds., Plenum Press, New York, 407, 1980.

3. **Kittler, J. M., Viceps-Madore, D., Cidlowski, J. A., and Thanassi, J. W.,** Immunoblot detection of pyridoxal phosphate binding proteins in liver and hepatoma cytosolic extracts, *Biochem. Biophys. Res. Commun.*, 112, 61, 1983.
4. **Chiu, J. -F., Fusitani, H., and Hnilica, L. S.,** Methods for selective extraction of chromosomal nonhistone proteins, in *Methods in Cell Biology*, Vol. 16, Stein, G., Stein, J., and Kleinsmith, J., Eds., Academic Press, New York, 1977, 283.
5. **Glass, W. F., Briggs, R. C., and Hnilica, L. S.,** Identification of tissue-specific nuclear antigens transferred to nitrocellulose from polyacrylamide gels, *Science*, 211, 70, 1981.
6. **Prince, L. O. and Campbell, T. C.,** Effects of sex difference and dietary protein level on the binding of aflatoxin B_1 to rat liver chromatin proteins *in vivo*, *Cancer Res.*, 42, 5053, 1982.
7. **Lowry, O. H., Rosebrough, N. J., Farr, A. L., and Randall, R. J.,** Protein measurement with the folin phenol reagent, *J. Biol. Chem.*, 193, 265, 1951.
8. **Wray, W., Boulikas, T., Wray, V. P., and Hancock, R.,** Silver staining of proteins in polyacrylamide gels, *Anal. Biochem.*, 118, 197, 1981.
9. **Towbin, H., Staehelin, T., and Gordon, J.,** Electrophoretic transfer of proteins from polyacrylamide gels to nitrocellulose sheets: procedure and some applications, *Proc. Natl. Acad. Sci. U.S.A.*, 76, 4350, 1979.
10. **Kittler, J. M., Meisler, N. T., Viceps-Madore, D., Cidlowski, J. A., and Thanassi, J. W.,** A general immunochemical method for detecting proteins on blots, *Anal. Biochem.*, 137, 210, 1984.
11. **Gershoni, J. M. and Palade, G. E.,** Electrophoretic transfer of proteins from sodium dodecyl sulfate-polyacrylamide gels to a positively charged membrane filter, *Anal. Biochem.*, 124, 396, 1982.
12. **Wojtkowiak, Z., Briggs, R. C., and Hnilica, L. S.,** A sensitive method for staining proteins transferred to nitrocellulose sheets, *Anal. Biochem.*, 129, 486, 1983.
13. **Laemmli, U. K.,** Cleavage of structural proteins during the assembly of the head of bacteriophage T4, *Nature (London)*, 277, 680, 1970.
14. **Grøndahl-Hansen, J., Huang, J. -Y., Nielsen, L. S., Andreasen, P. A., and Dano, K.,** General detection of proteins after electroblotting by trinitrobenzene sulphonic acid derivatization and immunochemical staining with a monoclonal antibody against the trinitrophenyl group, *J. Biochem. Biophys. Methods*, 12, 51, 1986.
15. **Gould, K. G. and Engel, P. C.,** The reactions of pyridoxal 5′-phosphate with the M_4 and H_4 isoenzymes of pig lactate dehydrogenase, *Arch. Biochem. Biophys.*, 215, 498, 1982.
16. **Morgan, E. M. and Kingsbury, D. W.,** Pyridoxal phosphate as a probe of reovirus transcriptase, *Biochemistry*, 19, 484, 1980.
17. **Griffith, M. J.,** Covalent modification of human α-thrombin with pyridoxal 5′-phosphate, *J. Biol. Chem.*, 254, 3401, 1979.
18. **Paech, C. and Tolbert, N. E.,** Active site studies of ribulose-1,5-bisphosphate carboxylase/oxygenase with pyridoxal 5′-phosphate, *J. Biol. Chem.*, 253, 7864, 1978.
19. **Kittler, J. M. and Thanassi, J. W.,** unpublished results.
20. **Kittler, J. M. and Thanassi, J. W.,** Immunochemistry and immunoassays of vitamin B_6 and derivatives, in *Coenzymes and Cofactors*, Vol. 2, *Pyridoxal Phosphate and Derivatives*, Dolphin, D., Poulson, R., and Avramovic, P., Eds., John Wiley & Sons, New York, 1986, 295.
21. **Galfre, G. and Milstein, C.,** Preparation of monoclonal antibodies: strategies and procedures, in *Methods in Enzymology*, Vol. 73, Part B, Langon, J. J. and Vunakis, H. V., Eds., Academic Press, New York, 1981, 3.

6.5 DESORPTION OF ANTIBODIES FOR REUSE OF BLOTS

Johan Geysen

INTRODUCTION

It is possible to visualize different antigens in the same specimen in two immunostaining sequences. Vandesande[1] has elaborated such a technique for tissue sections antigens using peroxidase-antiperoxidase (PAP) immunostaining.

The introduction of PAP immunostaining of vacuum-blotted antigens by Peferoen et al.[2] and the publication of a variety of desorption protocols for "erasing" first antibody and iodine-labeled protein A from blots[3-8] inspired us to explore the potential of double immunostaining of antigens vacuum blotted to nitrocellulose. The principle is simple: following PAP immunostaining of the first antigen with 3,3′-diaminobenzidine (DAB) as a chromogen, the primary antibody, link antibody, and PAP complex are desorbed, while the chromogen deposits are left in place. Subsequently, the other antigen can be visualized with the PAP method, in combination with 4-chloronaphtol, which yields a bluish instead of a brownish precipitate.

This desorption protocol eliminated cross-contamination, causing mixed colors as a result of peroxidase adsorbed during the first immunostaining, which precipitates 4-chloronaphtol or as a result of primary antibody, which will be detected by the link antibody in the course of the second immunostaining.

For blotted antigens, we could use a desorption procedure similar to one for tissue section antigens: it combines low pH, low ionic strength, and the presence of the chaotropic agent 2,2′ dimethylformamide with repeated changes from these "harsh" to "mild" conditions. The latter treatment proved to be a valuable alternative to electrophoretic removal of antibodies and immunochemicals.

EQUIPMENT AND REAGENTS

Materials

Incubations and desorption of blots were performed in heat-sealed soft plastic bags measuring 15 × 20 cm. Washing steps and chromogen reaction were performed in Petri dishes. Heat sealing can be performed with any available apparatus for sealing frozen foods. Throughout the procedure, blots were mechanically shaken on a "rocking table" (REAX 3, Heidolph, West Germany).

Chemicals

All chemicals were of analytical grade. 3,3′-Diaminobenzidine and 4-chloronaphtol were obtained from Sigma and Janssen Chimica, respectively. Nitrocellulose sheets (0.45 μm) and nitrocellulose disk filters (diameter 25 mm and 0.22 μm pore size) were from Millipore.

Immunochemicals

1. Normal goat serum was obtained by bleeding a goat from the carotid artery into vacuum tubes. After clotting and centrifugation, the serum was frozen and stored at −25°C.
2. Primary antibody was obtained by immunizing rabbits with either lipoprotein bands from polyacrylamide gradient gels or vitellogenin bands from SDS-PAGE (sodium dodecyl sulfate-polyacrylamide gel electrophoresis) gradient gels. The antigen sources were male hemolymph and egg homogenate of the blowfly *Sarcophaga bullata*, respectively.[9]

3. Link antibodies were produced in goat by injecting chromatographically pure rabbit immunoglobulin from Nordic.
4. Peroxidase-antiperoxidase complex was manufactured according to the method of Vandesande[10] and is commercially available from UCB Bioproducts, Gent, Belgium. All immunochemicals, except normal goat serum, were distributed in aliquots sufficient for one blot, frozen in liquid nitrogen and stored at −25°C.

Buffer Stocks

To be stored at 4°C.

1. Blot buffer: Dissolve 3.03 g Tris and 11.26 g glycine in distilled water and adjust volume to 1000 mℓ. Add 20% methanol, by volume.
2. Tris-buffered saline (TBS): Dissolve 1.21 g Tris and 9.00 g NaCl in 800 mℓ of distilled water and adjust pH to 7.60 with 1 N HCl. Add 1 mℓ Triton® X-100 and 0.04 g Merthiolate® and adjust volume to 1000 mℓ with distilled water.
3. Tris buffer: Dissolve 6.06 g Tris in 800 mℓ of distilled water, adjust pH to 7.60 with 1 N HCl, and add distilled water to a volume of 1000 mℓ.
4. Glycine buffer: Dissolve 15.02 g glycine in 800 mℓ of distilled water and adjust to pH 2.2 with 6 N HCl. Add distilled water to a volume of 1000 mℓ.

Prepare the following solutions immediately before use.

5. Blocking buffer: Mix 1.2 mℓ normal goat serum and 10.8 mℓ TBS.
6. Washing buffer: After blocking, the blocking buffer can be recuperated and diluted to a volume of 500 mℓ with TBS; this buffer can be used for washing.
7. Incubation solutions: The following dilutions in washing buffer are recommended: 1/1000 for primary antibody, 1/20 to 1/50 for link antibody, and 1/300 up to 1/1000 for PAP complex in a volume of 10 mℓ per blot.
8. 2,2′-Diaminobenzidine-peroxide solution: Dissolve 12 mg 2,2′ diaminobenzidine in 100 mℓ of Tris buffer, pH 7.6, force through a 0.22 μm disk filter with a 50 mℓ syringe and store in the dark. Add 10 μℓ of 33% (w/v) hydrogen peroxide just before use and shake.
9. Desorption medium: Mix glycine buffer, 2,2′-dimethylformamide and distilled water (1:2:4, by volume). For alternative buffers see Table 1.
10. 4-Chloronaphtol-peroxide solution: Dissolve 50 mg 4-chloronaphtol in 10 mℓ ethanol (100%), add 90 mℓ of Tris buffer, pH 7.6 and allow precipitate to form. Force through a 0.22 μm disk filter and keep in the dark. Add 10 μℓ of 33% (w/v) hydrogen peroxide immediately before use.

PROCEDURES

Electrophoresis and Blotting

Hemolymph of female *Sarcophaga bullata* blowflies was subjected to SDS-PAGE following the Laemmli system on 5 to 15% acrylamide gels.[11] Gels were packed in plastic bags, heat sealed, and frozen at −80°C on a copper plate measuring 16 × 16 × 1 cm^3. Prior to blotting, frozen gels were thawed by immersion in blotting buffer for 20 min. Protein patterns were transferred to nitrocellulose by means of vacuum blotting (see Chapter 4.2).

First Immunostaining

1. Wash the blot or moisten dried blot in TBS for a few minutes.
2. Quench blot for 30 min in blocking buffer.

Table 1
COMPOSITION OF BUFFERS AND BLOT TREATMENTS, DEVELOPED FOR DESORBING ANTIBODY FROM BLOTS

No.	Composition	Treatments	Ref.
1.	0.05 *M* Na-phosphate buffer, pH 7.6 2% (w/v) SDS 0.1 *M* 2-mercaptoethanol	60°C 60 min	3, (4)
2.	0.05 *M* Na-phosphate buffer, pH 7.6 0.154 *M* NaCl 10 *M* urea 0.1 *M* 2-mercaptoethanol	60°C 30 min	5, (6)
3.	0.01 *M* Tris/HCl, pH 7.6 3 *M* NaSCN	37°C 60 min	7
4.	0.04 *M* glycine/HCl, pH 2.2 30% (v/v) 2,2′-dimethylformamide	22°C Dipping procedure	9
5.	0.04 *M* glycine/HCl, pH 2.2 2% (w/v) SDS	60°C 30 min	—
6.	0.1 *M* glycine/HCl, pH 2.2	22°C 90 min	8

Note: References in parentheses refer to related techniques.

3. Transfer the blot to primary antibody appropriately diluted with washing buffer for 1 hr at room temperature.
4. Wash five times for 1 min with washing buffer.
5. Transfer the blot into link antibody solution (dilution 1/20 up to 1/50) for 30 min.
6. Wash as for step 4.
7. Transfer the blot into PAP solution diluted 1/300 up to 1/1000 for 30 min on the rocking table.
8. Wash five times 1 min with washing buffer, followed by five times 1 min with Tris buffer, pH 7.6.
9. Incubate with the 2,2′-diaminobenzidine-peroxide solution until the bands are clearly seen (after a few minutes).
10. Wash with distilled water and proceed with desorption or dry the blot.

Desorption Protocol

11. Wash the stained blot in TBS for 2 min.
12. Shake the blot on the rocking table in the desorption buffer for 2 min.
13. Repeat steps 11 and 12 twice and finish desorption by again performing step 11.

Second Immunostaining

14. Repeat steps 2 to 8 of the first immunostaining with the second primary antibody.
15. Incubate with the 4-chloronaphtol-peroxide solution.
16. Discard the chromogen solution and wash with distilled water.

The wet blot can now be photographed and then dried for storage.

USE AND LIMITATIONS

The original desorption protocol with three cycles of ''denaturing'' conditions has been applied succesfully in combination with antibody preparations and the PAP technique. The results are shown in Figure 1. Later, when we applied double immunostaining of blots with new antibody preparations against the same antigens but of higher avidity, we were unable to remove all the antibody from the blot with the described desorption technique. As a consequence, the 2,2′-diaminobenzidine-stained bands developed an intermediate brownish blue color, which can be distinguished from chloronaphtol-stained bands. However, we are not satisfied with this rebound. For this reason, in the following, we concentrate on alternative protocols and strategies which may help in finding the suitable protocol for the desorption.

STRATEGY FOR DESORPTION PROTOCOL

General Considerations

The avidity of the primary antibody is a crucial parameter for the potential of a given desorption protocol. On the other hand, a desorption protocol does not have to elute antibodies "quantitatively". As long as cross-contamination is reduced to a level below the detection limit of the immunostaining procedure, it can be applied successfully. It must be considered that desorption may exert negative effects such as: (1) denaturing of the epitopes of the immobilized compounds, (2) loosening of these compounds from the matrix, (3) destruction of or damage to the blotting matrix, and (4) dissolution of the chromogen deposits on the blot. The latter is particularly important for double immunostaining of blots. Whereas a negative effect on any of the latter three levels will be readily observed (e.g., staining of the blot negative; nitrocellulose dissolved or shrunken; chromogen deposits gone); the effects on epitopes must be evaluated separately. In general, the degree of preservation of antigenicity is on the expense of the quantitivity of desorption.

However, in many cases the blotted compounds have already been treated more harshly than during the desorption. The use of SDS and 2-mercaptoethanol in combination with elevated temperature for desorption is not likely to decrease the antigenicity of a sample separated with SDS-PAGE. Similarly, the use of urea and extreme of acidic and basic pH may be considered following isoelectric focusing or 2D-electrophoresis. Thus, there exists a reasonable margin for interfering with the immunological interactions without further affecting the antigenicity; however, the introduction of new treatments calls for control experiments.

Efficiency of Desorption and Effect on Antigenicity

An adequate control for cross contamination should include the following steps: IMMUNOSTAINING WITH OMISSION OF CHROMOGEN DEVELOPMENT → DESORPTION → IMMUNOSTAINING WITH OMISSION OF THE PRIMARY ANTIBODY.

When a desorption protocol is tested according to this principle and provided it is effective, the result will be comparable to a negative control, which is an undesorbed blot immunostained with omission of the primary antibody.

The principle outlay of an antigenicity test would be DESORPTION → IMMUNOSTAINING. A desorption protocol which has no significant effect on the epitopes will yield blots comparable to a positive control, which is an undesorbed blot stained for the given antigens.

Comparison of Desorption Protocols by Dot Immunobinding

Different buffers and protocols for desorption will now be tested by means of a dot immunobinding assay (Figures 2 and 3). As antigen, insect male hemolymph has been spotted onto nitrocellulose strips in different concentrations and pairs of strips were treated with a given desorption protocol. One strip was tested for the efficiency of desorption, while the second was screened for effects on antigenicity. A flowchart of the manipulations is presented in Figure 3, so that a maximum of steps can be performed for all strips simul-

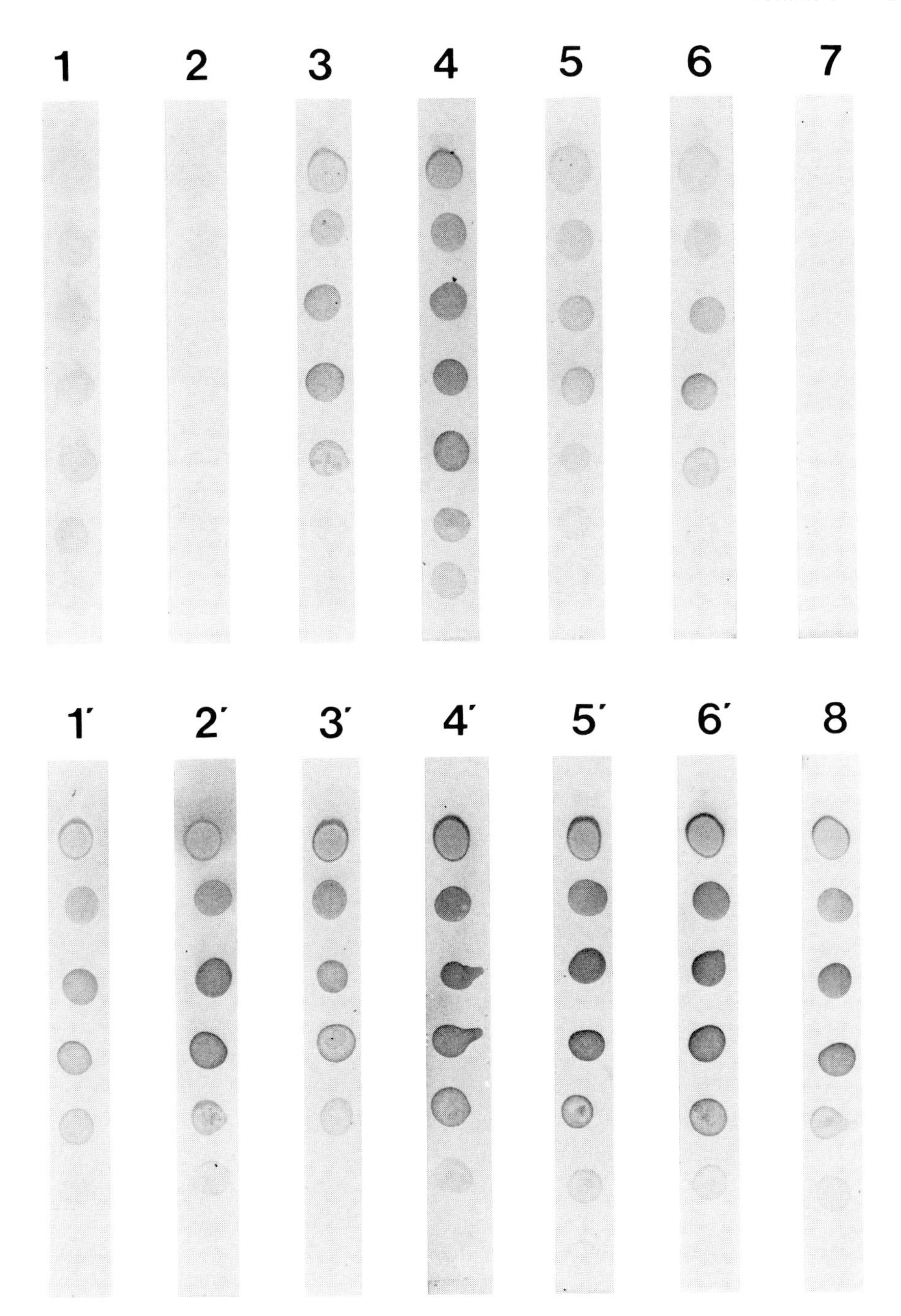

FIGURE 1. Single and double immunostaining by means of the PAP method and repeated change from "mild" to "denaturing" conditions. (A) Separation of female hemolymph proteins with SDS-PAGE and staining with Coomassie Brilliant Blue after vacuum blotting. (B) Single PAP immunostaining for vitellogenin with specific rabbit antibody diluted 1/1500, link antibody 1/20, PAP 1/600, and visualized with 2,2′-diaminobenzidine as chromogen. (C) Single immunostaining as in B, but now for lipoprotein, primary antibody diluted 1/750 and with 4-chloronaphtol as chromogen. (D) Double immunostained blot, first stained as in B, desorbed and subsequently stained as in C. Both antigens are clearly visible on the double-stained blot. (From Geysen, J., De Loof, A., and Vandesande, F., *Electrophoresis,* 5, 129, 1984. With permission.)

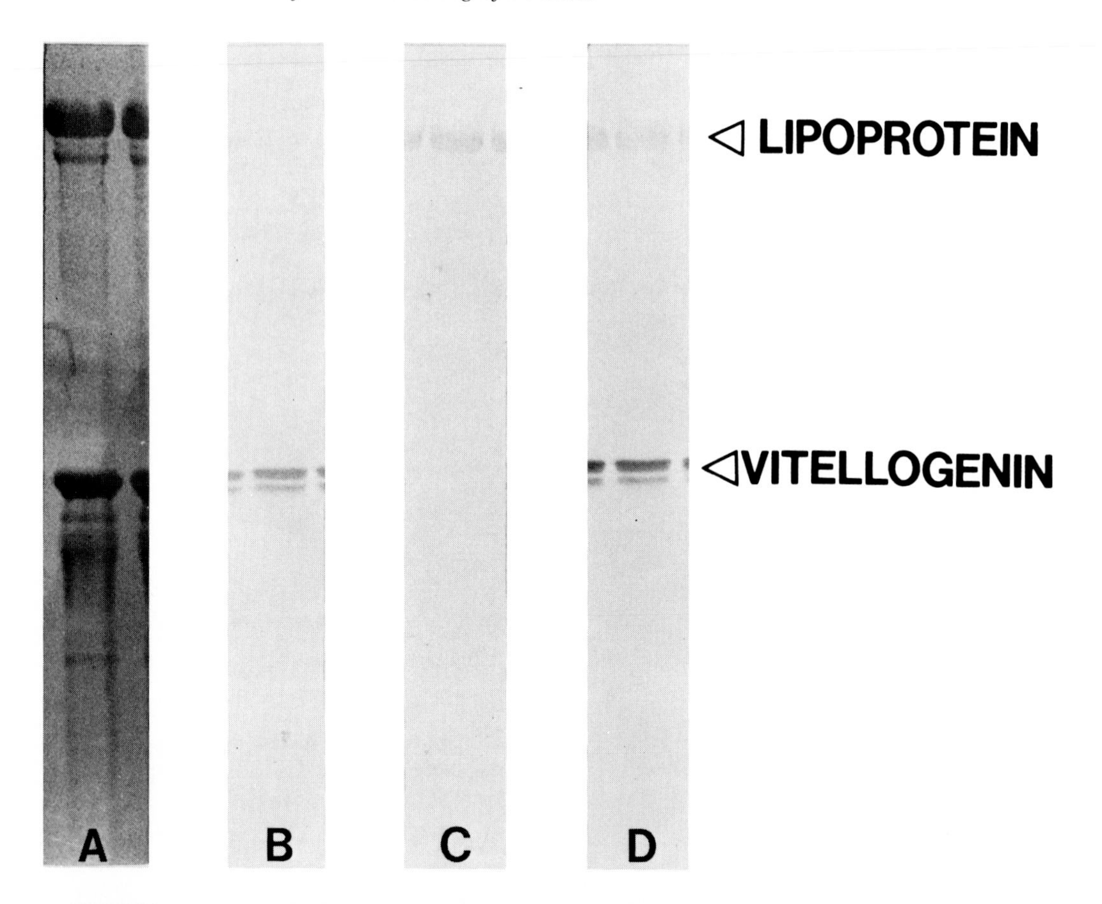

FIGURE 2. Dot immunobinding assay. Comparison of desorption protocols with respect to efficiency of desorption (upper half) and the effect on antigenicity (lower half). Insect male hemolymph was spotted in a fivefold dilution series (1/5 to 1/78125) in order to cover a wide range of antigen concentrations. Strips were treated as shown in Figure 3. High avidity antibody was applied in a dilution of 1/1000, link antibody 1/50, and PAP 1/1000. Spots were visualized with 2,2′-diaminobenzidine. Pairs of strips bearing the same number were treated (desorbed) with the buffer listed under the same number in Table 1. Strips 7 and 8 are the negative and positive controls, respectively.

taneously. The composition of the buffers is shown in Table 1 and the results are displayed in Figure 2.

The efficiency of desorption — All strips are stained less intensively than the positive control. This demonstrates that all buffers are able to remove at least part of the antibody and immunomarker from the spots. However, in comparison with the negative control, significant differences appear. Only buffer 2 desorps efficiently, while buffers 1 and 5 perform intermediately and the rest of the buffers poorly.

Effects on antigenicity — Except for strip 3, all strips of the lower half of Figure 3 are stained as intensely as the positive control; thus, the effect on the epitopes of the antigens investigated is negligible for all buffers except number 3.

Conclusion

All buffers preserve antigenicity well, however, the efficiency of desorption changes considerably from protocol to protocol. In combination with high avidity antibody and the PAP technique, only the buffer based on 10 *M* urea, 0.1 *M* 2-mercaptoethanol at 60°C was able to remove antibody and immunomarker nearly quantitatively.

TEST FOR EFFICIENCY OF DESORPTION	NEGATIVE CONTROL	POSITIVE CONTROL	EFFECT OF ELUTION ON ANTIGENICITY
MOISTEN			
QUENCH			
PRIMARY AB			
*			
LINK AB			
*			MOISTEN
PAP			QUENCH
* *			* *
DESORPTION	MOISTEN	MOISTEN	DESORPTION
*	QUENCH	QUENCH	*
*	*	PRIMARY AB	PRIMARY AB
*	*	*	*
LINK AB	LINK AB	LINK AB	LINK AB
*	*	*	*
PAP	PAP	PAP	PAP
* *	* *	* *	* *
CHROMOGEN REACTION WITH 3,3'-DIAMINOBENZIDINE			
COMPARE TO THE NEGATIVE CONTROL			COMPARE TO THE POSITIVE CONTROL

FIGURE 3. Flowchart of manipulations performed in a control assay for desorption of antibodies from nitrocellulose blots. The columns refer to experiments which can be performed as a test for desorption protocols. The principles set out in the text have been applied to PAP immunostaining. The sequence of steps which must be performed can be read vertically for each single set up. Washing steps have been represented by an asterisk. A double asterisk represents the two washing steps to be performed before desorption or chromogen reaction. Following desorption, the incubation with primary antibody has to be omitted in the test for efficiency of desorption and in the negative control. In an assay, simultaneously performed for efficiency of desorption and effect on antigenicity, this period can be bridged by treatment with washing buffer.

ALTERNATIVES AND PERSPECTIVES

There are protocols which can be performed in sequence without the risk of cross-contamination. A blot which has been treated with an iodine-labeled antibody against antigen A and autoradiographed can be reused for immunostaining of antigen B with a biotinylated antibody against B, avidin biotinylated-peroxidase, and 3,3′-diaminobenzidine. In combination with monoclonal antibodies, which can be easily labeled, this approach offers more perspectives than desorption.

Immunoenzyme and silver intensified immunometal[13] (see Chapter 6.4.4) methods both yield visible, stable, and water-insoluble deposits. These deposits tend to cover or overgrow the blotting matrix, the antigen, and the subsequent layers of immunoreagents. Thus, theoretically, deposit formation can be exploited as a means of eliminating cross-contamination. For example, when the density of the deposits makes immunochemicals irrecognizable and enzymes inactive, one can directly continue with the next immunostaining sequence.

Finally, other means of manipulation, which may improve the efficiency of desorption need attention. Besides applications of elevated temperature in combination with desorption buffers, electrophoresis might be used to remove desorbed antibody from blots as has been demonstrated for tissue sections.[1] Also, the technique of forcing a flow of desorption buffer through a blot as in vacuum blotting may be utilized.

ACKNOWLEDGMENTS

The author wishes to thank René Smisdom for technical assistance, Professors A. De Loof and F. Vandesande and M. J. Evans for critical reading and J. Puttemans for photography.

REFERENCES

1. **Vandesande, F.,** Peroxidase-antiperoxidase techniques, in *Immunohistochemistry,* Cuello, E., Ed., John Wiley & Sons, New York, 1983, 101.
2. **Peferoen, M., Huybrechts, R., and De Loof, A.,** Vacuum-blotting: a new simple and efficient transfer of proteins from sodium dodecyl sulphate polyacrylamide gels to nitrocellulose, *FEBS Lett.,* 145, 369, 1982.
3. **Symington, J., Green, M., and Brackman, K.,** Immunoautoradiographic detection of proteins after electrophoretic transfer from gels to diazo-paper: analysis of adenovirus encoded proteins, *Proc. Natl. Acad. Sci. U.S.A.,* 78, 177, 1981.
4. **Gullick, W. J. and Lindstrom, J. M.,** Structural similarities between acetylcholine receptors from fish electric organs and mammalian muscle, *Biochemistry,* 21, 4563, 1982.
5. **Renart, J., Reiser, J., and Stark, G. R.,** Transfer of protein from gels to diazobenzyloxymethyl-paper and detection with antisera: a method for studying antibody specificity and antigen structure, *Proc. Natl. Acad. Sci. U.S.A.,* 76, 3116, 1979.
6. **Erickson, P. F., Minier, L. N., and Lasher, R. S.,** Quantitative electrophoretic transfer of polypeptides from SDS polyacrylamide gels to nitrocellulose sheets: a method for their re-use in immunoautoradiographic detection of antigens, *J. Immunol. Methods,* 51, 241, 1982.
7. **Reiser, J. and Wardale, J.,** Immunological detection of specific proteins in total cell extracts by fractionating in gels and transfer to diazophenylthioether paper, *Eur. J. Biochem.,* 114, 569, 1981.
8. **Legocki, R. P. and Verma, D. P. S.,** Multiple immunoreplica technique: screening for specific proteins with a series of different antibodies using one polyacrylamide gel, *Anal. Biochem.,* 111, 385, 1981.
9. **Geysen, J., De Loof, A., and Vandesande, F.,** How to perform subsequent or "double" immunostaining of two different antigens on a single nitrocellulose blot within one day with an immunoperoxidase technique, *Electrophoresis,* 5, 129, 1984.
10. **Vandesande, F.,** Immunohistochemical double staining methods, in *Immunohistochemistry,* Cuello, E., Ed., John Wiley & Sons, New York, 257, 1983.
11. **Laemmli, U. K.,** Cleavage of structural proteins during assembly of the head of the bacteriophage T4, *Nature (London),* 227, 680, 1979.
12. **Geysen, J., De Loof, A., and Vandesande, F.,** The influence of link antibody quality and dilution on the appearance of false negativity in the PAP method, revealed by a two parameter immunospotting assay, *Histochemical J.,* in press.
13. **Moeremans, M., Daneels, G., Van Dijck, A., Langanger, G., and De Mey, J.,** Sensitive visualisation of antigen-antibody reactions in dot and blot immune overlay assays with immunogold and immunogold/silver staining, *J. Immunol. Methods,* 74, 353, 1984.

6.6 SUBSTRATE-SPECIFIC DETECTION ON NITROCELLULOSE OF SDS-PAGE SEPARATED ENZYMES EMPLOYING POLYSPECIFIC ANTIBODIES AND SECONDARY INCUBATION WITH NATIVE ENZYME-MIXTURE

Peter M. H. Heegaard

INTRODUCTION

The detection of enzyme activity associated with certain bands in SDS-PAGE (sodium dodecylsulfate-polyacrylamide gel electrophoresis) has not previously generally been possible since the denaturating conditions employed in SDS-PAGE destroy most types of enzyme activity. Therefore, a method was developed on the basis of the principles previously reported for the detection of phosphodiesterase[1] and RNA-polymerase activity.[2] It detects enzymes in complex mixtures after SDS-PAGE and electroblotting by way of their enzymatic activity, using a polyspecific antibody and a specific enzyme staining method.

The application given here concerns mouse serum α-1-esterase[3] and employs both purified and crude polyspecific rabbit antibodies against mouse serum proteins. Mouse serum was subjected to SDS-PAGE and blotting, whereafter blots were incubated with antibodies reacting with the α-1-esterase. After this incubation blots were further incubated with mouse serum, and the esterase activity bound to the intermediate antibody was then detected by a specific esterase stain.

REAGENTS

Fast Red TR salt (5-chloro-2-toluidinine diazoniumchloride-hemizincchloride) and 1-naphthyl acetate was obtained from Sigma and used in a 0.1 *M* phosphate buffer, pH 7.5, as described later. A Biorad 160/1.6 was used as a power supply for blotting. Nitrocellulose was from Schleicher & Schuell, BA 85, 0.45 μm pore size. Normal mouse serum was obtained from vena axillaris, collected in plastic tubes, clotted overnight at 4°C, centrifuged, and stored frozen until analysis. Antiserum against whole mouse serum was produced in rabbits and purified as described in Reference 4. The resulting immunoglobulin concentrations was 30 mg/mℓ. Antiserum against mouse serum α-1-proteins was also produced in rabbits.[5] This antibody preparation was used directly without further purification.

PROCEDURES

SDS-PAGE

Discontinuous SDS-PAGE was performed according to Reference 6 with modifications by Reference 7 in a homemade apparatus, according to Kerckaert.[8] Separation gels were homogenous with T = 15% and C = 1.5% and stacking gels, T = 5%, C = 1.5%. Sample buffer (5% (w/v) sucrose, 1% (w/v) SDS, 0.15 *M* Tris, 0.002 *M* EDTA, 2.5% (w/v) Pyronin Y, 0.32 *M* dithiothreitol) was diluted 1:4 with sample and boiled 2 min before the start of electrophoresis. A typical protein load was approximately 50 μg total serum proteins per lane.

Blotting Procedure

The transfer system was the semidry discontinuous system[9] (see Chapter 4.3.2). Transfer was performed at 0.8 mA/cm^2 for 1 to 1.5 hr corresponding to 10 to 30 V, the voltage increasing during transfer. In some experiments the dot immunobinding approach was used. After transfer, the strips were stained at room temperature by the following standard procedure: Blocking is performed for 2 min without shaking in 0.01 *M* Tris/HCl, 0.25 *M* NaCl, 2% (w/v) Tween® 20, pH 7.2, 10 mℓ per lane. The antibody incubation is done overnight with gentle shaking of nitrocellulose strips with polyspecific antibody (typical dilution: 1 + 50) in 10 mℓ of the above-mentioned buffer without Tween® 20. After washing three times for 10 min in the same buffer, the second incubation (2 hr with shaking) is performed in the same buffer containing a solution of the esterase (pure enzyme, crude mixture, etc.). After three washings as above, the blot is stained for 0.5 to 1 hr with shaking in 100 mℓ 0.1 *M* phosphate, pH 7.5, containing 1 mℓ 2.5% (w/v) 1-naphthylacetate and 50 to 100 mg of Fast Red TR salt.[10]

COMMENTS ON THE PROCEDURE

Of key importance is that a surplus of antibodies is ensured in the second incubation. The stain was originally developed for the detection of esterase activity in crossed immunoelectrophoresis (limit of sensitivity in this system is 0.1 ng cholinesterase[10]), but is also suitable here, as the colored diazoproduct of the enzyme reaction precipitates on the nitrocellulose.

EVALUATION OF THE PROCEDURE

Dot Immunobinding Assay

Since esterase activity was not detectable by direct staining of blots from SDS-PAGE of normal mouse serum, conditions for detecting esterase activity was established using dot immunobinding. When the serum was spotted directly on nitrocellulose, the presence of 2% (v/v) Tween® 20 in the blocking buffer drastically reduced the intensity of the subsequent esterase staining (Figure 1A).

When antibodies were spotted on the paper and subsequently incubated with antigen in different buffers not containing Tween® 20, a 2-hr incubation at pH 7.2 was found to result in far better staining than incubation at the pH value normally used for antibody incubations, which is pH 10.2 (see Chapter 1; Figure 6).

When antigen spotted on nitrocellulose was followed by antibody and antigen incubations at different pH values (pH 7.2 or 10.4) the pH value of the first incubation, however, was found to have essentially no influence on the final staining intensity (not shown). Also, blocking by Tween® 20 did not interfere with the final enzyme staining after antigen and antibody incubations, indicating that the direct effect observed with Tween® 20 (Figure 1A) is a reversible, inhibiting effect rather than a detergent displacement of the protein from the paper.

Spotting antigen on the paper followed by antibody and antigen incubations was also used to establish optimal concentrations of the agents (Figure 1C). The fine dots on the figure are caused by Fast Red TR salt particles.

These experiments established the general procedure — no Tween® 20 in the incubation buffers, which should be of neutral pH, at least for the second incubation. A surplus of antibody in the first incubation must be ensured, while the concentration of the second incubation antigen should be as low as possible.

Electroblotting

Blotting in the discontinuous transfer system and staining for esterase as above show the presence in normal mouse serum of a band with an apparent M_r of 70,000 under reducing conditions (Figure 2). The negative controls show no staining. Crude antiserum (lane 2) is seen to give rise to a higher background than purified immunoglobulins, probably due to unspecific binding of esterase activity in the rabbit serum. The M_r of the enzyme was verified analyzing purified mouse α-1-esterase in a conventional SDS-PAGE.

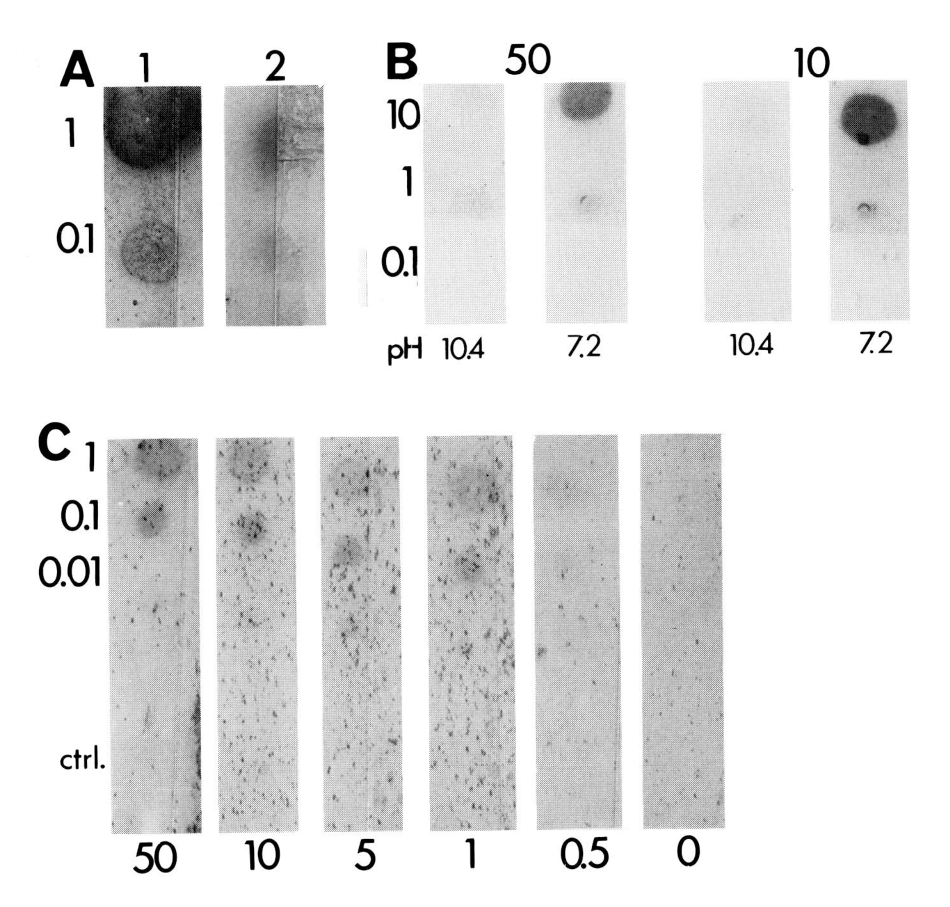

FIGURE 1. Dot immunobinding experiments. Panel A: 1 and 0.1 μℓ mouse serum was spotted on nitrocellulose. Strip 1 was stained directly for esterase. Strip 2 was blocked for 2 min in 2% (w/v) Tween® 20 in pH 10.4 buffer before staining for esterase. Panel B: 10, 1, and 0.1 μℓ of rabbit immunoglobulins against mouse serum proteins were spotted on nitrocellulose as indicated and incubated for 2 hr at 20°C with 50 and 10 μℓ unfractionated mouse serum, respectively, in either pH 10.4 (0.05 *M* Tris, 0.150 *M* NaCl) or pH 7.2 (0.05 *M* Tris, 0.5 *M* NaCl) buffer as indicated and stained for esterase activity. Panel C: On each strip are spotted (from top to bottom) 1, 0.1, and 0.01 μℓ of rabbit immunoglobulins against mouse serum proteins plus 1 μℓ of control rabbit immunoglobulins (ctrl.). After blocking in 2% Tween® 20, strips were incubated with 50, 10, 5, 1, 0.5, and 0 μℓ, respectively, of mouse serum in 10 mℓ incubation buffer (pH 7.2) without Tween® 20, 2 hr at 20°C. Staining for esterase was then performed after washing three times with incubation buffer (pH 7.2).

DISCUSSION

The method provides a general way to detect enzyme activity on nitrocellulose blots after SDS-PAGE. The only demands are that the activity in question shall be soluble and recognizable by an antibody after SDS-PAGE and that a specific nonsoluble stain for the activity should be known. Other methods for reactivating special enzymes after SDS-PAGE exist,[11,12] but are difficult to practice. The principle presented here was also successfully employed in the detection of antigens by otherwise nonbinding monoclonal antibodies (the clono-Glad technique).[13]

Even if the enzyme after SDS-PAGE blotting was still recognizable by antibodies, its enzyme activity was lost. This loss was due to the denaturing conditions of the SDS-PAGE procedure, as direct dot immunobinding of enzyme, in the absence of Tween® 20, retained activity (Figure 1A). The inhibiting effect of detergents was also noted by Muilerman et al.[1] in their system where a change of detergent from Triton® X-100 to Tween® 20 solved

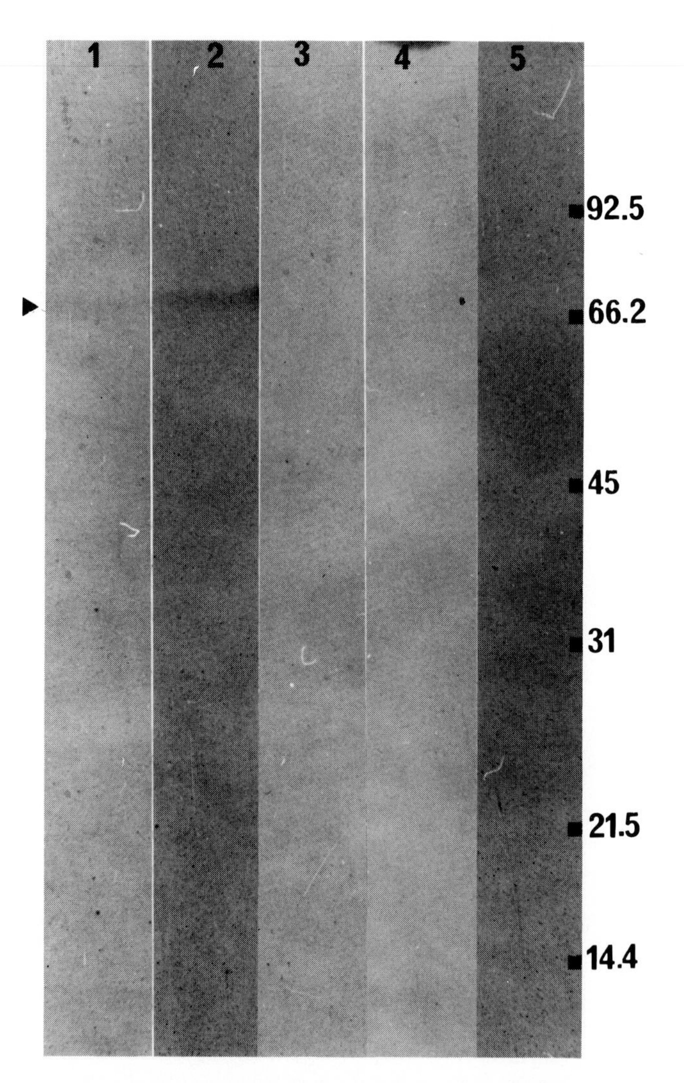

FIGURE 2. Esterase staining after SDS-PAGE and electroblotting of 20 $\mu\ell$ normal mouse serum (diluted 1 + 20). After blocking for 2 min in 2% Tween® 20 at pH 7.2, the strips were incubated overnight at 20°C as follows: (Lanes 1 + 4) Rabbit immunoglobulins against mouse serum proteins (1 + 100), (lanes 2 + 5) antiserum against mouse serum α-1-proteins (1 + 300). Lane 3 contains no antibody. After washing three times for 10 min with incubation buffer, the second incubation was performed for 2 hr as follows: (Lanes 1 − 3) Mouse serum (1 + 1000) and lanes 4 − 5 without protein. After washings as above, the blots were stained for esterase as described. Molecular weights × 10^{-3} are indicated to the right and the specifically stained esterase by the arrow to the left.

the problem. Due to the specificity of the staining, background caused by the absence of Tween® 20 does not occur.

The data presented here on the effect of the second incubation pH on final enzyme staining indicate that buffer ions influence the microenvironment of the nitrocellulose even after several washings and change of buffer.

Generally, the approach can be used in the following way: (1) Is staining directly on the blot possible? (2) If not, optimize conditions (pH, detergents — types, concentrations), concentration of reagents, time of incubations in dot immunobinding experiments. Buffer conditions should be evaluated very carefully. (3) Use the optimized system in the final experiment with appropriate controls (leaving out either antibody or antigen in first and second incubations, respectively).

REFERENCES

1. **Muilerman, H. G., Ter Hart, H. G., and van Dijk, W.,** Specific detection of inactive enzyme protein after polyacrylamide gel electrophoresis by a new enzyme-immunoassay method using unspecific antiserum and partially purified active enzyme: application to rat liver phosphodiesterase I, *Anal. Biochem.*, 120, 46, 1982.
2. **van der Meer, J., Dorssers, L., and Zabel, P.,** Antibody-linked polymerase assay on protein blots: a novel method for identifying polymerases following SDS-polyacrylamide gel electrophoresis, *EMBO J.*, 2(2), 233, 1983.
3. **Heegaard, P. M. H. and Bøg-Hansen, T. C.,** Transferrin and α-2-macroglobulin I are circulating acute-phase reactants in the mouse, in *Marker Proteins In Inflammation*, Vol. 3, Laurent, P., Grimaud, J. A., and Bienvenu, J., Eds., Walter de Gruyter, Berlin, 1986, 275.
4. **Harboe, N. and Ingild, A.,** Immunization. Isolation of immunoglobulins, estimation of antibodytiter, *Scand. J. Immunol.*, 2(Suppl. 1), 161, 1973.
5. **Bøg-Hansen, T. C., Krog, H. H., and Back, U.,** Plasma lipoprotein-associated arylesterase is induced by bacterial lipopolysaccharide, *FEBS Lett.*, 93(1), 86, 1978.
6. **Laemmli, U. K.,** Cleavage of structural proteins during the assembly of the head of bacteriophage T4, *Nature (London)*, 227, 680, 1970.
7. **Nielsen, C. S. and Rose, C.,** Separation of nucleic acids and chromatin proteins by hydrophobic interaction chromatography, *Biochim. Biophys. Acta*, 696, 323, 1982.
8. **Kerckaert, J. P.,** Highly simplified analytical or preparative slab gel electrophoresis, *Anal. Biochem.*, 84, 354, 1978.
9. **Kyhse-Andersen, J.,** Electroblotting of multiple gels: a simple apparatus without buffer tank for rapid transfer of proteins from polyacrylamide to nitrocellulose, *J. Biochem. Biophys. Methods*, 10, 203, 1984.
10. **Brogren, C.-H. and Bøg Hansen, T. C.,** Enzyme characterization in quantitative immunoelectrophoresis, *Scand. J. Immunol.*, 4(Suppl. 2), 37, 1975.
11. **Lacks, S. A. and Springhorn, S. S.,** Renaturation of enzymes after polyacrylamide gel electrophoresis in the presence of sodium dodecyl sulfate, *J. Biol. Chem.*, 255, 7467, 1980.
12. **Chang, L. M. S., Plevani, P., and Bollum, F. J.,** Evolutionary conservation of DNA polymerase beta structure, *Proc. Natl. Acad. Sci. U.S.A.*, 79, 758, 1982.
13. **Scopsi, L., Bock, E., and Larsson, L. -I.,** Monoclonal antibody immunocytochemistry: novel method extending usefulness of monoclonal antibodies for antigen visualization, *Eur. J. Cell Biol.*, 41, 97, 1986.

Section 7

NON-SPECIFIC BINDING AND ARTIFACTS

SPECIFICITY PROBLEMS AND TROUBLESHOOTING WITH AN ATLAS OF IMMUNOBLOTTING ARTIFACTS

Ole J. Bjerrum, Kurt Pii Larsen, and Niels H. H. Heegaard

INTRODUCTION

Immunoblotting experiments can, in principle, be disturbed in two ways. First, disturbances may result from inappropriate events taking place when using apparatus and materials. This is what we call artifacts or technical errors. Secondly, they may result from antigen-antibody interactions confusing the outcome because of the very high resolution and sensitivity of the technique permitting detection of weak cross-reactions not seen with other immunochemical systems. In contrast to the real artifacts which often are observed as peculiarities on the finished blots, this latter group of problems are not due to faulty work, i.e., they are not eliminated by technically optimal experiments.

These considerations are responsible for the division of this chapter into two main parts. One, which can be applied on all types of immunoblotting, deals with specificity problems while the other part deals with artifacts connected with immunoblotting on nitrocellulose with immunoenzymatic detection systems. The artifacts are systematically presented in a troubleshooting guide including figures illustrating characteristic artifactual manifestations.

SPECIFICITY PROBLEMS

Controls

The appearance of a stained band in an immunoblotting experiment does, in fact, only reflect the presence of some enzymatic activity or radioactivity in other detection systems at that particular place on the blotting membrane. Further, the binding pattern of the detecting antibody is a more or less valid reflection of the distribution of primary bound ligands which again might possess binding characteristics not easily elucidated by other immunochemical methods. Finally, the antigenic properties might change after, e.g., SDS-treatment, reduction, boiling, transfer, or because of contamination from utensils and reagents where albumin, keratins,[1] immunoglobulins, bacterial products, or thiols[2] may give unwanted bands not visible on the SDS-PAGE gel in conventional Coomassie brilliant blue staining. Figure 1 shows such an artifact deriving from presence of dithiothreitol in the sample buffer. Alkylation with iodoacetamide of the sample buffer remove the bands.[36] Accordingly, problems with "nonspecificity",[4] "multiple bands", and "novel cross reactions"[5-7] are frequently encountered and emphasize the requirement of adequate controls for the interpretation of any immunoblotting experiment since unspecific bands cannot simply be identified by their morphology and staining intensity (see Chapter 6.3.1).

In analyzing more or less harshly treated antigens on a solid phase by means of variable immunochemical approaches, the immunoblotting technique and the specificity problems encountered in many ways resemble immunocytochemical techniques and the designs of the specificity controls are also very similar. It is convenient to let the concept of "unspecific staining" include genuine unspecific staining of nonimmunological origin, which is not always sharply distinguishable from artifacts. These false positive bands can be due to nonimmunological binding of antibodies by Fc-receptors, e.g., from bound complement[8] or

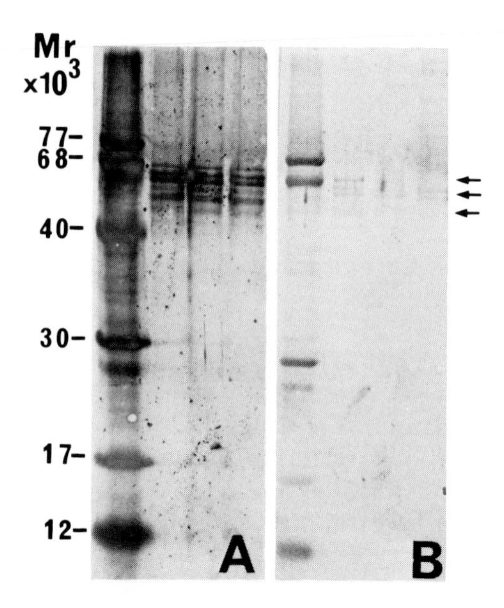

FIGURE 1. Irrelevant bands with M_r around 66,000 deriving from mercaptoethanol or dithiotreitol observed on nitrocellulose in (A) gold staining and (B) after immunoblotting (arrows). *Experiment:* Gold staining and immunoblotting was performed as described in Appendix to Chapter 1. Semidry electrotransfer to nitrocellulose of M_r markers (LKB) in sample buffer containing 40 m*M* dithiothreitol in lane 1 (see Chapter 9.2, Volume II). The three other lanes only contain sample buffer. Immunostaining was performed with corresponding polyspecific antibody.[5] Alkaline phosphatase conjugated swine antirabbit IgG antibody (Dakopatts, Copenhagen) (1 + 2000, 2 hr) was used as secondary antibody. Staining: Nitroblue tetrazolium and 5-bromo-4-chloro-3-indolyl phosphate.

by unspecific protein absorption to the blotting matrix itself and also to endogenous peroxidase activity of the sample analyzed.[9] When this is the case the blocking substance (see Chapter 6.2) and buffer conditions for incubation[10] (see Chapter 6.3.1) should always be carefully examined. Thus, blocking in the presence of immunoglobulins of the same species as the secondary antibody may be useful. Also, a change of blocking substance from Tween® 20 to nonfat dry milk may have a significant effect (see Chapter 8.3).

The other group of "unspecific staining" is unwanted but specific staining[8] resulting from antisera containing antibodies of different specificities because of preimmune antibody populations (antibodies existing before immunization) or induced by impurities in the immunogen preparation. This is the problem of monospecificity, which might even be encountered with monoclonal antibodies which react with identical epitopes on different proteins.[5] In some cases it may be necessary to raise new antibodies reacting with a more unique part of the molecule.

The problem of unwanted specificities is often related to the high sensitivity of immunoblotting which visualizes antigen-antibody reactions not detected by other immunochemical methods currently in use for characterization of antigen preparations. Thus, for polyclonal antibodies, monospecificity is an operational term delineated by the method in question (Chapter 6.3.1). For monoclonals, some specificity problems were mentioned earlier, but properties established by other methods seem to be well preserved in immunoblotting.[5]

In Table 1 the principles of the detection controls are given. It is convenient to deal with 1st level (specificity) and 2nd level (staining) controls.[8] It should be noted that the nature of the detecting antibodies (xeno-, allo-, or auto-antibodies) also implicates different controls. Specificity problems connected with autoantibodies are also mentioned in Chapter 9.3 (Volume II). First level or absorption controls ensure the specificity of the primary antigen-antibody interaction. As mentioned in Table 1, solid-phase absorption controls are preferred.

Table 1
CONTROL EXPERIMENTS FOR EVALUATION OF A STAINED BAND IN IMMUNOBLOTTING

First level controls	Comments
Liquid phase absorption with antigen	Absorb over broad range of antigen concentrations
Solid phase absorption with antigen	Preferable as soluble antigen-antibody complexes may bind to blotted antigen
Preimmune (nonimmune) or hyperimmune serum (against unrelated antigen) instead of primary antibody.	Control for presence of preimmune antibodies and for genuine unspecific interaction between blotted antigen and primary antibodies
Second level controls	
General	
Omission of antigen (blotting of blank gel or of irrelevant antigen)	Control for impurities
Omission of primary antibody	Control for presence of immunoglobulin in antigen preparation and/or binding of 2nd antibody to the antigen
Omission of all antibodies	Control for endogenous enzyme activity
Omission of linking antibodies	In multilayer techniques; control for cross-reactions between antigens on blots and 2nd or 3rd layer antibodies or nonimmunological interaction between detecting system and primary antibodies
Antibody-type dependent	
Alloantibodies: homozygotic antisera applied as primary antibodies	Specificity control; should react identically; no reactions with autologous antigen-preparation
Autoantibodies: sera from normal and diseased animals or individuals applied as primary antibodies	Different reaction patterns

Second level controls evaluate all other factors possibly involved in unwanted but specific or genuine unspecific staining such as the performance of the detection system and the pitfalls of the antigen preparation. The result of a 1st level control absorption is inactivation of the antiserum, making bands disappear on the blots, and confirming the specificity of the antigen-antibody interaction in these cases. Controls at both the 1 and 2 levels are imperative for the evaluation of any immunoblotting experiment.

Monospecificity

The controls can clarify the nature of unspecificity and thus give some hint as to how to obtain more monospecific experiments. Some practical approaches are outlined in Table 2.

Before an antiserum is subjected to an absorption procedure, some simple precautions can be performed. However, it should be stressed that the appearance of several bands may be an observation of significance (polymerization, degradation, various post–translatory forms). The business of the controls is to elucidate this question. Routinely, the antibodies should be diluted and be of as high avidity as possible. Dilution, adjustment of antigen load, and incubation at elevated pH make only high-affinity interactions survive,[11,12] (see Chapter 1) while optimal blocking conditions,[4] decomplementation procedures, and blocking of SH-groups are directed against nonimmunological absorption. The third group of measures aim at developing fewer possibilities for antibody specificities in the antibody preparation by the various absorption procedures mentioned in Table 2 including employment of region-specific antibodies (see Volume II, Chapter 8.2) or, alternatively, by selecting the wanted antibody specificities during the incubation by means of the double-antigen techniques (see Chapter 6.8).[11] Figure 2 illustrates how it has been possible by means of many different approaches to obtain apparent monospecificity against the individual polypeptides of the human erythrocyte membrane. As some of the antibodies were raised against SDS-treated material it turned out to be necessary to absorb the antibody with SDS-treated proteins as shown for

Table 2
HOW TO OBTAIN APPARENT MONOSPECIFICITY IN IMMUNOBLOTTING

Precaution	Reasoning
I. Quenching of weak affinity interactions	
Adjustment of antigen load	Use of an appropriate amount of the antigen reduces conc. of unwanted antigens
Dilution of primary antibody	Specific titer of wanted antibody is often so much higher than contaminating antibodies that these will disappear by simple dilution
Incubation at elevated pH	Increase, e.g., to pH 10.2 allows only binding of antibodies with high affinity (see Figure 6, Chapter 1).
II. Blocking of nonimmunological interactions	
Presence of carrier proteins	Addition of 1% albumin, gelatin, or serum of same species as secondary antibody to antibody solutions prevents nonspecific immunoglobulin binding
Blocking of thiol groups	Unspecific binding via disulfide-bridge formation may be avoided by treatment of blot with, e.g., *N*-ethyl-maleimide or iodoacetamide
III. Selection of specific antibodies	
Absorption with native materials	Absorption with immobilized materials is preferable to liquid phase absorption as soluble antigen-antibody complexes may bind to blotted antigen
Material to be analyzed	Since titer of the wanted antibody is high, it is sometimes possible to absorb directly with material to be analyzed
Material depleted for antigen in question	This could be material obtained from deficient individuals, an extracted residue, or neighboring fractions from a fractionation procedure; however, it may be necessary to use complete antigen (see above)
Absorption with SDS-treated material	Some epitopes are first exposed after SDS-treatment; absorption with immobilized SDS-treated material (e.g., on nitrocellulose) is an advantage as the influence of SDS on antibody is avoided
Absorption with region-specific peptides	In case of cross-reactions with identical epitopes in different proteins
Employment of double antigen techniques	Select specificities wanted by incubation with a surplus of primary polyclonal antibody reacting with both native and SDS-denatured epitopes followed by addition of a native antigen preparation and a monoclonal antibody against the protein[11]

the anti-band 3 antibody. The absorption procedure is facilitated by the fact that very dilute solutions of the antibody can be used in the immunoblotting procedure.

Conclusively, every immunoblotting experiment should be evaluated carefully on the basis of control experiments performed simultaneously as outlined in Table 1. Both level of controls are needed and experiments without these are not conclusive. The amount of unwanted bands — immunological or nonimmunological — can in most instances be reduced considerably by employing some of the measures proposed in Table 2 guided by the results of the control experiments.

ARTIFACTS

In contrast with the above-mentioned specificity considerations artifacts are technical errors often observed as peculiarities on the developed blots. Familiarity with such artifacts makes their identification and subsequent correction easier and prevents misinterpretation of the resulting staining patterns. Some artifacts have been briefly mentioned by other authors.[9,16]

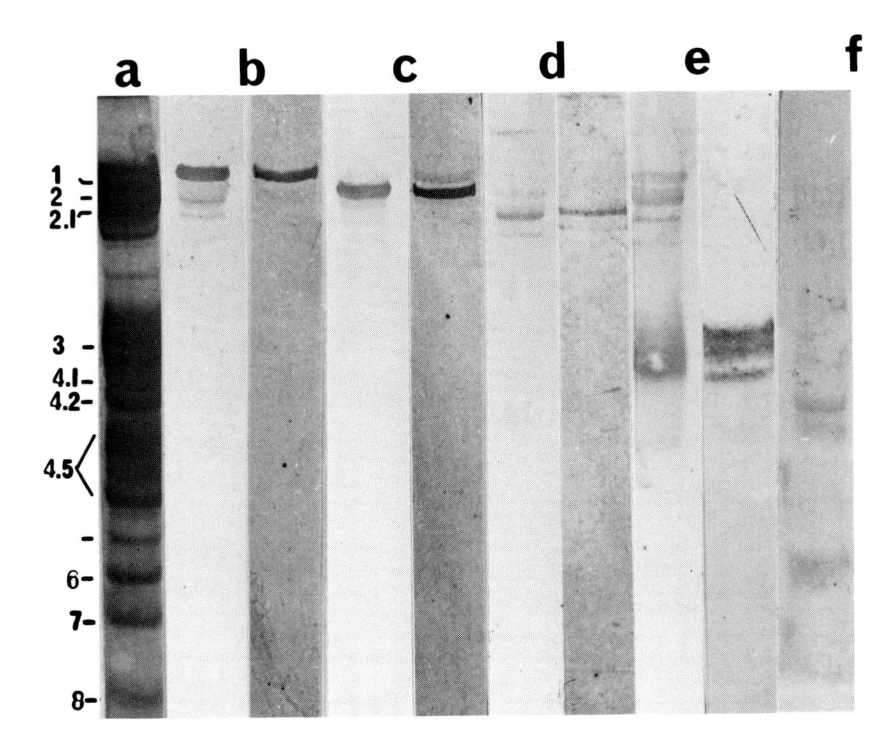

FIGURE 2. Preparation of monospecific rabbit antibodies against selected human erythrocyte membrane polypeptides by various absorption procedures.[13] *Experiment:* Erythrocyte membrane proteins (10 μg) were separated and designated (to the left) according to Fairbanks et al.,[14] gel stained with (a) Coomassie Brilliant Blue and immunoblotted with nonabsorbed and absorbed (b) antispectrin (band 1), (c) antispectrin (band 2), (d) anti-ankyrin (band 2.1), (e) antiband 3, and (f) and antihemoglobin rabbit antibodies. Secondary antibody was peroxidase-conjugated swine antirabbit IgG antibodies (Dakopatts) with 3-amino-9-ethyl carbazole as a substrate (see Appendix to Chapter 1.) Antispectrin antibodies were raised in rabbits with excised line-immune precipitates of plasmin degraded spectrin.[15] Absorption was performed with alkali-stripped (pH 11) erythrocyte membranes and small amounts of washed normal ghosts not sufficient to cause an appreciable reduction in the antispectrin titer. Anti-ankyrin antibodies were raised against precipitates excised from crossed immunoelectrophoresis.[15] Antiband 3 antibodies were raised against the proteins eluted from the band 3 region after SDS-PAGE. Absorption was performed with SDS-treated EDTA-extractable proteins of erythrocyte membrane proteins and finally with small amounts of purified glycophorin. A 6 *M* urea extract could be used instead of the EDTA extract. Antihemoglobin antibody (Dakopatts) was absorbed with hemoglobin-free erythrocyte ghosts.[13] Anode at the bottom.

A more systematic description of artifacts seen in connection with immunoblotting can be found in some of the information sheets that are enclosed with commercially available immunoblotting kits.[17] By analogy with an earlier published troubleshooter for electroimmunoprecipitation artifacts,[18] Table 3 is subdivided according to the observed phenomenon with indication of the cause followed by a short description of how it can be remedied. The figures illustrating some characteristic artifacts are numbered according to this subdivision. The account is for most of the experiments based on a "standard" immunoblotting employing nitrocellulose and immunoenzymatic detection.

ACKNOWLEDGMENTS

Miss Kirsten Olesen is thanked for her skillful technical assistance. For the supply of artifactual blots thanks are due to Peter Hinderson (Figures VI.3.3., VIII.1 and X.3), Jørgen Vinten (Figure I.4), and Peter M. H. Heegaard (Figure III.1). Financial support has been provided by the Danish Medical Research Council (grant no. 12-5257) and Harboefonden.

Table 3
TROUBLESHOOTER FOR ARTIFACTS OBSERVED IN IMMUNOBLOTTING ON NITROCELLULOSE WITH ENZYME-LABELED DETECTING ANTIBODIES

	Observation		Cause	Precaution/correction
I.	No or reduced staining with general protein stains	I.1.	No or wrong antigen applied	Always stain the gel and one of the nitrocellulose lanes for protein after transfer
		I.2.	Insufficient transfer; no current or reversed current applied; the voltage or transfer time too low (Figure I.2.A)	Check polarity and measure voltage or current; use timer
			Transfer of high M_r components from polyacrylamide gels of high concentration including gradient gels and thick gels (Figures I.2.B and C) is protracted	Buffer change;[19,20] addition of SDS to the transfer buffer[19,21,22] or proteolytic cleavage of antigen *in situ* may help;[23] change gel conc. of crosslinking[24] or use reversibly crosslinked gels;[25] prolong elution time or apply a gradient electric field[19]
		I.3.	Insufficient binding to nitrocellulose; low M_r antigens (<20.000) may pass the membranes (Figures I.3.A and B)	In such cases membranes with pores of 0.10—0.22 μm are recommended;[26] use of methanol[27] or salt[28] in transfer buffer and lower voltage gradient may also help[21]
		I.4.	Overloading with protein may displace antigens present in lower conc. (Figure I.4)	Binding capacity of nitrocellulose for protein is 15—80 μg/cm²;[26,29] apply reasonable amount of antigen and perform titrations
		I.5.	Dissociation of antigen from nitrocellulose (Figure I.5.A and B)	Use smaller pore size for the nitrocellulose;[24] prolonged blocking and storage of blotted nitrocellulose in presence of nonionic detergent should be avoided as it displaces a fraction of bound antigen;[30,31] fixation with 50% ethanol may help also chemical crosslinking has been performed[31,33]
		I.6.	Digestion of antigen on nitrocellulose	Blocking substance may contain proteolytic activity, e.g., gelatin
		I.7.	Nitrocellulose blocked or inactivated	Store nitrocellulose protected below 20°C out of direct sunlight and away from chemical vapors
II.	Impaired color development	II.1.	Antigen denatured by exposure to heat; the epitopes are SDS-sensitive or thiol-dependent (Figure II.1)	Heat accumulation during transfer should be avoided: reduce current, lower ionic strength, use short electrode-to-electrode distance, use large buffer volume with stirring, introduce cooling; monoclonal antibodies are most sensitive to denaturing of epitopes; test selective elimination of denaturants (SDS, methanol) from electrophoretic transfer buffer; ensure that SDS has been completely removed from blot, i.e., by treatment with Tween® 20
		II.2.	Antigen covered by blocking substance	Too-high conc. of blocking substance,[30] e.g., 1% gelatin may cover antigen[37]
		II.3.	Nonsaturating binding of primary antibody in assay; unfavorable binding conditions: pH, NaCl, and presence of SDS; titer too low (Figure II.3)	Increase the antibody conc.; extend incubation time (24 hr); use agitation and incubate at elevated temp., at pH 7.4, and at ionic strength of 0.15

Table 3 (continued)
TROUBLESHOOTER FOR ARTIFACTS OBSERVED IN IMMUNOBLOTTING ON NITROCELLULOSE WITH ENZYME-LABELED DETECTING ANTIBODIES

	Observation		Cause	Precaution/correction
		II.4	Secondary antibody of wrong specificity applied, e.g., anti-rabbit IgG antibodies instead of antimouse IgG or antihuman IgG antibodies; conc. too low (see also IV.4)	Use correct detecting antibody; do not reuse the solution
		II.5.	Enzyme inactivation (Figure II.5)	Test activity of the enzyme-labeled antibody by direct dotting onto nitrocellulose; tap water deionized by polystyrene resins and presence of low quality (impure) methanol or peroxidase may inactivate enzymes;[17] azide is a potent inhibitor;[17] presence of Pb^{2+} and impure H_2O_2 also inhibit (see Chapter 6.1.2); alkaline phosphatase has a pH optimum approx. 10 and presence of phosphate inhibits the reaction (see Chapter 6.4.2)
		II.6.	Staining/color development solution is inactive (Figure II.6)	Test activity of the mixed solution by direct addition of enzyme; store reagents and solutions properly; keep out of direct sunlight; to avoid reagent precipitation mix stock solutions and buffer at room temperature. For peroxidase, use fresh, stabilized H_2O. Too high concentration of H_2O_2 inhibits the enzymatic reaction. In case of *0*-dianisidine, plastic trays may absorb the dye. 4-chloro-1-naphthol: blocking with Tween 20 interferes with the staining. (Cf. Chapter 8.3)
		I.7.	Substrate solution does not correspond to employed enzyme conjugated antibody	Use correct substrate solution
III.	High background staining			
III.1.	Uniformly distributed white-appearing protein bands	III.1.	Insufficient blocking step and/or maintenance of blocking (Figure III.1)	Use large vol of fresh blocking solution; increase conc. of blocking substance, e.g., from protein to Tween® 20 or vice versa (see Chapter 6.2); incorporate 1% normal serum during incubation with primary antibody[9] Addition of low conc. of blocking substrate to buffers for incubation and washing is important for keeping nitrocellulose blocked during various incubation steps (see Chapter 6.2); especially Tween® 20 tends to dissociate
III.2.	Uniformly distributed	III.2.1.	Reaction of primary or secondary antibody or conjugated protein A with proteins present in blocking solution (Figure III.2.1)	Instead of protein mixture use pure protein (e.g., albumin) or Tween® 20 for blocking (see Chapter 6.2)

Table 3 (continued)
TROUBLESHOOTER FOR ARTIFACTS OBSERVED IN IMMUNOBLOTTING ON NITROCELLULOSE WITH ENZYME-LABELED DETECTING ANTIBODIES

	Observation		Cause	Precaution/correction
		III.2.2.	Improperly cleaned incubation trays	Use clean incubations trays
		III.2.3.	Insufficient washing step after 1st or 2nd antibody incubation	Increase no. of washes to 4—5 × 10 min each; a washing buffer of pH 10.2 may be advantageous
		III.2.4.	Nitrocellulose left in color development solution too long (Figure III.2.4)	Remove nitrocellulose once reaction appears to be complete; avoid evaporation; for stronger reaction restain with fresh solution
III.3.	Randomly distributed	III.3.1.	Antigen contamination of upper buffer reservoir of SDS-PAGE apparatus (Figure III.3.1)	Use fresh buffer; clean apparatus properly
		III.3.2.	Antigen contamination of gel, transfer buffer, sponges, or blotting cell (Figure III.3.2, see also IV.4)	Discard transfer buffer after use; use properly cleaned or fresh sponges; clean blotting cell properly; performance of a "blind" blotting experiment in 50 m*M* sodium carbonate is effective; antibody-containing agarose gel should be pressed and washed properly (see Chapter 8.5) Antigen contamination can also be avoided by application of dialysis film or nitrocellulose on both sides of polyacrylamide gel/nitrocellulose sandwich while blotting
		III.3.3.	Overdevelopment in staining solution (Figure III.3.3)	See III.2.4
III.4.	Following banding pattern	III.4.1.	Too high conc. of primary antibody (polyclonal) causes binding of antibodies present in low conc. (see Figure 1)	Reduce antibody conc.; absorb antibody (see sect. on specificity problems)
		III.4.2.	Presence of SDS in primary antibody solution causes unspecific binding to immobilized protein bands (Figure III.4.2)	Wash blots carefully before incubation with primary antibody
		III.4.3.	Conc. of secondary antibody/antiserum too high	Reduce conc.
IV.	Stained spots			
IV.1.	Sharply stained irregular spots and anomalies	IV.1.	Contamination with antigen or immunoglobulin reacting with primary or secondary antibody before blocking has taken place, e.g., from nondissolved antigen left over in the application well, squirts, fingerprints, sneezing, drops of blood, sweat, and tears (Figure IV.1)	Work cleanly and carefully; use gloves
IV.2.	Sharply stained areas	IV.2.	Contamination with antigen present on the grids, sponges, electrodes, or blotting cell (Figure IV.2, see also III.3.2)	Clean blotting cell, grids, and sponges carefully (see also III.2.2)
IV.3.	Blurred stained squirt-like streaks	IV.3.	Secondary antibody has been squirted directly into nitrocellulose through incubation buffer (Figure IV.3)	Mix antibody and buffer before strips are incubated

Table 3 (continued)
TROUBLESHOOTER FOR ARTIFACTS OBSERVED IN IMMUNOBLOTTING ON NITROCELLULOSE WITH ENZYME-LABELED DETECTING ANTIBODIES

	Observation		Cause	Precaution/correction
IV.4.	Scattered and distinct grains	IV.4.	Precipitated grains of stain (Figure IV.4, see also Figures II.4. and VI.3.3)	Filter solution and avoid evaporation; do not incubate strips in same color solution too long
IV.5.	Fluffy confluent colonies	IV.5.	Microbial growth due to prolonged incubation at elevated room temp. (Figure IV.5)	Add preservative to incubation solution (e.g., 5 m*M* sodium azide)
V.	Blank areas			
V.1.	Partly developed pattern	V.1.1.	Insufficient transfer from gel to nitrocellulose	Ensure good contact between gel and nitrocellulose; Gradient gels are more difficult to elute than homogeneous gel[16]
		V.1.2.	During incubation strips have not been totally covered with buffer (Figure V.1.2)	Use agitation
		V.1.3.	The strips have covered each other; this is especially seen during substrate incubation (Figure V.1.3)	Incubate only one strip per ditch or tray
		V.1.4.	Lanes have been wrongly cut so that empty space between tracks appears in middle of strip (Figure V.1.4)	Mark position of lanes with pyronin G, which binds to nitrocellulose; it can be used as marker dye and also applied just before electrophoresis is terminated (see Chapter 1, Appendix)
		V.1.5.	Incubation on agitator with ditches perpendicular to horizontal movements may cause standing waves to appear in ditches; this gives rise to uneven distribution of antibodies and stain (Figure V.1.5)	Place incubation rack with ditches parallel with longitudinal movements
V.2.	Bald areas and spots	V.2.1.	Presence of air bubbles or dirt between nitrocellulose membrane and gel slab (Figure V.2.1)	Membrane is best wetted by being floated on top of buffer; squeeze out accidentally caught air bubbles with finger
		V.2.2.	Any contamination of the nitrocellulose before use (e.g., by squirting, sneezing, and fingerprinting), nonionic detergent, oil, or dirt (Figure V.2.2)	Store nitrocellulose protected; work carefully; use gloves (see also IV.1)
		V.2.3.	Stain can be removed by rubbing (Figure V.2.3)	Protect strips after drying, e.g., by mounting them below Magic Tape (3M Company) or in a folder
VI.	Disturbed pattern			
VI.1.	Irregular patterns	VI.1.1.	SDS-PAGE has failed	Control stain a lane of SDS-PAGE
VI.2.	Jagged banding pattern	VI.2.1.	Contact between gel and nitrocellulose has only partly been established during blotting; Rubber band holding the sandwich together broken (Figure VI.2.1)	Never move gel placed on nitrocellulose; constrict blotting sandwich carefully
		VI.2.2.	Polyacrylamide gel has swelled during the transfer step	Methanol conc. in transfer buffer affects size of polyacrylamide gel
VI.3.	Blurred banding pattern	VI.3.1.	Too heavy load of protein applied for electrophoresis (see Figure II.5)	Normally load should be 1/5 of that applied for gel staining with Coomassie brilliant blue

Table 3 (continued)
TROUBLESHOOTER FOR ARTIFACTS OBSERVED IN IMMUNOBLOTTING ON NITROCELLULOSE WITH ENZYME-LABELED DETECTING ANTIBODIES

	Observation		Cause	Precaution/correction
		VI.3.2.	Diffusion of separated protein bands took place before blotting (Figure VI.3.2)	If gel cannot be blotted same day as electrophoresis, store it frozen
		VI.3.3.	Dissemination of antigen from original band or dot (Figure VI.3.3)	Surplus of antigen in dot dissolves in blocking buffer and binds to nitrocellulose before blocking is effective
		VI.3.4.	Wrong side of blotting matrix turned against gel	Nonlustrous side of nitrocellulose sheet is usually protein binding site
VII.	Brittle nitrocellulose	VII.1.	Temp. 75°C + make nitrocellulose brittle (Figure VII.1)	Avoid exposure to elevated temp.
		VII.2.	When nitrocellulose dries out, it becomes brittle	Store nitrocellulose protected from dirt, avoid sunlight and drying out; drying of strips is best performed at room temp. in air stream
VIII.1.	Curled nitrocellulose	VIII.1.	Exposure to vapors from organic solvents (Figure VIII.1)	Do not store nitrocellulose together with organic solvents
IX.	Shrunken nitrocellulose	IX.1.	Methanol/ethanol containing solution causes shrinking	Use colloidal staining (see Chapter 6.1.3, and VI.2.2)
X.	Discolored nitrocellulose	X.1.	Electrode products from nonprecious metals (Figure X.1)	Use platinum electrodes, at least for anode
		X.2.	Carbon particles from graphite electrodes; liberated from cathode at extreme alkaline pH values, presence of SDS strongly accentuates process[35] (Figure X.2, see also Figure VII.1)	Wash electrodes properly
		X.3.	Adherent polyacrylamide gel (Figure X.3)	Avoid drying polycrylamide gel on nitrocellulose
		X.4.	At temp. 140°C + nitrocellulose becomes brownish (Figure X.4)	Avoid exposure to heat
XI.	Faded patterns	XI.1.	Exposure to light	Developed prints should stored protected from light
		XI.2.	Oxidation	Fading may be reduced by washing prints in antioxidants (e.g., 25 m*M* sodium pyrosulfite[12] or tochopherol) before storage; developed prints can be sealed in vacuum packs designed for storage of salami Succession of substrates indicates their sensitivity to fading with most resistant mentioned 1st: 5-bromo-4-chloro-3-indohyl phosphate, 4-chloro-1-naphthol; diaminobenzidine, tetramethyl benzidine; 0-dianisidine; 3-amino-9-ethylcarbazole (see Chapter 1). When working with fading substrates it is advisable to photograph wet nitrocellulose strips at day of development, otherwise keep in water
		XI.3.	During drying overlayed strips may give rise to different fading (Figure XI.3)	Place the strips neatly when drying

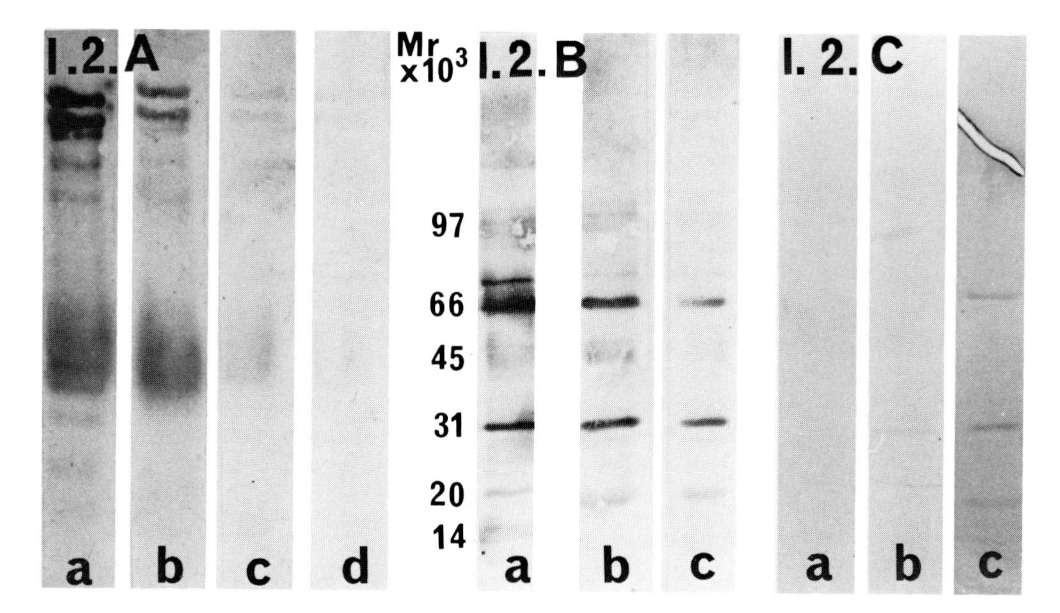

FIGURE I.2. Insufficient transfer. Immunoblotting. (A) The transfer time is reduced from 60 (a), to 6 (b), to 3, (c) and to 1 min (d). (B) Gel thickness is increased from 0.75 (a), to 1.5, (b), and 3 mm (c). (C) Gel staining for retained protein with Coomassie brilliant blue of the 0.25- (a), 1.5- (b), and 3-mm (c) thick gels. *Experiment:* (A) 10 μg human erythrocyte membrane proteins were separated as in Figure 2. (B and C) Low M_r marker mixture: phosphorylase b 97k, bovine serum albumin (BSA) 66k, ovalbumin 45k, carbonic anhydrase 31k, soybean trypsin inhibitor 20k, and α-lactalbumin 14k (1 μg) Kem-En-Tec, Hellerup, Denmark) was separated on a 7 to 15% gradient polyacrylamide gel in presence of 0.1% (w/v) SDS. Dots indicate position of bands. Immunostaining was performed with corresponding polyspecific rabbit antibodies (A) Dakopatts (1 + 2000, 18 hr) and (B and C) (1 + 400, 8 hr) (see Chapter 10). Staining was as in Figure 1, otherwise the conditions were as for Figure 2.

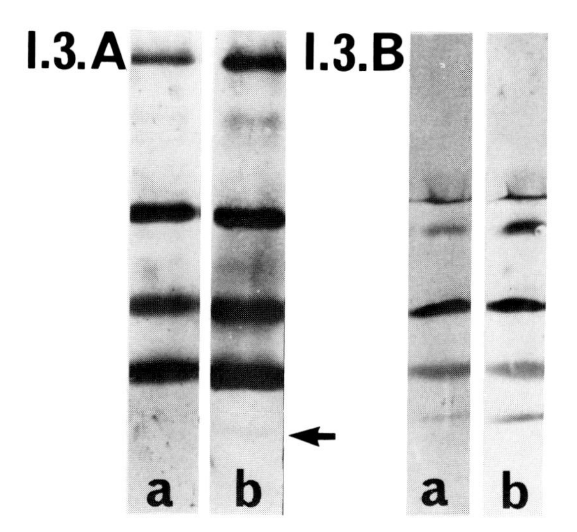

FIGURE I.3. Insufficient binding to nitrocellulose. (A) A membrane with a pore size of 0.45 μm (a) does not bind a component with M_r = 10,000, which binds to nitrocellulose with a pore size of 0.22 μm (b, arrow). (B) Less binding is observed without methanol in the transfer buffer (a) compared to the control (b) with 20% (v/v) methanol. *Experiment:* Conditions as for Figure I.2.B. Load in (A) 2 μg; in (B) 1 μg.

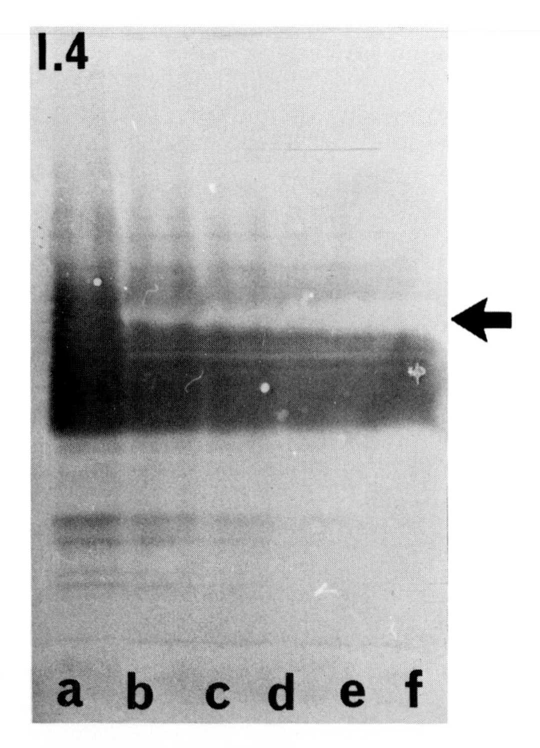

FIGURE I.4. Competition for binding sites on nitrocellulose. Arrow points to position of BSA which has displaced the examined antigen. *Experiment:* Immunoblotting of 80 μg human erythrocyte membrane protein (a) and subsequent 2-fold dilutions with 10 mg/mℓ of bovine albumin (b-e) using rabbit antihuman erythrocyte glucose transporter antiserum (1 + 300, 18 hr). Conditions otherwise as in Figure I.2.B.

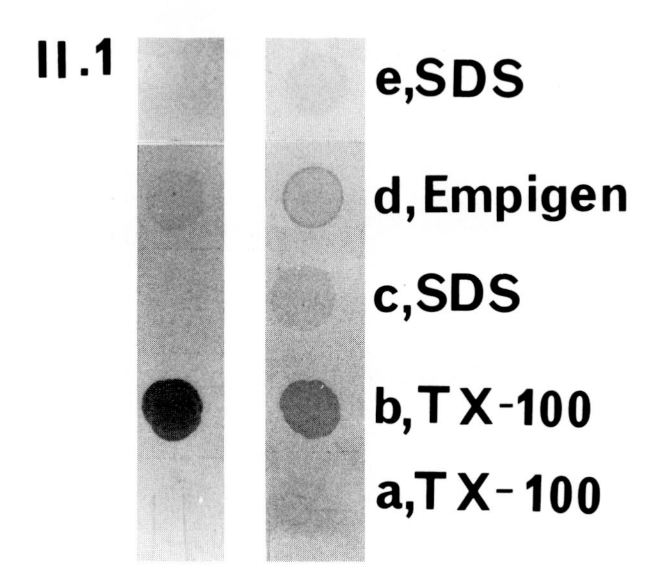

FIGURE II.1. Denaturing of epitopes by exposure to SDS. *Experiments:* (a, c, d) Dots of human erythrocyte membranes and (b, e) purified brain acetylcholinesterase solubilized in 0.5% (v/v) Triton® X-100 (a), 0.2% (v/v) Triton® X-100) (b), 1% (w/v) SDS (c, e), and 0.5% (w/v) Empigen BB (d) were reacted with two different (I, II) hybridoma culture supernatants positive against acetylcholinesterase (1 + 20, 18 hr). Note the impaired binding (a) of acetylcholinesterase to nitrocellulose in presence of 0.5% (v/v) Triton® X-100. Secondary antibody: alkaline phosphatase conjugated rabbit antimouse IgG antibodies. Conditions were otherwise for Figure I.2.B.

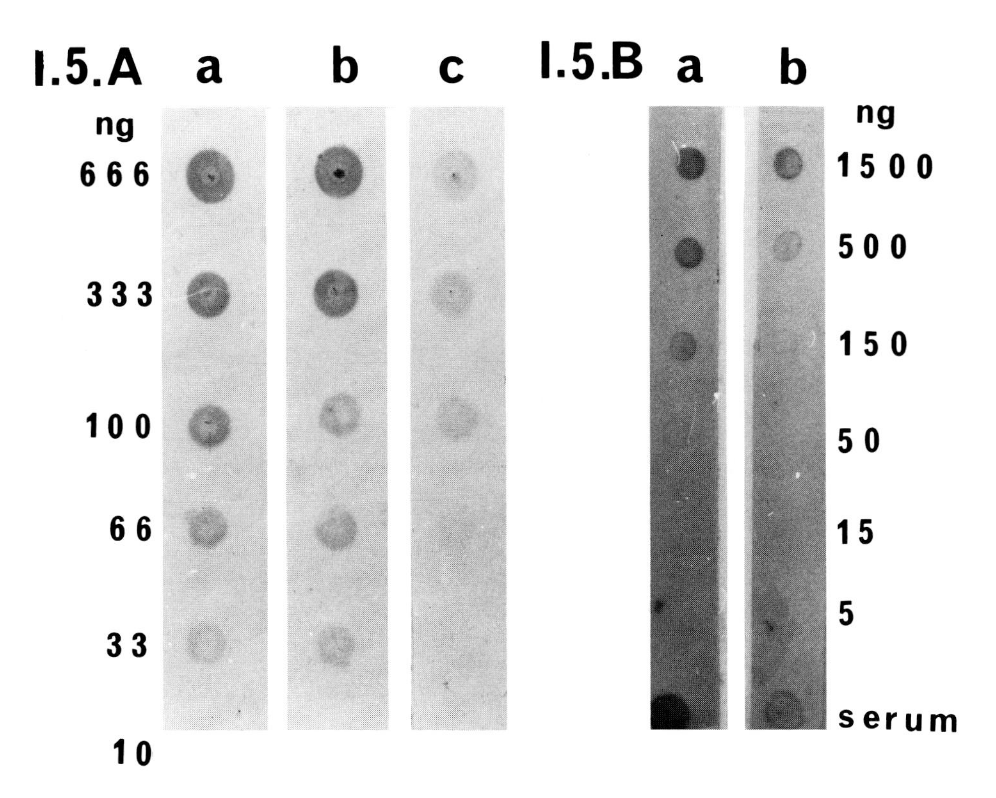

FIGURE I.5. (A) Displacement in the blocking step of antigen from nitrocellulose by Tween® 20 applied in a concentration of (a) 0.1%, (b) 0.5%, and (c) 2% (v/v). (B) Fixation of antigen on nitrocellulose by treatment with 45% (v/v) ethanol for 5 min at 20°C (lane a) before blocking with Tween® 20. *Experiment:* (A) Dot immunobinding of human orosomucoid diluted in 0.154 *M* NaCl to the amount (ng) stated (to the left) and reacted with rabbit anti-orosomucoid antibodies (Dakopatts) (1 + 1000, 18 hr). The strips are blocked in Tween® 20 for 3 min. Staining as for Figure I.2. (B) Dots of human IgA diluted in 0.154 *M* NaCl to the amount (ng) stated (to the right). (a) Treated with ethanol before blocking with Tween® 20 2%, (v/v), 2 min, 20°C. (b) Control. Both were reacted with a human serum containing anti-IgA antibodies (1 + 333, 18 hr). Secondary antibody: peroxidase conjugated rabbit antihuman IgA antibodies (Dakopatts) (1 + 2000, 2 hr). Substrate: 3-amino-9-ethyl carbazole. Conditions are otherwise as described in Appendix to Chapter 1.[32]

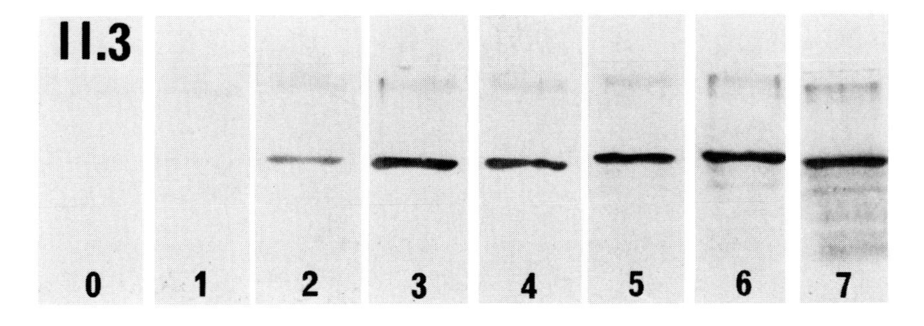

FIGURE II.3. Strengthened color development due to increased concentration of primary antibody. *Experiment:* Human transferrin (5 μg) immunoblotted with consecutive bleedings from a rabbit during an immunization course with 100 μg transferrin given at weeks 0, 2, 4, and 6. Immunization time before bleeding is indicated (weeks).[34] Conditions were otherwise as for Figure I.2.B.

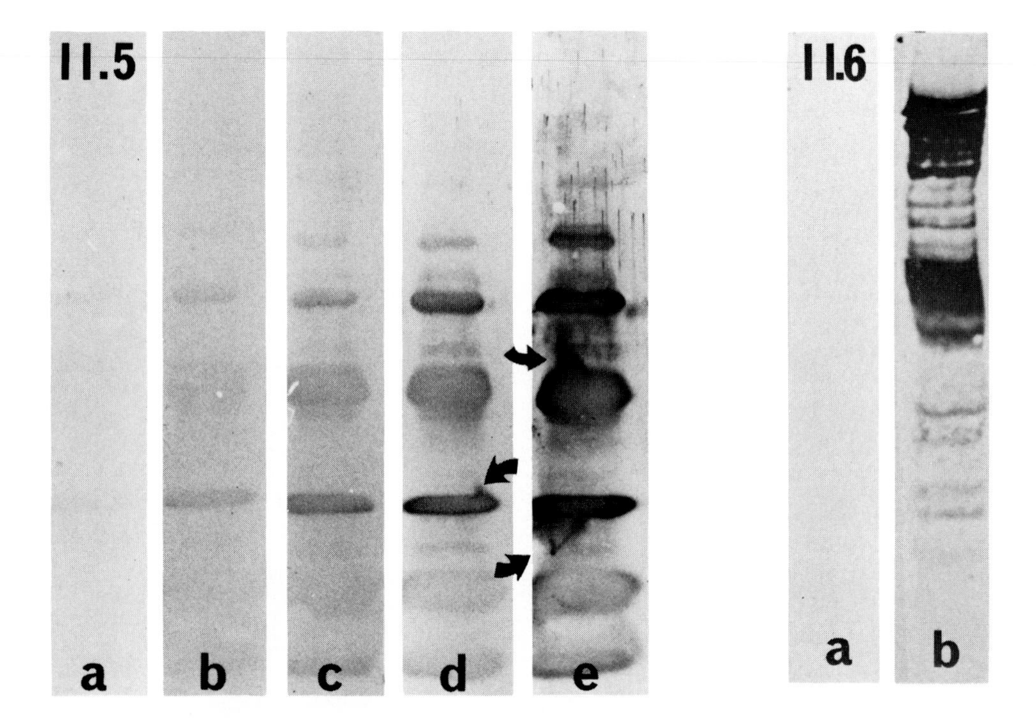

FIGURE II.5. Inactivation of alkaline phosphatase during staining due to presence in the staining buffer of phosphate (a) 30 m*M*, (b) 10 m*M*, (c) 3 m*M*, (d) 1 m*M*, (e) control. Note the broadness and dissimination of the protein bands because of a heavy load (arrows). *Experiment:* Immunoblotting of low M_r mixture (40 μg). Conditions were otherwise as in Figure I.2.B.

FIGURE II.6. Inactivation of 4-chloro-1-naphthol staining due to presence of Tween® 20. *Experiment:* Immunoblotting of human erythrocyte membrane proteins (5 μg). After transfer to nitrocellulose, blocking was performed (a) with 2% (v/v) Tween® 20 (3 min), (b) gelatin 3% (w/v) for 30 min (1 + 1000, 2 hr). Substrate: 4-chloro-1-napthol (0.5 mg/mℓ). Conditions were otherwise as for Figure 2.

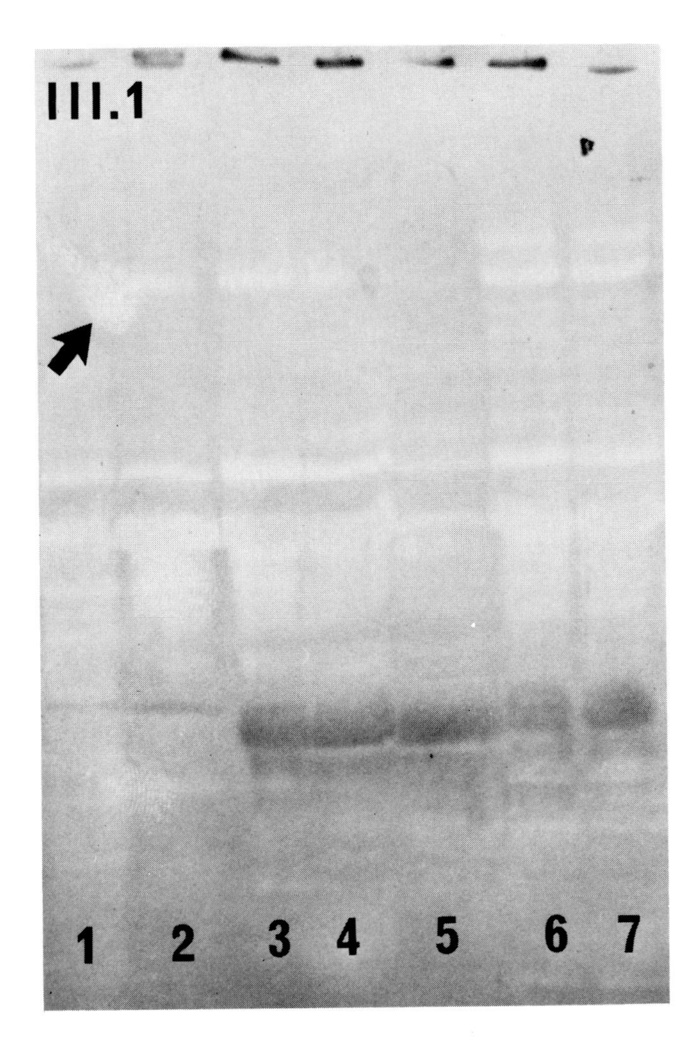

FIGURE III.1. White protein bands on a stained background. Tween® 20 has been omitted from the washing and staining buffers. Blocking has taken place only at the protein bands and fingerprinting (arrow). *Experiment:* Immunoblotting of 120 μg serum protein from normal (lanes 1, 2) and acute phase (lanes 3—7) mouse with antihuman haptoglobin antibodies (Dakopatts) (1 + 200, 40 hr). Staining: 3-amino-9-ethyl carbazole. Conditions were otherwise as in Figure 2.

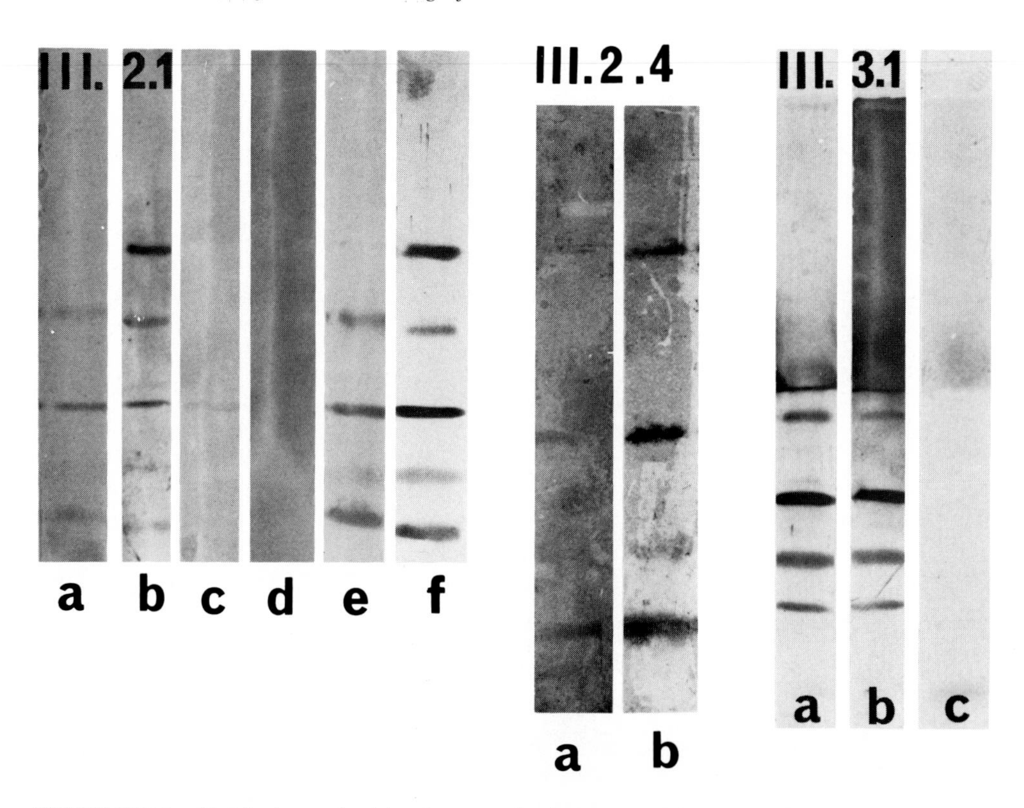

FIGURE III.2.1. High background staining due to reaction between secondary antibodies and proteins uniformly bound to nitrocellulose. Blocking performed with (a) 3% (w/v) BSA and (b) 3% (w/v) human serum albumin (HSA). Primary antibody contains anti–BSA antibodies. Blocking performed with rabbit IgG (c) 30 μg/mℓ for 2 min and (d) 5 μg/mℓ for 18 hr. (d) No blocking performed. (White bands do not appear because all the blotted proteins reacts with primary antibody). (e) Control. The white line in the middle of the strips probably is a result of standing waves; see Figure V.1.5. *Experiments:* Low M_r markers (2 μg) inclusive BSA analyzed with corresponding rabbit antibody.[3] Conditions were otherwise as for Figure I.2.B.

FIGURE III.2.4. High background staining due to extended incubation in color development solution in combination with some evaporation. Note the loose application of stain. *Experiment:* Immunoblotting of low M_r markers (1 μg) stained with peroxidase conujugated secondary antibody using (a) 3-amino-9-ethyl carbazole and (b) tetramethyl benzidine as substrate (see Chapter 9.1.2.)

FIGURE III.3.1. High background staining on the upper part of the blot due to albumin contamination of the buffer reservoirs of the SDS-PAGE apparatus. *Experiment:* Immunoblotting of low M_r markers (1 μg) with (a) corresponding antiserum (1 + 400, 18 hr),[3] (b) corresponding antibody plus antihuman albumin antibody (1 + 10,000, 18 hr), (c) electroblotted nitrocellulose strip in Amido Black staining. Buffer reservoir was contaminated with 0.05% (w/v) HSA.

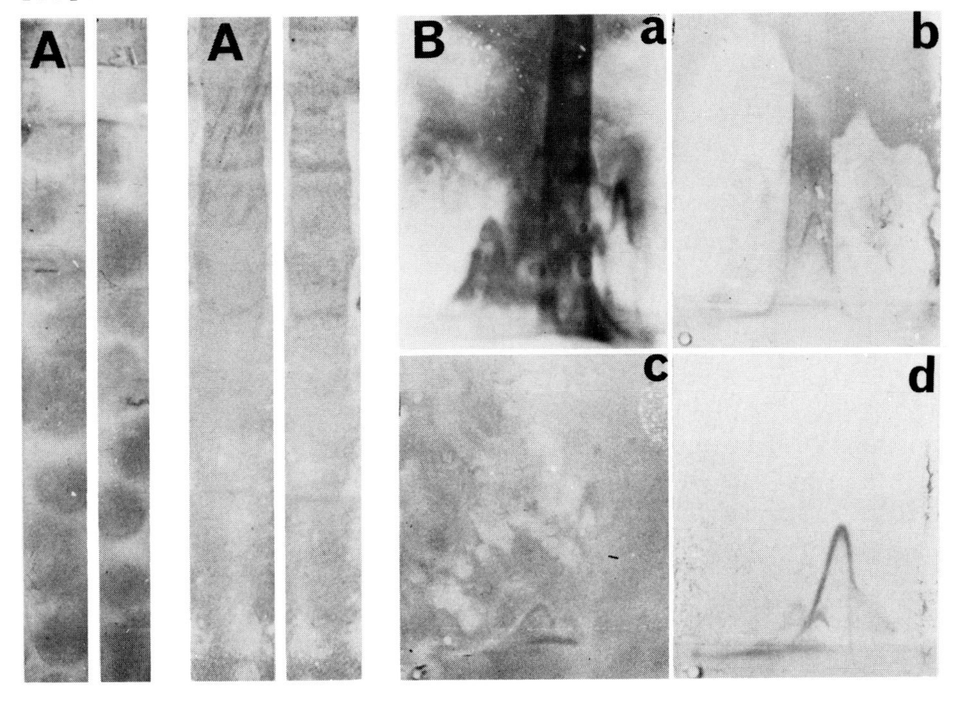

FIGURE III.3.2. Different types of background staining due to (a) antigen contamination of blotting cell and (b) presence of residual antigen in the gel. *Experiments:* (a) Immunoblotting of SDS-PAGE gels by a conventional sandwich in a contaminated buffer vessel. (b) Immunoblotting of crossed immunoelectrophoresis of (a, b) human serum and (c, d) human erythrocyte membrane proteins. (a-c) shows varying degree of contamination, and (d) shows a clean control. After washes and 20 min treatment at pH 11.5 semidry transfer was performed and the blots developed with alkaline phosphatase conjugated swine antirabbit IgG antibodies (1 + 2000, 18 hr; see Chapter 8.5).

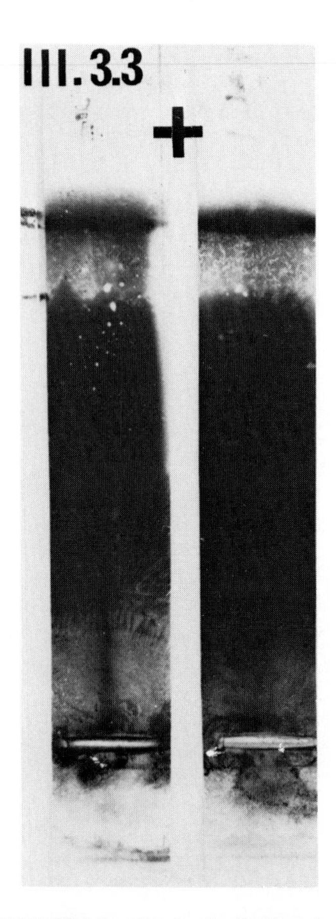

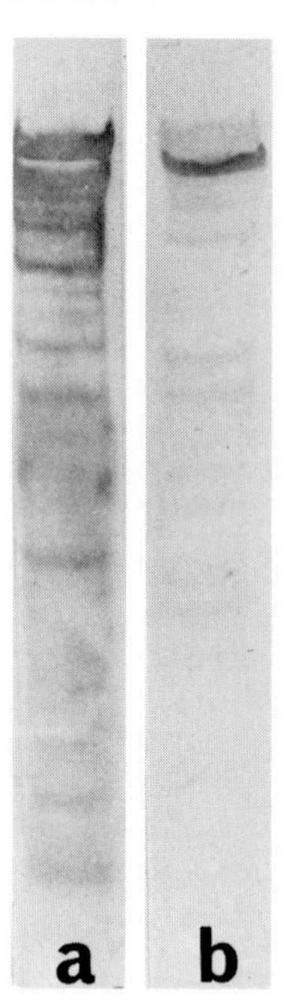

FIGURE III.3.3. Randomly distributed staining due to overdevelopment in staining solution. *Experiment:* Immunoblotting of 2 lanes 20 μg human Ig*M* crosslinked with glutardialdehyde, separated by agarose gel electrophoresis and developed with rabbit antihuman Ig*M* antibodies (1 + 1000, 18 hr). Staining was as in Figure I.2.B. Anode at the top.

FIGURE III.4.2. Unspecific staining following the banding pattern. (a) 0.5% (v/v) SDS has been added to the primary antiserum. (b) is the control without SDS. *Experiment:* Immunoblotting of human erythrocyte membrane proteins (20 μg) using human serum (1 + 330, 18 hr) as primary antibody and alkaline phosphatase conjugated rabbit antihuman IgG antibodies (Dakopatts) as secondary antibody (1 + 2000, 2 hr; see Figure 2).

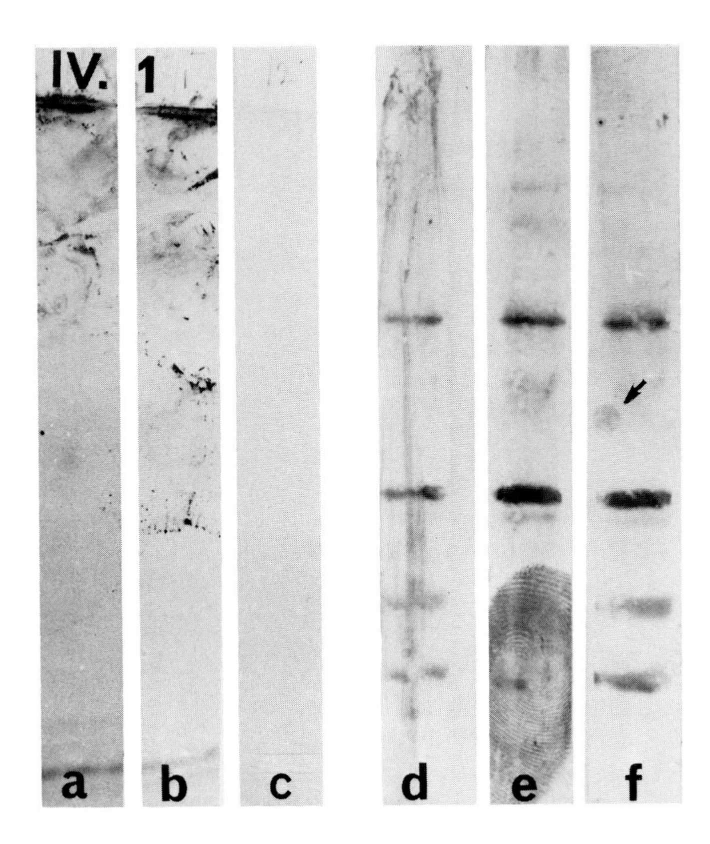

FIGURE IV.1. Various types of irregular spots and anomalies due to contamination of the nitrocellulose. (a and b) Antigen contamination derived from nondissolved material precipitated in the application well, (c) a control lane without load, (d) smeared nasal secretion, (e) fingerprint, and (f) drop of albumin solution 0.05% (w/v) (arrow). *Experiment:* Lanes a and b shows SDS-solubilized bovine spermatozoal proteins incubated with human serum (1 + 100, 18 hr), and developed as in Figure III.4.2. Lanes (d-f) correspond to Figure I.2.B.

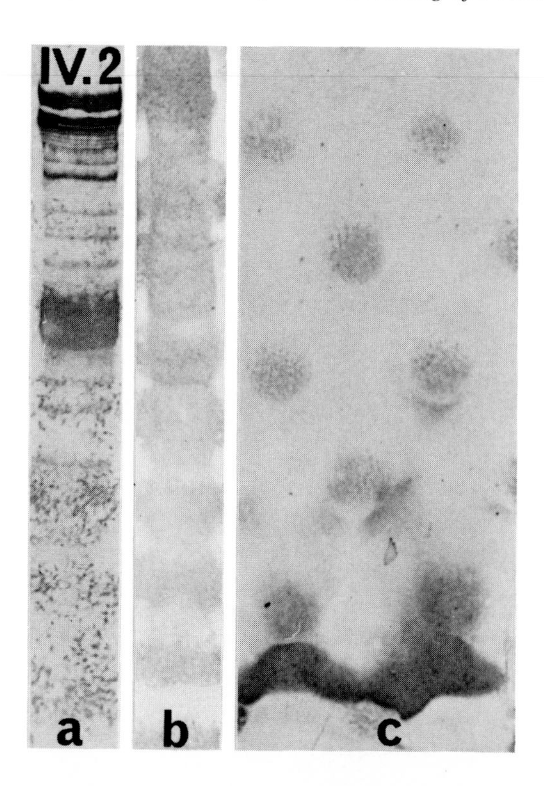

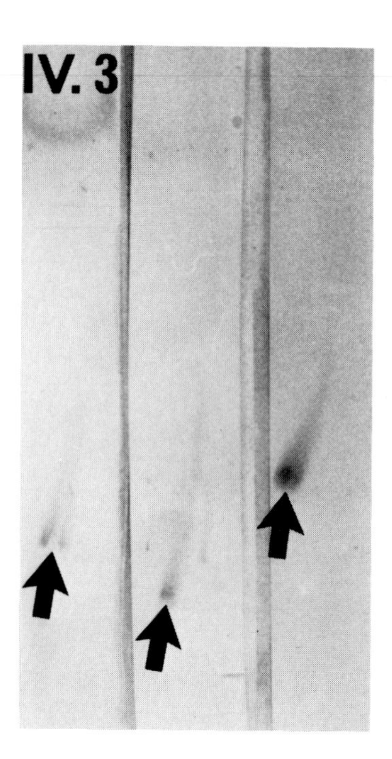

FIGURE IV.2. Various types of sharply stained spots deriving from contamination of the blotting cell. Antigen contamination (a) located corresponding to the sponges, (b) to the electrodes, and (c) to the perforations of the grids. *Experiments:* The electroblotting took place in a conventional sandwich in a buffer vessel between a set of platinum electrodes. Staining: 3-amino-9-ethyl carbazole.

FIGURE IV.3. Blurred stained streaks (arrows) due to direct squirting of secondary antibodies on the nitrocellulose through the incubation buffer. *Experiment:* As in Figure I.4.

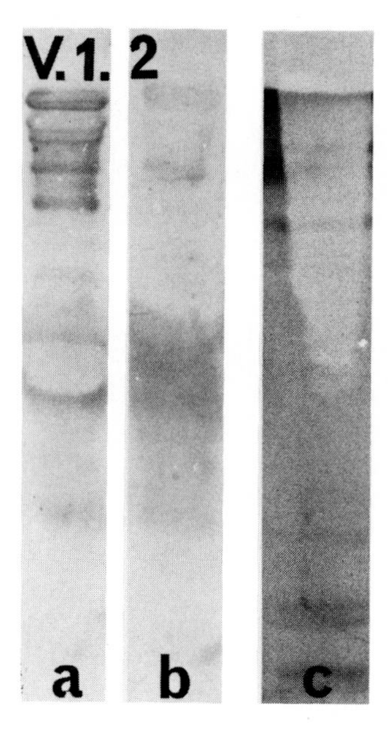

FIGURE V.1.2. Partly developed pattern due to insufficient immersion of nitrocellulose strip in primary (a) and secondary antibody (b, c). *Experiment:* (a, b) As for Figure I.2.B and (c) as for Figure IV.1.

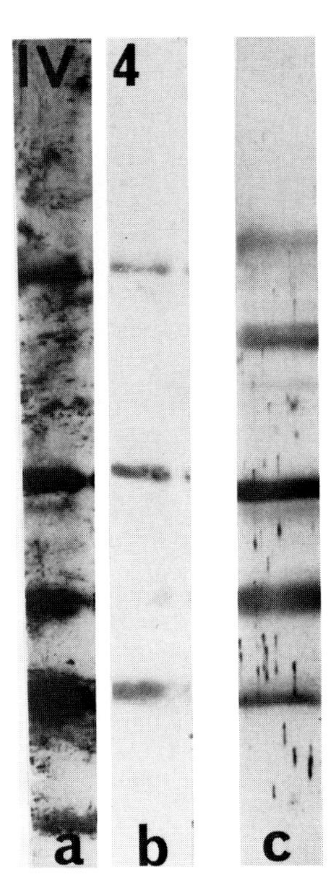

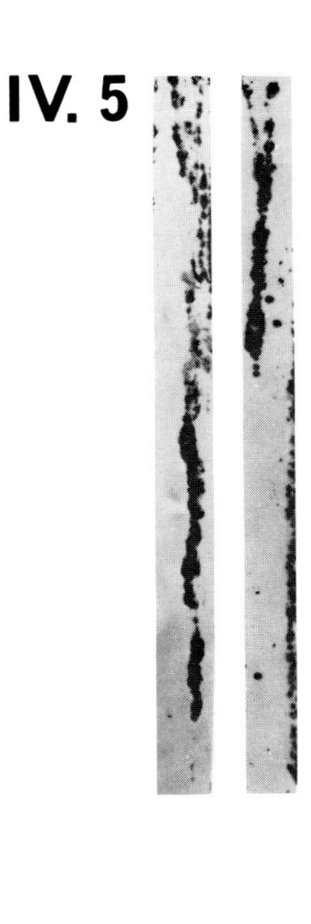

FIGURE IV.4. Scattered and distinct grains of stain due to (a) precipitation on nitrocellulose after prolonged development in substrate solution with evaporation, (b) normal staining, and (c) precipitation of stain from a nonfiltered solution. *Experiment:* As in Figure I.2.B.

FIGURE IV.5. Fluffy and confluent, highly stainable colonies due to microbial growth. *Experiment:* As in Figure IV.1.

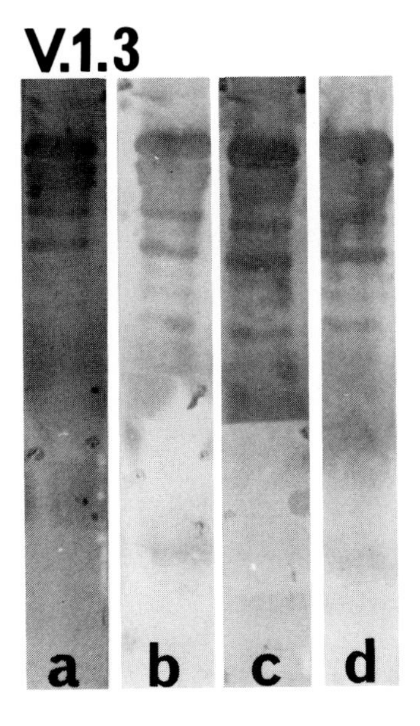

FIGURE V.1.3. Partly developed patterns due to covering of nitrocellulose blots during incubation in (a) primary antibody, (b) secondary antibody, (c) staining solution, and (d) control lane. *Experiment:* Nitrocellulose pieces have been placed on top of the original blotting strips and sewed together by a thread where the holes are visible. Conditions were otherwise as for Figure I.2.A.

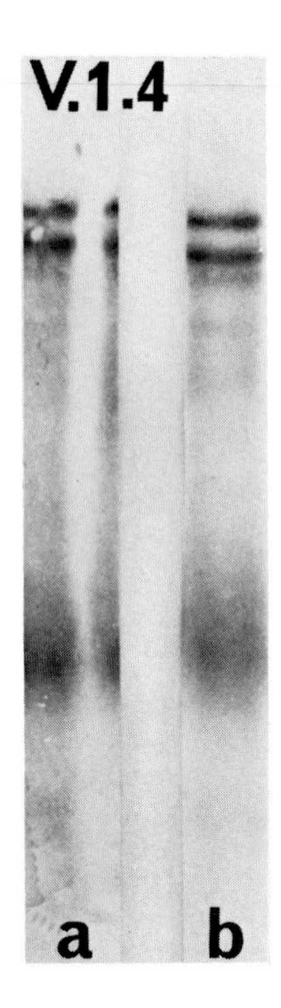

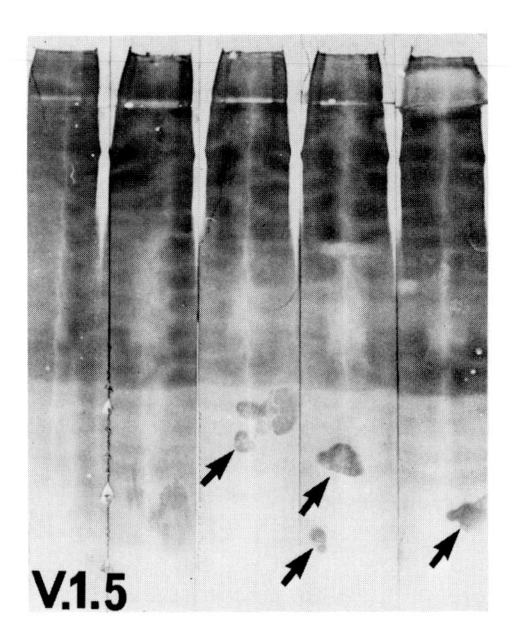

FIGURE V.1.4. Partly developed pattern. Lane (a) has been wrongly cut so the empty space between the tracks appears on the blotting strip. (b) Control. *Experiment:* As in Figure I.2.A.

FIGURE V.1.5. Uneven staining caused by standing waves in the antibody solution induced by the agitation. Note the light line in the middle of the strips; see also Figure III.2.1. Antigen contaminations are indicated by arrows, see also Figure IV.1 (b). *Experiment:* As in Figure IV.1 (a).

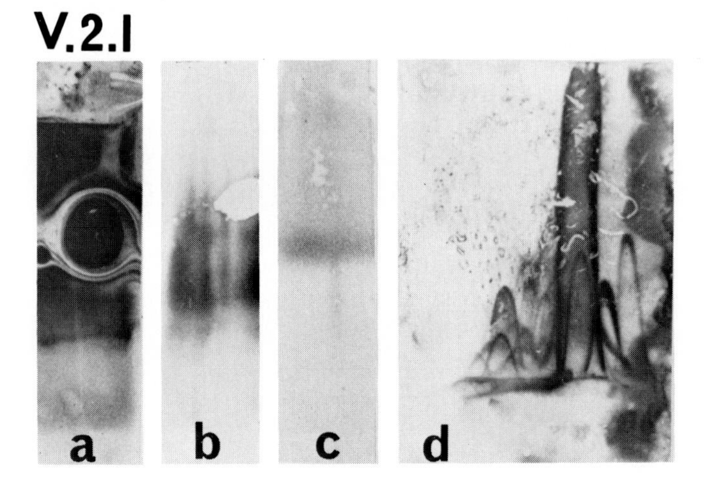

FIGURE V.2.1. Bald areas and spots due to air or dirt trapped in the sandwich. *Experiment:* (a) agarose gel electrophoresis (b, c) SDS-PAGE, and (d) crossed immunoelectrophoresis. (a) See Figure III.3.3, (b, c) see Figure I.2, and (d) see Figure III.3.2.

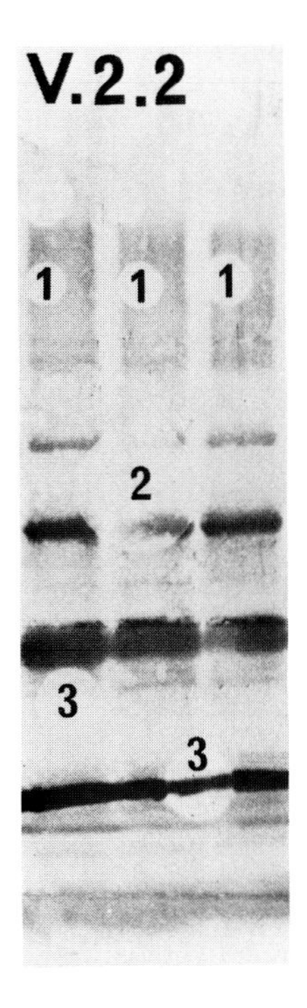

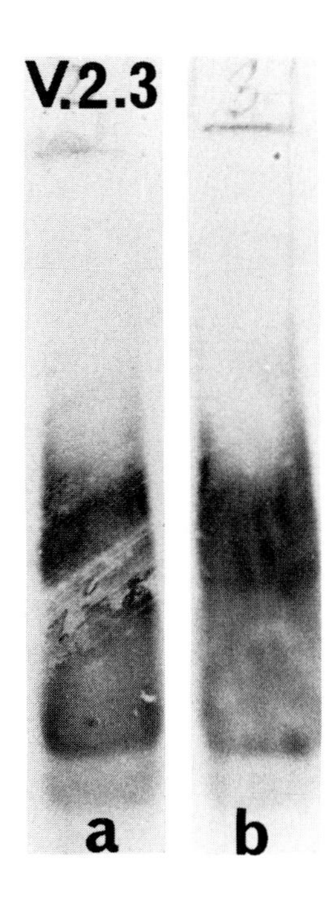

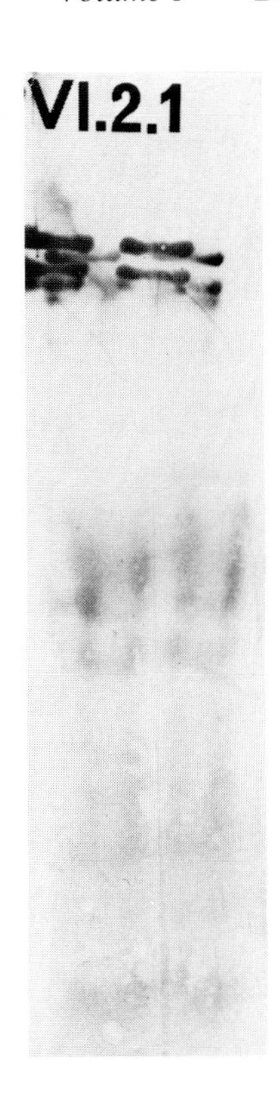

FIGURE V.2.2. Bald spots due to contamination of the nitrocellulose before electroblotting. *Experiment:* The following solutions have been spotted onto the nitrocellulose sheet before normal processing: (1) human transferrin (0.5% (w/v)), (2) Tween® 20 (1% (v/v)), and (3) human hemoglobin (0.5% (w/v)). Conditions correspond to those of Figure I.2.B.

FIGURE V.2.3. Empty areas where stain has been erased. *Experiment:* Immunoblotting of high M_r mixture pharmacin fine chemicals with corresponding antibodies. (a) Erased. (b) Control. Electrophoretic separation took place in a composite gel consisting of 2% (w/v) polyacrylamide and 0.5% (w/v) agarose. Conditions otherwise as for Figure I.2.B.

FIGURE VI.2.1. Double imprint due to displacement of nitrocellulose and gel during electrotransfer. *Experiment:* Gold-stained blots of 10 μg human erythrocyte membrane proteins; see Appendix to Chapter 1.

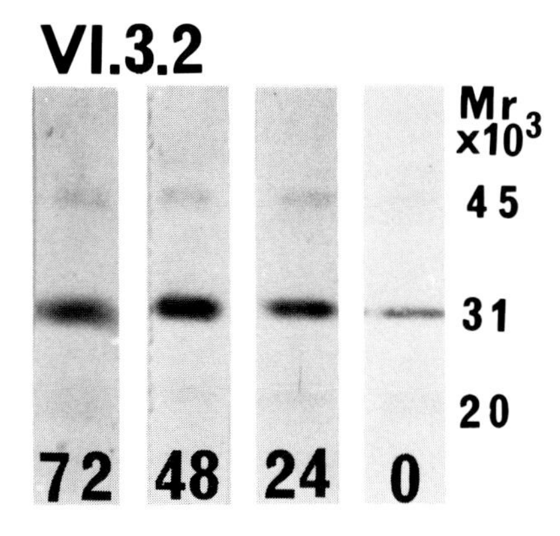

FIGURE VI.3.2. Blurred banding pattern because the proteins of the SDS-PAGE gel have been allowed to diffuse. Diffusion time at room temperature in hours is indicated on the figure. *Experiment:* As for Figure I.2.B.

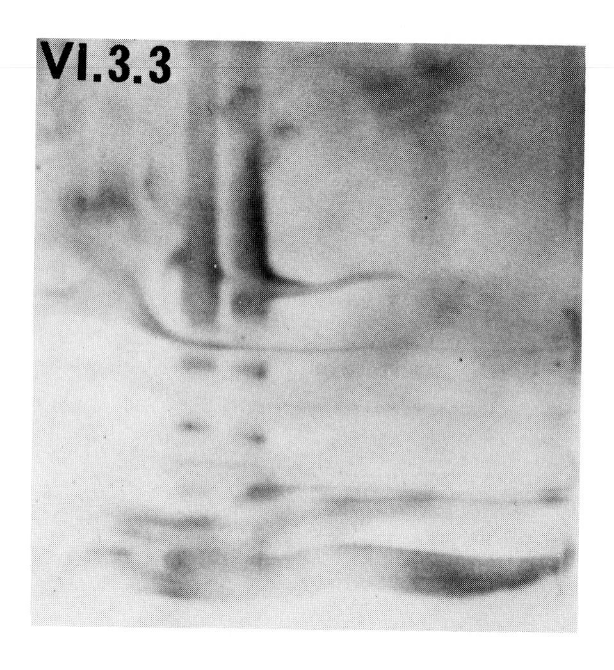

FIGURE VI.3.3. Surplus of antigen disseminates to the surroundings before blocking is effective (see also Figure II.5). *Experiment:* Autoradiography of an electroblot of radioactive labeled antigen from *Treponema reiteri*.

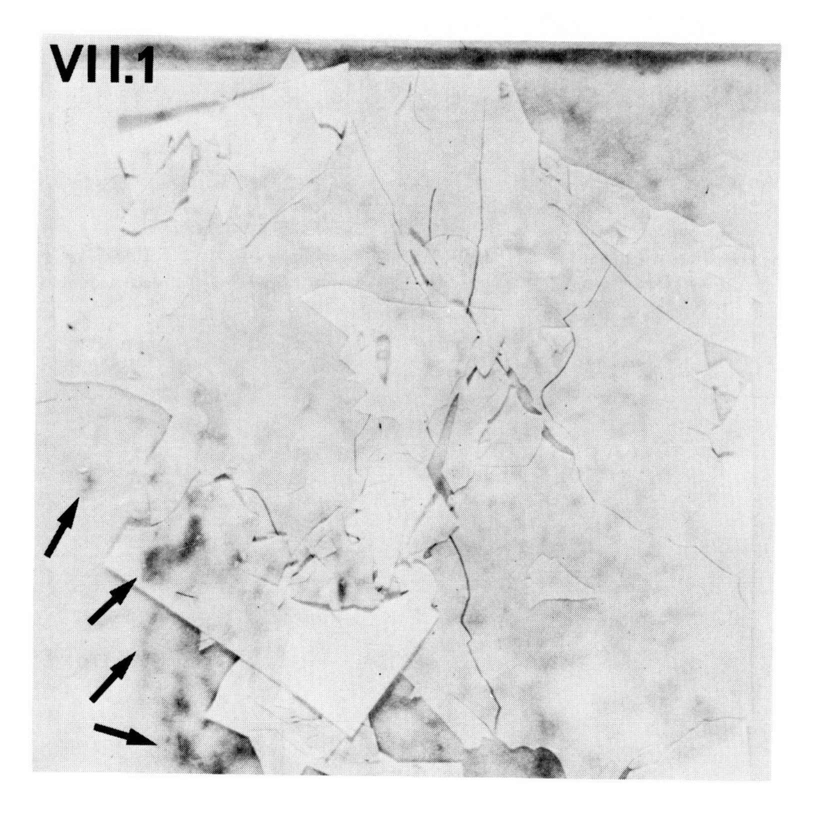

FIGURE VII.1. Nitrocellulose becomes brittle upon heating. Here seen after harsh semidry electroblotting. Note the presence of carbon particles from the graphite electrodes (arrows).

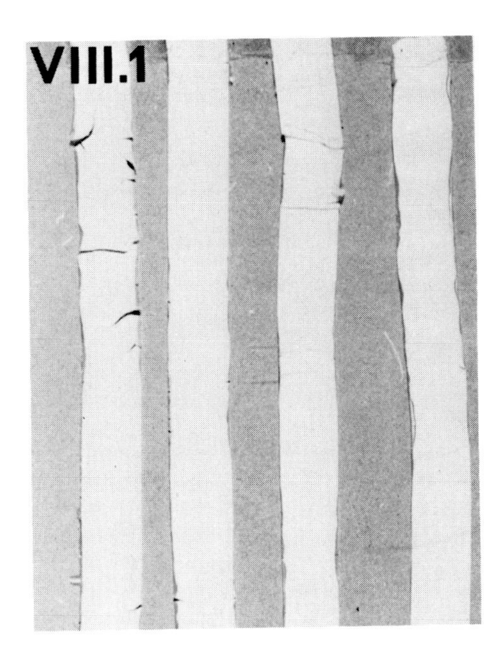

FIGURE VIII.1. Nitrocellulose becomes curled after exposure to vapor from some kind of organic solvents.

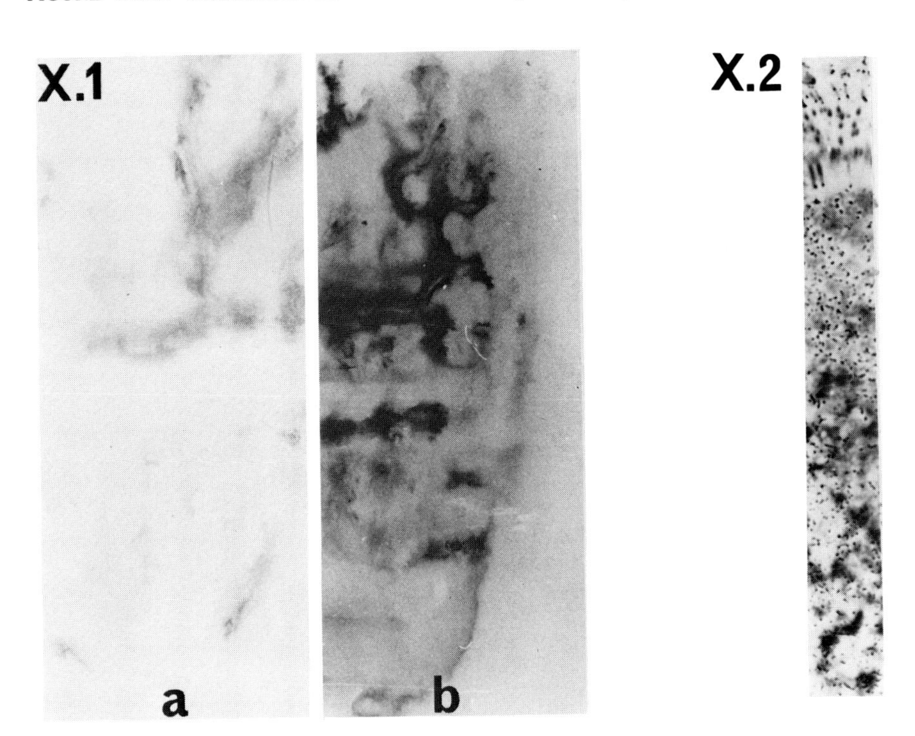

FIGURE X.1. Discolored nitrocellulose due to binding of electrode products: (a) unstained, (b) stained with Amido Black. *Experiment:* Electrotransfer took place for 48 hr in a buffer tank with electrodes of stainless steel.

FIGURE X.2. Nitrocellulose impregnated with carbon particles from graphite electrodes in semidry electroblotting. *Experiment:* Semidry electroblotting performed in 50 m*M* Tris, 0.1% (v/v) SDS, 20 V for 1 hr.

FIGURE X.3. Nitrocellulose with adherent polyacrylamide gel (arrows). *Experiment:* The gel was dried down on the nitrocellulose. Staining: Amido Black.

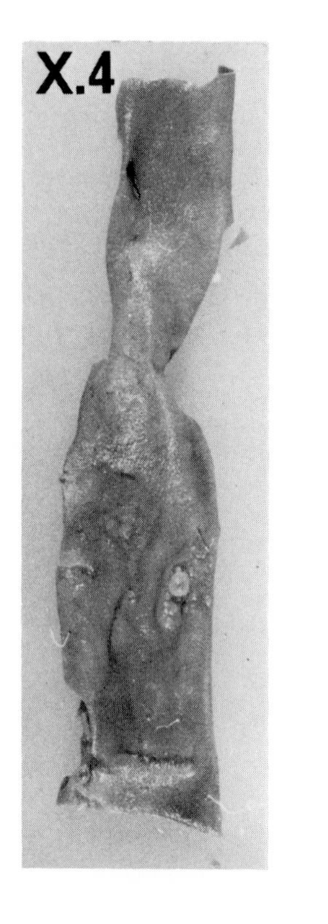

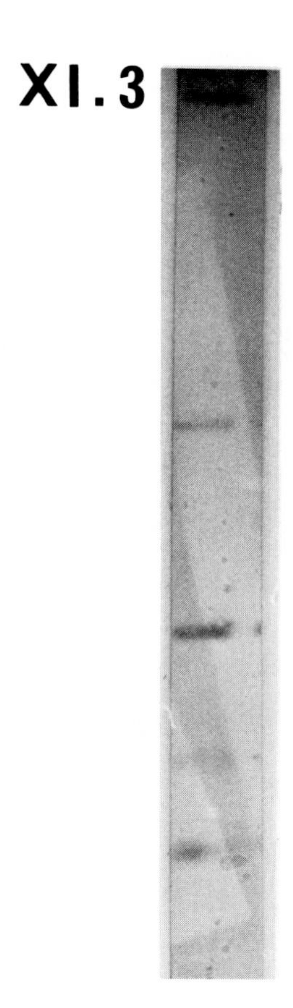

FIGURE X.4. Brownish, curled, and brittle nitrocellulose. *Experiment:* The nitrocellulose has been heated in an oven for 15 min at 140°C.

FIGURE XI.3. Fading during drying made visible by overlaying strips. *Experiment:* As in Figure III.2.4 (a).

REFERENCES

1. **Ochs, D.,** Protein contaminants of sodium dodecyl sulfate-polyacrylamide gels, *Anal. Biochem.,* 135, 470, 1983.
2. **Tasheva, B. and Dessev, G.,** Artefacts in sodium dodecyl sulfate-polyacrylamide gel electrophoresis due to 2-mercaptoethanol, *Anal. Biochem.,* 129, 98, 1983.
3. **Bjerrum, O. J., and Hinnerfeldt, F. R.,** Visualization of molecular weight standards after electroblotting: Detection by means of corresponding antibodies, *Electrophoresis,* 9, 1987, in press.
4. **Saravis, C. A.,** Improved blocking of nonspecific antibody binding sites on nitrocellulose membranes, *Electrophoresis,* 5, 54, 1984.
5. **Lampson, L. A. and Fisher, C. A.,** Immunoblot specificity of monoclonal antibodies assayed against complex extracts, *Anal. Biochem.,* 144, 55, 1985.
6. **Nigg, E. A., Walter, G., and Singer, S. J.,** On the nature of crossreactions observed with antibodies directed to defined epitopes, *Proc. Natl. Acad. Sci. U.S.A.,* 79, 5935, 1982.
7. **Berzofsky, J. A. and Schechter, A. N.,** The concepts of crossreactivity and specificity in immunology, *Mol. Immmunol.,* 18, 751, 1981.
8. **Larsson, L. I.,** Peptide immunocytochemistry, *Prog. Histochem. Cytochem.,* 13, 4, 1981.
9. **Polvino, W. J., Saravis, C. A., Sampson, C. E., and Cook, R. B.,** Improved protein analysis on nitrocellulose membranes, *Electrophoresis,* 4, 368, 1983.
10. **Wedege, E. and Svenneby, G.,** Effects of the blocking agents bovine serum albumin and Tween 20 in different buffers on immunoblotting of brain proteins and marker proteins, *J. Immunol. Methods,* 88, 233, 1986.
11. **Larsson, L.-I.,** Simultaneous ultrastructural demonstration of multiple peptides in endocrine cells by novel immunocytochemical methods, *Nature (London),* 282, 743, 1979.
12. **Bjerrum, O. J., Larsen, K. P., and Wilken, M.,** Some recent developments of the electroimmunochemical analysis of membrane proteins. Application of Zwittergent, Triton X-114 and Western blotting technique, in *Modern Methods in Protein Chemistry,* Tschesche, M., Ed., Walter de Gruyter, Berlin, 1983, 79.
13. **Bjerrum, O. J., Selmer, J., Larsen, F., and Naaby-Hansen, S.,** Exploitation of antibodies in the study of membranes, in *Methodological Surveys in Biochemistry and Analysis. Investigation and Exploitation of Antibody Combining Sites,* Vol. 15, Reid, E., Cook, G. M. W., and Morré, D. J., Eds., Plenum Press, New York, 1985, 231.
14. **Fairbanks, G., Steck, T. L., and Wallach, D. F. H.,** Electrophoretic analysis of the major polypeptides of the human erythrocyte membrane, *Biochemistry,* 10, 2606, 1971.
15. **Bjerrum, O. J.,** Detergent-immunoelectrophoresis. General principles and methodology, in *Electroimmunochemical Analysis of Membrane Proteins,* Bjerrum, O. J., Ed., Elsevier, Amsterdam, 1983, 3.
16. **Gershoni, J. M. and Palade, G. E.,** Protein blotting principles and applications, *Anal. Biochem.,* 131, 1, 1983.
17. Information sheet to BIO-RAD IMMUN-BLOT (Gar-HRP) ASSAY KIT, Biorad Laboratories, Richmond, Calif., 1982.
18. **Bjerrum, O. J.,** A trouble shooter and an atlas for electroimmunoprecipitation artefacts, *Electrophoresis,* 6, 209, 1985.
19. **Gershoni, J. M., Davies, F. E., and Palade, G. E.,** Protein blotting in uniform or gradient electric fields, *Anal. Biochem.,* 144, 32, 1985.
20. **Sutton, R., Wrigley, C. W., and Baldo, B. A.,** Detection of IgE- and IgG-binding proteins after electrophoretic transfer from polyacrylamide gels, *J. Immunol. Methods,* 52, 183, 1982.
21. **Svoboda, M., Menris, S., Robyn, C., and Christophe, J.,** Rapid electrotransfer of proteins from polyacrylamide gel to nitrocellulose membrane using surface-conductive glass as anode, *Anal. Biochem.,* 151, 16, 1985.
22. **Nielsen, P. J., Manchester, K. L., Towbin, H., Gordon, J., and Thomas, G.,** The phosphorylation of ribosomal protein diabetes and after denervation of diaphragm, *J. Biol. Chem.,* 257, 316, 1982.
23. **Gibson, W.,** Protease-facilitated transfer of high molecular-weight proteins during electrotransfer to nitrocellulose, *Anal. Biochem.,* 118, 1, 1981.
24. **Elkon, K. B., Jankowski, P. W., and Chu, J.-L.,** Blotting intact immunoglobulins and other high-molecular-weight proteins after composite agarose-polyacrylamide gel electrophoresis, *Anal. Biochem.,* 140, 208, 1984.
25. **Renart, J., Reiser, J., and Stark, G. R.,** Transfer of proteins from gels to diazobenzyloxymethyl-paper and detection with antisera. A method for studying antibody specificity and antigen structure, *Proc. Natl. Acad. Sci. U.S.A.,* 76, 116, 1979.
26. **Burnette, W. N.,** Western blotting. Electrophoretic transfer of proteins from sodium dodecyl sulfate-polyacrylamide gels to unmodified introcellulose and radiographic detection with antibody and radionated protein A, *Anal. Biochem.,* 112, 195, 1981.

27. **Lin, W. and Kasamatsu, H.,** On the electrotransfer of polypeptides from gels to nitrocellulose membranes, *Anal. Biochem.,* 128, 302, 1983.
28. **Rochette-Egly, C. and Daviaud, D.,** Calmodulin binding to nitrocellulose and zeta pore membranes during electrophoretic transfer from polyacrylamide gels, *Electrophoresis,* 6, 235, 1985.
29. **Towbin, H., Staehelin, F., and Gordon, J.,** Electrophoretic transfer of proteins from polyacrylamide gels to nitrocellulose sheets. Procedure and some applications, *Proc. Natl. Acad. Sci. U.S.A.,* 76, 4360, 1979.
30. **Spinola, S. M. and Cannon, J. G.,** Different blocking agents cause variation in the immunologic detection of proteins transferred to nitrocellulose membranes, *J. Immunol. Methods,* 81, 161, 1985.
31. **Van Eldik, L. J. and Wolchok, S. R.,** Conditions for reproducible detection of calmodulin and S100B in immunoblots, *Biochem. Biophys. Res. Commun.,* 124, 752, 1984.
32. **Bjerrum, O. J.,** Dot-immunobinding as a simple test for demonstration of human IgA-antibodies, unpublished result.
33. **Kakita, K., O'Connell, K., and Permutt, M. A.,** Immunodetection of insulin after transfer from gels to nitrocellulose filters, *Diabetes,* 31, 648, 1982.
34. **Harboe, N. M. G. and Ingild, A.,** Immunization, isolation of immunoglobulins and antibody titre determination, *Scand. J. Immunol.,* 17(Suppl. 10), 345, 1983.
35. **Bjerrum, O. J. and Schäfer-Nielsen, C.,** Analysis of buffer systems and transfer parameters for semi-dry electroblotting, in *Electrophoresis '86,* Dunn, M. J. Ed., VCH Publishers, Weinheim, West Germany, 1986, 315.
36. **Ipsen, H. and Bjerrum, O. J.,** unpublished results.
37. **Hald, J., Vinten, J., and Bjerrum, O. J.,** Quantitative aspects of immunoblotting, employing alkaline conjugated secondary antibodies, *Elektrophorese Forum '86,* Rodola, B. J., Ed., Technische Universität München gedruckt, Munich, West Germany, 66, 1986.

INDEX

A

B

C

D

E

F

G

H

I

L

M

N

O

P

Q

R

S

T

U

V

W

Y

Z